Environmental Governance of Common-Pool Resources

This book analyses the way in which public and private regulation address environmental problems related to fishing and forestry by establishing and maintaining property rights and by overcoming the limitations of property rights in coping with these common resource problems. It compares the environmental governance of fishery and forestry in nine jurisdictions.

Jing Liu is Associate Professor of Law in the Research Institute of Environmental Law, Wuhan University, China.

Michael Faure is Professor of Comparative Private Law and Economics at the Erasmus School of Law, Erasmus University Rotterdam, and Professor of Comparative and International Environmental Law at Maastricht University.

Peter Mascini is Professor of Empirical Legal Studies at the Erasmus School of Law and Associate Professor of Sociology at the Erasmus School of Social and Behavioural Sciences, Erasmus University Rotterdam.

Routledge Research in International Environmental Law

A full list of titles in this series in available at:
www.routledge.com/

Responsibilities and Liabilities for Commercial Activity in the Arctic
The Example of Greenland
Edited by Vibe Ulfbeck, Anders Møllmann and Bent Ole Gram Mortensen

Ecological Integrity and Global Governance
Science, Ethics and the Law
Laura Westra

Environmental Law and the Ecosystem Approach
Maintaining Ecological Integrity Through Consistency in Law
Froukje Maria Platjouw

Stratospheric Ozone Damage and Legal Liability
US Public Policy and Tort Litigation to Protect the Ozone Layer
Lisa Elges

Ecological Restoration in International Environmental Law
Anastasia Telesetsky, An Cliquet and Afshin Akhtar-Khavari

Sustainable Development Principles in the Decisions of International Courts and Tribunals
1992–2012
Marie-Claire Cordonier Segger and H.E. Judge C.G. Weeramantry

Culture and International Economic Law
Valentia Vadi and Bruno de Witte

Natural Resources Law, Investment and Sustainability
Shawkat Alam, Jahid Hossain Bhuiyan and Jona Razzaque

Protecting Traditional Knowledge
The WIPO Intergovernmental Committee on Intellectual Property and Genetic Resources, Traditional Knowledge and Folklore
Edited by Daniel F. Robinson, Ahmed Abdel-Latif and Pedro Roffe

Environmental Governance of Common-Pool Resources

A Comparison of Fishery and Forestry

Jing Liu, Michael Faure and Peter Mascini

Routledge
Taylor & Francis Group
LONDON AND NEW YORK

First published 2018 by Routledge

2 Park Square, Milton Park, Abingdon, Oxfordshire OX14 4RN
52 Vanderbilt Avenue, New York, NY 10017

Routledge is an imprint of the Taylor & Francis Group, an informa business

First issued in paperback 2019

British Library Cataloguing in Publication Data
A catalogue record for this book is available from the British Library

Library of Congress Cataloging in Publication Data
Names: Liu, Jing, 1985 May 1-, author. | Faure, Michael (Michael G.), author. | Mascini, Peter, 1968-, author.
Title: Environmental governance of common pool resources / By Jing Liu, Michael Faure, and Peter Mascini.
Description: New York : Routledge, 2017. | Series: Routledge research in international environmental law
Identifiers: LCCN 2017018034 | ISBN 978-1-138-04976-5 (hbk) | ISBN 978-1-315-16939-2 (ebk)
Subjects: LCSH: Environmental law. | Environmental policy–International cooperation. | Natural resources–Management. | Right of property.
Classification: LCC K3585 .L585 2017 | DDC 344.04/6–dc23
LC record available at https://lccn.loc.gov/2017018034

ISBN: 978-1-138-04976-5 (hbk)
ISBN: 978-0-367-88873-2 (pbk)

Typeset in Baskerville
by FiSH Books Ltd, Enfield

Contents

Abbreviations

ABC	Allowable Biological Catch
ACE	Annual Catch Entitlement
AFCC	Area Fisheries Coordination Committees (Japan)
AFL	Sea Lord and Aotearoa Fisheries Limited (New Zealand)
ANP	Protected Natural Area (Mexico)
APCL	Act on Preservation and Control of Living Marine Resources (Japan)
APT	Forest and Land Inspection and Social Authority (Bolivia)
ASL	Local Community Associations (Bolivia)
BAL	Basic Agrarian Law (Indonesia)
BANPESCA	Fishery Development Bank (Mexico)
BFL	Basic Forestry Law (Indonesia)
BMP	Best Management Practice
BPN	National Land Agency (Indonesia)
C&Is	Criteria & Indicators
CANAIPESCA	National Fisheries and Aquaculture Industry Chamber (Mexico)
CAR	Corrective Action Requirements
CB	Certification Body
CBFM	Community-Based Forest Management
CCP	Collaboration with Companies (Bolivia)
CFMP	Coastal Fisheries Management Plan (Japan)
CFV	Certificación Foresta Voluntaria (Bolivian Council for Voluntary Forest Certification) (Bolivia)
CGF	Consumer Goods Form
CNCP	National Confederation of Fishing Cooperatives (Mexico)
COBI	Comunidad y Biodiversidad (Community and Biodiversity) (Mexico)
CONABIO	Mexican Biodiversity Commission
CONAPESCA	National Commission for Aquaculture Fisheries
CSA	Canadian Standards Association
CWA	Clean Water Act (US)

DAFF	Department of Agriculture, Forestry and Fisheries (South Africa)
DoC	Department of Conservation (New Zealand)
DWG	Deep Water Group
EEZ	Exclusive Economic Zone
ENGO	Environmental Non-Governmental Organization
FAO	Food and Agriculture Organization of the United Nations
FCAs	Fisheries Cooperative Associations (Japan)
FCAL	Fisheries Cooperative Association Law (Japan)
FEDECOOP	Regional Federation of Cooperative Societies of Baja California (Mexico)
FES	Economic and Social Function (Bolivia)
FIP	Fishery Improvement Project
FL	Fishery Law (Japan)
FMO	Fishery Management Organization (Japan)
FOS	Friend of the Sea Certification Scheme
FPIC	Free and Prior Informed Consent
FSC	Forest Stewardship Council
GDP	Gross Domestic Product
GHQ	General Headquarters of the Supreme Commander for the Allied Powers
HD	Historically Disadvantaged
HFMC	Hoki Fishery Management Company (New Zealand)
IFCC	Indonesian Forestry Certification Cooperation
IFF	Intergovernmental Forum on Forests
IMP	National Institute for Fisheries Research (Mexico)
INE	Institute of Ecology (Mexico)
INRA	National Institute for Agrarian Reform Law (Bolivia)
IPF	Intergovernmental Panel on Forests
IQ	Individual Quota
ISO	International Standard Organization
ITQ	Individual Transferable Quota (System)
ITTA	International Tropical Timber Agreement
ITTO	International Tropical Timber Organization
IUU	Illegal, Unreported and Unregulated (Fishing)
LEI	Eco-Label Institute (Indonesia)
LP	Independent auditors (Indonesia)
LV	Independent Verifier (Indonesia)
MAF	Ministry of Agriculture and Fisheries (New Zealand)
MAFF	Ministry of Agriculture, Forestry and Fisheries (Japan)
MCM	Department of Marine and Coastal Management (South Africa)
MEL	Marine Eco-Label Japan Programme
MLRA	Marine Living Resources Act (South Africa)
MoF	Ministry of Fisheries (New Zealand)

MoF	Ministry of Forestry (Indonesia)
MOU	Memorandum of Understanding
MPI	Ministry for Primary Industries (New Zealand)
MSC	Marine Stewardship Council
MSY	Maximum Sustainable Yield (New Zealand)
NGO	Non-Governmental Organization
NOM	Official Mexican Standard (Mexico)
NPS	Non-Point Source Water Pollution
NZRFC	New Zealand Recreational Fishing Council
OMP	Operational Management Procedure (South Africa)
P&C	Principles and Criteria
PEFC	Programme for the Endorsement of Forest Certification
PHAPL	Sustainable Production Natural Forest Management Certification (Indonesia)
PHBML	Sustainable Community-Based Forest Management System (Indonesia)
PHPL	Mandatory Certification Scheme (Indonesia)
PHTL	Sustainable Plantation Forest Management Certification (Indonesia)
PROFEBA	Environmental Enforcing Agency (Mexico)
PROPEMEX	Mexican State Fishery Company
QMA	Quota Management Area (New Zealand)
QMS	Quota Management System
QOA	Quota Owner Association (New Zealand)
RFBPS	Royal Forest and Bird Protection Society (New Zealand)
RMWG	Resource Management Working Group (South Africa)
RRP	Resource Recovery Plan (Japan)
RSB	Round Table on Sustainable Biofuels
RSP	Round Table on Sustainable Palm Oil
RTRS	Round Table on Responsible Soy
RTRWP	Provincial Spatial Plants (Indonesia)
SA	Agrarian Superintendence (Bolivia)
SADSTIA	South Africa Deep Sea Trawling Industry Association
SAGARPA	The Secretariat of Agriculture, Livestock, Rural Development, Fisheries and Food (Mexico)
SECIFA	South-East Coast Inshore Fishing Association (South Africa)
SF	Forestry Superintendence (Bolivia)
SFA	State Forest Agency (Sweden)
SFI	Sustainable Forest Initiative
SLIMF	Small and Low Intensity Management Forests Standards
SSF	Small-Scale Fisheries
SSNC	The Swedish Society for Nature Conservation
SWGs	The Scientific Working Group (South Africa)
TAC	Total Allowable Catch

TACC	Total Allowable Commercial Catch (New Zealand)
TAE	Total Allowable Effort
TCOs	Tierras Communitarias de Origen (Native Community Lands) (Bolivia)
TLAS	Legal Verification Scheme (Indonesia)
TOKM	Treaty of Waitangi Fisheries Commission (New Zealand)
TUR	Territorial User Right
UFMs	Unidades Forestales Municipales (Municipal Forestry Unit) (Bolivia)
UNCLOS	United Nations Convention on the Law of the Sea 1982
UNFF	United Nations Forum on Forests
US	United States
US$	United States Dollars
USFS	United States Forest Service
VMES	Vulnerable Marine Eco System
VMS	Vessel Monitoring System
VPA	Voluntary Partnership Agreement
WAFCC	Wide-Area Fisheries Coordinating Committee (Japan)
WGI	Worldwide Governance Indicators
WWII	Second World War
WWF	World Wildlife Fund

Foreword

This book is one of the results of a project subsidized by the Netherlands Royal Academy of Arts and Sciences on smart mixes in relation to transboundary environmental harm. The project was sponsored within the framework of the Academy project 'Beyond borders' (*Over grenzen*). It consisted of a collaboration between the Erasmus School of Law in Rotterdam and the Faculty of Law of the University of Amsterdam, with the collaboration from the Law Faculty of Maastricht University as well. The project work was largely executed by two postdocs financed by the Academy, Dr Markos Karavias and Dr Liu Jing and supported by an expert team consisting of: Professor Michael Faure (Erasmus University Rotterdam and Maastricht University); Professor André Nollkaemper (University of Amsterdam); Professor Peter Mascini (Erasmus University Rotterdam); Professor Judith van Erp (Utrecht University); Professor Marjan Peeters (Maastricht University); Professor Niels Philipsen (Erasmus University Rotterdam and Maastricht University).

The project team benefited from the expertise of an advisory committee consisting of: Jan van den Broek (VNO-NCW); Professor Joyeeta Gupta (University of Amsterdam); Dr Veerle Heyvaert (London School of Economics); Gerald Kok (Shell); Professor Catherine Redgwell (University of Oxford); Dr Peter Sand (University of Munich); Professor Jaap Spier (Supreme Court of the Netherlands); Jasper Teulings (Greenpeace International); Ludo Veuchelen (Erasmus University Rotterdam).

Within the framework of this project research was carried out on the governance of Common-Pool resources, more particularly forests and fishery resources. Two workshops were organized: one on smart mixes in relation to forest and climate governance on 4–5 February 2015 and one workshop on smart mixes in relation to fishery and oil pollution governance on 7–8 October 2015 both at the Facility of the Netherlands Royal Academy of Arts and Sciences in the Trippenhuis in Amsterdam.

As authors of this book which largely builds on the mentioned research project we are grateful to our colleagues, members of the expert team, to the members of the advisory committee, to the participants in the two mentioned workshops and especially to the Netherlands Royal Academy of

Arts and Sciences for their invaluable input and support without which this book could not have been published.

Jing Liu, Michael Faure, Peter Mascini
Wuhan/Maastricht/Rotterdam, March 2017

1 Introduction

1.1 Background

1.1.1 Governing environmental problems: an intertwined system of public and private regulation

Apart from the intrinsic value of the earth's systems, human beings are dependent on their natural environment for a respectable and prosperous life.[1] The environment, however, has been threatened, for example by industrialization, urbanization and population growth. Many environmental problems, such as the diffusion of chemicals and other pollutants and the depletion of forests and fisheries have come into the public's attention and have aroused broad concerns since the mid-twentieth century.

The prominence of these environmental problems has been accompanied by rising environmental public regulation. The last century has witnessed a rapid increase of international environmental agreements and international organizations governing various environmental problems. Many countries have also adopted domestic regulation. Nonetheless, despite the ongoing proliferation of international agreements and domestic regulation, environmental problems have persisted.

In the 1980s and 1990s, neoliberal ideas on free trade, liberalization, and privatization diffused globally.[2] As a result, for example, attempts of developed countries to prevent illegal logging in tropical countries via import bans lost support. As governmental and intergovernmental options for addressing environmental issues were blocked or failed by the invocation of legislation that enabled free trade, states, NGOs and social movement groups put more energy and resources into non-governmental programmes – like nascent certification and monitoring efforts.[3] For example, forest certification started to develop when the international community failed to conclude a legally binding forest agreement, while

1 Hey 2016, p. 2.
2 Dobbin, Simmons & Garrett 2007, pp. 449–472.
3 See Bartley 2003.

fishery certification emerged as a response to the perceived failure of international law to address overfishing.[4]

Public and private regulation are not functioning in isolation, but are intricately linked. Besides, domestic public regulation promotes or constrains the development of private and hybrid regulation via specific governance instruments.[5] Moreover, public regulation and private regulation do not only interact, but also create multiple layers of rules regulating potential polluters.[6] To what extent the interaction between public and private regulation induces more environmental friendly behaviour and hence contributes to mitigating environmental problems still remains a complicated question to be understood and answered.

1.1.2 Common-pool resources problems

There are countless environmental problems, caused by numerous drivers, which call for different regulatory regimes and instruments. This research focuses on one prominent problem: common-pool resources problems. Many natural resources, such as forests and fish stocks are common-pool resources: the exclusion of other users is very costly if not impossible (non-excludability) and exploitation by one user diminishes the availability for others (subtractability).[7] Because exclusion is costly, when no institutions restrict access, resource users may tend to free-ride and to overexploit resources. In this case, 'the tragedy of the commons' scenario materializes.[8] Scholars have identified basically three types of property rights to restrict access to common-pool resources: public, private, and communal.[9] Extensive academic attention has been paid to how common-pool resources are protected under these different property rights and to the preconditions for each property right to function effectively.[10] The effective functioning of property rights means that the overexploitation of common-pool resources is prevented by either excluding so many users that the subtraction of resources does not exceed levels of sustainability or making sure that present users do not exceed these levels.

4 See Gulbrandsen 2010.

5 E.g. many European countries have public procurement policies requiring timber products under public procurement to be legal and sustainable and they have recognized products under many certification schemes to be sustainable. Governments can also regulate private certification schemes via publishing (inter)governmental guidelines, having access to market standards and so on, see Gulbrandsen 2014.

6 Bartley 2011, p. 517.

7 Feeny *et al.* 1990.

8 Hardin 1968, p. 1243.

9 Feeny *et al.* 1990; Ostrom 2010.

10 For example, Ostrom has identified eight design principles for the success of communal property rights regimes. The limitations of private property rights and public property rights have been discussed extensively in Law and Economics literature. For details, see *infra* Chapter 2.

In order to be able to be effective, property rights need to be affirmed by the state. For example, public regulation can affirm private ownership of forests via registration or the creation of (quasi-) private property rights[11] on state-owned forests via granting concessions. Traditional community rights on forests also require the recognition in domestic law. States can also intervene, for example, by correcting market failures in case property rights are privatized.

Rather than operating separately, public and private regulation often interact. Their interaction influences the existence and performance of various property rights and can help to overcome their limitations. However, how such influence plays out in practice and contributes to the management of common-pool resources problems, has not been extensively studied yet. The main goal of this study is to explore this.

1.2 Research questions and problem definition

This research aims at exploring the question: how do public and private regulation interact to address common-pool resources problems via influencing the establishment and functioning of property rights? As mentioned before, basically three types of property rights can be used to address common-pool resources in order to avoid 'the tragedy of the commons': public, private, and communal. This research hence analyzes first how public and private regulation influences the establishment and functioning of public, private, and communal property rights. However, since all property rights have their limitations, the effectiveness of addressing the overexploitation of common-pool resource, also depends on whether public and private regulation compensate these limitations. This research will also analyse whether public and private regulation help to overcome the limitations of property rights in addressing common-pool resources problems.

Our theoretical framework will be applied to forestry and fishery governance. Problems associated with these two types of common-pool resources have caused broad attention: many countries have suffered deforestation and forest degradation and overfishing is a well-recognized problem in both high seas and areas falling under domestic sovereignty. Various types of property rights have been established for forestry and fishery in different jurisdictions. Forests are immobile resources growing on a certain area of land. Usually ownership can be established on a determined area of forests or the land the forests grow on. The forests or land can either be

11 It is debatable whether such rights should be classified as 'property rights' or 'quasi-property rights'. This research will not engage in this legal discussion, but uses the term 'property rights' broadly to denote 'a set of rights to control assets', or in other words the form of power where 'a sanction and authority for decision-making' over resources have been established. See Cole 2010; Denman 1978; Dasgupta 1982.

publicly owned, privately owned or communally owned. Fish resources are mobile and are usually claimed as state resources. Nevertheless, rights to access and harvest can be vested in private actors and communities.[12] The property rights of forestry and fishery have traditionally been primarily regulated under domestic law. However, in recent decades, many private regulatory regimes, such as certification schemes, have also been implemented and have covered a substantial proportion of forests and fisheries.[13] Such privatization schemes often rely on clear definitions of ownership.

Although it is the goal of this study to examine the effectiveness of addressing common-pool problems, it cannot make claims regarding the direct effects of property rights and the indirect effects of regulation on the extent to which there is deforestation, forest degradation or overfishing. These common-pool problems are caused by so many conjunctive factors that the empirical studies on which this analysis is based cannot isolate the impact that property rights and regulation exert on these problems. For example, natural fluctuations and border crossing human activities can impact deforestation, forest degradation, and overfishing at the same time as property rights and regulation do. However, what this study can and will try to answer is the question in what ways the preconditions are met that can induce the compliance and behavioural change that are necessary in order to be able to prevent the overexploitation of common-pool resources. More specifically this study focuses on the five following preconditions: the clarity with which property rights are defined, the monitoring and enforcement of property rights, the collection and sharing of information on which basis property rights can be based, the coordination of the actions of the stakeholders, and the scaling of the management of property rights. In the next chapter the choice for these five preconditions will be accounted for.

1.3 Methodology

An interdisciplinary approach is used in this study. Economic analysis is followed both in the theoretical set-up and in evaluating the practice in governing forest and fishery resources. Traditional economic analysis of property law has examined when and how different types of property rights are suitable.[14] In the empirical part, this research will evaluate whether and

12 Huppert 2005.

13 For example, by December 2015, the two largest certification programmes, the Forest Stewardship Council (FSC) and the Programme for the Endorsement of Forest Certification (PEFC) have certified 188 million hectares and 268 million hectares respectively. See https://ic.fsc.org/en/facts-figures; www.pefc.org/about-pefc/who-we-are. The Marine Stewardship Council (MSC) covers ten per cent of the global fish products. See www.msc.org/business-support/key-facts-about-msc.

14 E.g. Alchian 1965, pp. 816–829; Alchian 1979, pp. 233–252; Demsetz 1967, pp. 347–359; Bouckaert 2010.

how public and private regulation have contributed to the realization or jeopardizing of the necessary preconditions for the effective functioning of different types of property rights.

The governance literature is also essential for this research. The emergence and rapid prevalence of private and hybrid forms of regulation are comparatively new phenomena. Governance literature broadens the traditional understanding of the forms of regulation and helps to explain how and why private regulation emerges. Recent literature starts to analyse how public and private regulation interact, via the architecture of regulatory regimes. The various approaches in analyzing interaction inspire our analysis of public and private interaction in overcoming the limitations of property rights in addressing forest and fishery problems.

Another equally important approach is the doctrinal legal research. Property rights are a product of legal rules. They include a bundle of rights which 'describe what people may and may not do with the resources they own'.[15] In essence, property rights entails the rights to exclude all others from the use of these resources.[16] Legal texts and case law provide the basis to understand the different types of property rights used in different jurisdictions.

In the empirical part of forest and fishery governance, nine purposely selected countries will be analysed as case studies. These case studies will show in what respects public and private regulation satisfy the necessary preconditions for property rights to be able to effectively address common-pool resources problems.

1.4 Structure

After this introduction, Chapter 2 will first present the types of property rights which can be used for common-pool resources and the preconditions under which they can function effectively. This will be followed by an analysis of the interaction of public and private regulation, as well as of the influence of this interaction on the preconditions of property rights' proper functioning. After the theoretical framework the selection of the case studies will be accounted for.

The theoretical framework will be applied to forest governance in Chapter 3. Forests in different countries suffer from deforestation or forest degradation. In order to address these problems, in addition to public regulation, many private regulatory regimes have emerged and have spread rapidly in the last few decades, especially forest certification schemes. In this chapter, public, private and communal forests will be defined according to the ownership of forests or the land on which forests grow. Case studies of the following five countries will be presented:

15 Cooter & Ulen 2010, p. 73.
16 Blomley 2016, pp. 226–227.

Indonesia, Bolivia, Canada, the United States, and Sweden. A comparison between them will be conducted to show how the interaction of public regulation and private certification influences the preconditions determining property rights' contribution to manage forest degradation or deforestation in different national contexts and how it affects property rights limitations.

Chapter 4 will examine how property rights function and ways to overcome their limitations by the interaction of public regulation and private certification affect overfishing. Three main types of fishing rights will be discussed: limited entry, quotas, especially individual transferable quotas (ITQs), and territorial user rights for fisheries (TURFs). As explained later, these fishing rights subsequently show similarity with public, private, and communal property rights. Case studies of the following four countries will be presented: Mexico, South Africa, New Zealand and Japan. The comparison will show the role of public and private regulation to address overfishing under different property rights regimes.

Chapter 5 compares forest and fishery governance by following the theoretical structure discussed below. It first compares the individual elements: property rights, public regulation and private regulation concerning forestry and fishery. Then the similarities and differences of the interactions between public and private regulation concerning forest and fishery are explored. How such interactions influence the functioning of property rights is also examined. After the comparison of individual elements, this part provides an integrated picture of the case studies and tries to understand the situated patterns of the different cases. Based on this comparison, this chapter tries to provide some policy implications.

Chapter 6 formulates general conclusions based on the theoretical framework and the case studies. Subsequently, it outlines avenues for further research examining smart mixes between legal and policy instruments in the fight against transboundary harm.

2 Common-pool resources, property rights and public and private regulation

2.1 Different property rights and institutional arrangements

As discussed earlier, common-pool resources have two important characteristics: subtractability and non-excludability. For such resources, a mismatch between the benefits and the costs of consuming the commons can occur: resource users capture the full benefits and share the costs with others.[1] Common-pool resources such as forests or the oceans are so-called open-access resources in the sense that in the absence of any legal rule or social norm anyone could in theory take for example the fish in the ocean or cut the trees in a forest.[2] Multiple resource users may compete with each other to harvest the resources disregarding their ecological limits. There is therefore a great likelihood that overharvesting or overfishing will take place leading to forest depletion[3] or to an extinction of species.[4] This phenomenon that an open-access resource will in the end lead to a depletion of resources has been described in the seminal paper by Garrett Hardin as 'the tragedy of the commons'.[5] It results from the fact that no one is excluded from the use of the particular resource. Private ownership is de facto simply established by the person who takes first access, for example of the fish in the ocean or the tree in the forest. This first mover advantage consequently leads to a situation where competing users will have an incentive to engage in a race whereby all try to harvest as much as possible as quickly as possible from the resource in order to avoid others doing so. Without rules aiming at the regulation of the common-pool resources extinction may be the result. The goal of the creation of property rights is to provide exclusive control and hence to internalize the externalities that follow from open access. Various property rights can be used to internalize the externalities,[6] including public, private and communal property rights.[7]

1 Hardin 1968; Alexander & Peñalver 2012.
2 Cooter & Ulen 2008, p. 154.
3 Ibid.
4 Faure & Skogh 2003, p. 40.
5 Hardin 1968.
6 Demsetz 1967, p. 347.
7 Feeny *et al.* 1990, p. 1.

These different types of property rights are often related to different institutions: the state, the market and self-governing communities.[8] Traditional literature regards external interventions as necessary to help resource users to get out of the trap of collective action by resorting to either the state or the market.[9] The first approach is often associated with claiming state ownership over resources and relying on state regulation to decide on how to harvest and manage resources. The second approach is to establish private property rights over common-pool resources and hence to mimic the market.[10] In addition to these two institutions, later literature indicates that sometimes user groups manage to get themselves out of the collective action trap via self-governing communities.[11] In this case, property rights are vested in a group of resource users rather than in individuals.

These property rights and institutions are not established without costs. Property rights need to be defined; properties need to be policed and when disputes arise, dispute settlement mechanisms are necessary. These various activities associated with the establishment and maintenance of property rights create transaction costs.[12] Therefore, it is not always desirable to establish property rights for common-pool resources. If the resource is uncongested and the transaction costs for establishing property rights are high, it is desirable to leave the resource in open access.[13] When congestion increases as time passes, or technological innovations reduce transaction costs, such as boundary maintenance costs, establishing property rights becomes more desirable. This also implies, the best ways to balance environmental protection and minimalize transaction costs are not fixed but dynamic and evolving.

Transaction cost is a concept that is often vaguely defined or only illustrated with examples.[14] The property rights literature usually uses it in a broad sense to incorporate 'costs associated with the transfer, capture and protection of rights'[15] or 'the resources used to establish and maintain property rights'.[16] Cole differentiates transaction costs into two types: exclusion costs and coordination costs. 'Exclusion costs are the costs of drawing and enforcing boundaries to restrict access to and use of the resource to the owner(s) of the property'.[17] Coordination costs are the costs

8 Ostrom 1990; Ostrom 2010; Imperial & Yandle 2005, pp. 493–509.
9 The famous public choice scholar, James Buchanan, referred to the state as a leviathan. See Buchanan 1975.
10 Ostrom 1990, pp. 8–13.
11 Ibid., pp.13–21.
12 Allen 2000, pp. 898–899; Allen 1991, pp. 1–18.
13 Cooter & Ulen 2010, p. 148.
14 Allen 2000, pp. 898–899. For such examples, see Barzel 1985, p. 8; Alchian & Woodward 1988, p. 66.
15 Barzel 1989, p. 122.
16 Allen 1991.
17 Cole 2002, p. 131.

associated with solving collective action problems.[18] To reduce exclusion costs hence requires a clear definition of property rights and the capacity to enforce the rights, including policing, dispute settlement and sanctioning. For private property, decision-making power is vested in individuals. Sometimes, however, the individual owners need to negotiate with external parties in using and managing resources. Multiple individual fishery quota holders using the same fishery are such an example. For both communal and public property, collective decision-making institutions are established and coordination takes place inside these institutions. These three types of transaction costs (definition, enforcement and coordination) are connected to two other factors: information and scale. Collecting and distributing information is crucial in both exclusion and coordination activities. Scale concerns the level to which property rights are managed and possibly the links between the units that may manage property rights at different levels. The scale of governance can influence how property rights are enforced and also how rights holders coordinate with other stakeholders. These five factors set the necessary preconditions for property rights to be able to effectively address common-pool resources problems: clear definition of property rights, enforcement, coordination, information creation and sharing, and scaling. In fact, these elements which are referred to in the literature as transaction costs will be used in this study as the preconditions that need to be fulfilled for the proper functioning of property rights. Proper functioning, or effectiveness, is used in a broad sense in this study. It can either refer to compliance, actions that aim to reduce common-pool resources problems or the actual reduction of common-pool resources problems. The five preconditions will be the crucial elements to be used in our case studies to analyse the way in which public and private regulation affect the functioning of property rights.

Those preconditions are determined by diverse factors, including the nature of the resources, ecological, technological and institutional circumstances, and even culture and ideology.[19] This research focuses on the influence of institutional factors on the preconditions for a proper functioning of property rights, more particularly on public and private regulation. In other words, this research will examine to what extent public and private regulation interact in facilitating the necessary preconditions for effectively addressing common-pool resources problems by clearly defining property rights, enhancing enforcement capacity, smoothing coordination, generating and sharing information and operating the right scale of governance.

The establishment of property rights may contribute to solving common-pool resources problems. However, all types of property rights

18 Ibid.
19 Ibid., p. 135.

have their limitations in internalizing externalities. Public and private regulation can aim to overcome these limitations. The following section discusses the preconditions for a proper functioning of property rights and the limitations of each type of property rights respectively.

2.1.1 Private property rights and the market

To establish private property rights and to mimic the market is one approach to governing common-pool resources. Private property 'typically denotes property owned by individuals holding rights to use (in socially acceptable, i.e., non-nuisance-causing ways), dispose of, and exclude others from resources'.[20] With the clear definition of the resource boundaries and its right holders, private property rights attempt to align the cost benefit analysis of the resource users with that of society, hence to overcome the externalities caused by individual decision-making.[21] Individual property rights determine which individuals can decide how to access, harvest and manage the resources. They also establish who can patrol their resources, negotiate and settle disputes with outsiders and the way for individuals to resort to justice if private settlement fails. Private enforcement can create costs for both the rights holders and the public (via the court system).

Although individuals have some decision-making power concerning their resources, activities of externals also influence the status of the resource. For example, a fishery is often exploited by multiple individual quota holders. Although each fisherman has his own (quasi) property rights, the interaction of individuals influences the health of the overall fish stocks. The protection of one piece of privately owned land may also be influenced by a public road, or visitor services nearby.[22] In such cases, substantial coordination costs exist if the individuals involved want to protect their resources.

To enforce their rights and to coordinate with external parties, individual rights holders need to possess sufficient information. On the one hand, the rights holders may be close to the resources and hence have intimate knowledge about the resources they are using. On the other hand, individuals may have limited capacity in producing and processing information about the bigger picture.

Private property rights can be established on both large and small-scale resources. Establishing private property rights on small-scale resources may run the risk of fragmenting larger resources and misaligning the cost benefit analyses of private actors. Larger scale resources may, on the other hand, create challenges in terms of coordination costs.[23]

20 Cole 2010, p. 229.
21 Demsetz 1967, pp. 348–349.
22 Cole 2002, p. 147.
23 Cole 2002, pp. 150–151.

The creation of private property rights can hence be an important tool to remedy the risk of depletion which would occur in an open-access regime. However, even if private property rights have been clearly established, market failures may still occur. For example, although private property rights are created for common-pool resources to internalize externalities, not all benefits and costs of the resource are automatically internalized. The protection of forests provides such a case in point. When private property rights have been established over forests and only the market value of the timber is internalized, other non-market values, such as the ecological and aesthetic value of the forest, are not considered by the self-interested private actors. In other words, a private actor may still use his private property in a way which creates externalities for third parties. This results in market failure which may therefore justify external intervention, including both public and private regulation.

2.1.2 Public property rights and the state

Public property rights may also be chosen to solve common-pool resources problems. Though theoretically public property is owned by all citizens, it is usually the state that exercises the rights on their behalf. The state decides how resources are accessed, harvested and managed. The decisions that the state makes concerning the protection of resources usually take place via public regulation, such as prohibiting the commercial use of forests without permits, requiring the establishment of riparian buffer zones, mandating biodiversity protection during forest harvest and management. 'Such state regulations may be characterized as exercises in sovereignty (imperium) rather than ownership (dominium). But this makes little practical difference. Whether the state is purporting to act as sovereign or owner, the rights it asserts are in the nature of property'.[24]

The enforcement of public property rights relies on public agencies. For example, the competent authority needs to decide on daily operational rules to manage their resources, to patrol the public forests or fisheries to exclude poachers and to issue sanctions in cases of non-compliance. Such enforcement may cause substantial costs and create challenges for many countries with limited governmental capacity but large areas of public resources to protect.

Designating the state as the owner and regulator of public property does not imply that the state is a unitary body. Resource management is usually an outcome of collective decision-making. Unlike private property, public property can keep the costs of negotiating and coordinating with external agents low. However, internal coordination costs can be high, including the costs of public administration and collective bargaining and decision-making.[25]

24 Cole 2010, p. 236.
25 Cooter & Ulen 2010, p. 104.

Again, enforcement and exclusion are not possible without information. On the one hand, compared to private property, the state can generate a high economy of scale in generating and spreading information. On the other hand, however, centralized information gathering can also be costly because of the large distance between daily management and central decision-makers. Therefore, whether public or private property is preferable in terms of information costs is an empirical question, depending on the balance between the advantages and disadvantages of both options in specific situations.

Public property rights can also be established over resources with various economies of scale. Economic analysis has especially propagated the advantages of public property rights over large-scale resources, such as protecting larger species in larger habitats, due to lower coordination costs compared to private property rights.[26]

Resorting to public property rights assumes that the state will manage the resources and regulate behaviour on behalf of the public interest.[27] However, the actual decision-makers and enforcers are politicians and bureaucrats. 'Like private owners, the politicians and bureaucrats who are de facto owners of de jure "public" property presumably manage resources so as to maximize their self-interests'.[28] Politicians face re-election in a cycle of several years. They seldom have the long-term view to go beyond their term of office. Therefore, they may consider the short-term market value of the natural resources but not the long-term benefits from resources conservation.[29] Bureaucrats, although they do not face re-election, may 'maximize budget allocations and administrative turf' rather than maximizing social welfare.[30] In addition, although public officials may want to serve the public interest, they do not invest in the resources personally and hence do not bear the full costs of their management.[31] Therefore, the cost benefit calculation of politicians and bureaucrats may not coincide with that of the general public. Moreover, the politicians and bureaucrats may be influenced by lobby groups. Often, business is well-organized and exerts more influence than countervailing interest groups such as NGOs or the public.[32] In summary, like market failures which may prevent private property rights from executing their full potential to achieve environmental goals, public property rights may also suffer governmental failures. In order to overcome these limitations, public and private regulation may be warranted.

26 Lueck 1989, p. 291.
27 Prakash & Potoski 2007.
28 Cole 2010, p. 252; see also Anderson & Leal 1991; Stroup & Baden 1983.
29 Cole 2010, p. 253; Stroup & Baden 1983; Stroup & Goodman 1992.
30 Cole 2010, p. 253; Stroup & Baden 1983; Orzechowski 1977.
31 Anderson & Leal 1991.
32 Buchanan & Tollison 1984; Mueller 2004; Olson 1973; Olson 1986; Revesz 2001.

2.1.3 Communal property rights and self-governing institutions

The adoption of private property rights and public property rights assumes that resource users cannot get out of the collective action dilemma themselves and that external interventions are required, either via the market or the state. Literature, however, has identified a third approach, where property rights are vested in a group of resource users rather than in individuals or the state. The group of resource users can be self-organized to set up resource use rules as well as to monitor and to enforce the rules themselves.[33] After examining the characteristics of many successful self-governing institutions, Ostrom summarizes eight design principles.[34] These principles are in line with the above-mentioned preconditions of property rights' effectiveness.

The first two principles discussed by Ostrom concern the definition of property rights. They require establishing clear boundaries of resources and users, and suitable appropriation and provision rules, which define the content of the property rights. Principle three, the collective-choice arrangement, influences the coordination costs of the communal property. The communal property rights are imposed on a group of resource users and the decisions regarding resource use and management are usually made via self-governing institutions. Providing broad participation rights to resource users in the community is crucial to reduce coordination costs. Otherwise, the marginalized actors may have no incentives to comply with the rules. Principles four to six concern the enforcement of property rights. Enforcement relies on the self-governing community and the resource users themselves. The users are close to the resources and can monitor the resources' status and each other's performances. In case of violation, communities can impose graduated sanctions in proportion to the extent of the violation. Local conflict-settlement mechanisms can also provide cheap and low threshold solutions for the resource users to solve their disputes. The last two principles are linked to the institutional embedding of communal property rights: these rights need to be recognized under domestic law and a larger social-ecological system needs to support the usually small-scale communal property system.

As with the two other types of property rights, collecting and sharing information and scaling the management of community property rights are also important. Under communal property rights, resource users can participate in the decision-making, enforcement of and benefiting from the resource use. This creates incentives for them to collect and share information related to the status and management of the resource.[35]

33 Ostrom 1990, 2010.
34 Ostrom 2010, p. 422.
35 Wilen, Cancino & Uchida 2012.

Literature has shown that communal property rights are most suitable for managing resources on a small scale. To scale up small governance regimes several institutional adjustments need to be made. 'It is not just a matter of nesting local arrangements within a supportive institutional context at the national and international levels. Equally important is nesting political institutions within supportive and complementary institutions from other sectors'.[36]

In comparison to individual property rights, the rights granted to a community can be more comprehensive. Communal property rights are often meant to safeguard a traditional lifestyle of a particular community in a geographically defined area rather than to allocate a particular resource to a community. For example, when fishing rights are vested in individuals, usually multiple individuals hold the rights to fish for specific species in the same fishery. One type of fishing rights commonly used is individual quotas for a certain species. Such a scheme creates incentives for fishermen to protect the targeted fish stocks but does not provide incentives to protect by-catch.[37] When one type of communal property rights, namely territorial user rights, is established, the rights to fish for resources in specific areas is granted to a group of resource users. In this situation, the externality concerning by-catch can be internalized, since the value of by-catch becomes internal to the users' group.[38] However, some externalities may still not be internalized. For example, the influence of fishing activities on the biodiversity and benthic habitats may still be beyond the consideration of self-interested user groups.

Different from a public property rights regime, the decision-makers under communal property rights are resource users themselves, rather than politicians and bureaucrats. Hence, on the one hand, they can have more intimate knowledge of the resources and their users' behaviour and they bear the costs and benefits of resource use directly. As such, the users of communal property may have better incentives to protect common-pool resources. On the other hand, local elites may have more influence on the self-governing institutions than other resource users. Such elites may act like politicians and bureaucrats, trying to maximize their own interests rather than the welfare of the whole resource user group.[39]

2.2 Refinements

The trichotomy of property rights and accompanying institutional arrangements – market, state, community – is a simplification of a complicated

36 McGinnis & Ostrom 2008, pp. 189–211.
37 Wilen, Cancino & Uchida 2012.
38 Ibid.
39 For example, research on community-managed forests in Indonesia shows that local elites have manipulated the decision-making process and have failed to reach a desirable outcome for the whole group. See Urano 2014.

practice. Property concerns the right to exclude all others from an asset. Ostrom distinguishes five components: the rights of access, withdrawal, management, exclusion and alienation.[40] These different elements can be granted to the same actor or different ones, leading to different types of property rights. Although the differentiation of public, private and communal property rights is a useful heuristic, it cannot accommodate hybrid forms of property rights. Taking into consideration that ownership and less complete types of property rights can be established separately on the same resource, some overlap may exist between the different categories. For example, many countries claim state ownership over forests but allow private actors to harvest and manage a certain unit of the forests via permits or concessions. Such property rights possess mixed characteristics of public and private property rights. As Cole argues, 'Most property regimes governing environmental goods are admixtures of individual private ownership, private (non-state) common property management, state ownership and management (i.e., regulation)'.[41]

Similarly, a firm distinction between the three types of institutional arrangements does not exist. Though private property and communal property are usually linked with the institutional arrangements of the market and the community, public regulation also plays a role in these two types of property rights. Public regulation can alter the 'actual or presumed property rights among independent agents'.[42] For example, public regulation can confirm private ownership of forests via registration or (quasi-)private property rights over state-owned forests via concessions. The existence of traditional community rights over forests requires domestic law to recognize traditional rights and not to jeopardize traditional self-governing institutions. The existence of a competitive market for private property also needs the support of public regulation.[43]

The above analysis implies that the trichotomy of institutional arrangements cannot accommodate all varieties of property rights adopted in practice and the actual property rights are often mixtures of the different institutions. However, no matter which categories the actual property rights fall into, in order for them to be effective the necessary preconditions need to be fulfilled and potential market or governmental failures need to be overcome. Clear definitions, sufficient enforcement capacity and coordination, collection and distribution of information and establishing the right scale of governance are also necessary preconditions for hybrid property rights' effectiveness.

Our rudimentary model concerning the design of property rights does not suggest that a best solution to overcome common-pool resources

40 Schlager & Ostrom 1992; Cole 2010, p. 231; see also Honoré 1961, pp. 107–147.

41 Cole 2010, p. 235.

42 Bromley 1991.

43 Ostrom 1990, p. 15.

problems can always be established. The above discussion shows that the choice of property rights involves a comparison of the various transaction costs between different alternatives. However, an accurate estimation of the various elements of transaction costs is extremely difficult, especially ex ante.[44] These elements are dynamic and vary with economic, ecological, technological and institutional settings.[45] For example, technological innovations can reduce exclusion costs, or, conversely, can increase over-exploitation when technological innovations render common-pool resources more accessible. Such complexities may challenge the validity of theory-based solutions. Moreover, even if effective solutions can be established, property rights and their corresponding institutions are socially embedded. Although findings may suggest that alternative property rights and institutions are more effective than current ones, making changes may be costly. The benefits of shifting to more effective property rights and institutions therefore need to be balanced with the costs incurred by institutional changes.[46] The above-mentioned complexity, however, does not make a comparative analysis useless. It simply means one has to be cautious in drawing conclusions and formulating policy implications.

2.3 The interaction between public and private regulation and their influence on the functioning of property rights

As discussed earlier public and private regulation can influence the preconditions of property rights' effectiveness. The de jure property rights need to be recognized by law and the state can also formalize, invalidate or weaken the de facto property rights. The prevalence of private regulation can also influence the transaction costs of different property rights. Moreover, public and private regulation have the potential to address the limitations of different property rights. For example, the existence of market failures can result in external interventions, including both public and private regulation. Governmental failures may prevent public property rights from achieving an optimal outcome. In this situation, private regulation may complement or substitute public regulation. Before exploring the influence of public and private regulation on property rights and their limitations in addressing common-pool resources problems, it is worth noting that public regulation and private certification often interact, rather than operate separately. Therefore, before discussing the joint effect of public and private regulation on property rights of common-pool resources, this research first examines how public regulation may interact with private regulation.

44 Cole 2002, p. 136.
45 Ibid., p. 135.
46 Ostrom 1990, Chapter 4.

2.3.1 The interaction between public and private regulation

Traditionally, the state plays a central role in social control. Through regulatory instruments such as zoning and spatial planning the state will decide which activities are allowed in specific areas. These instruments often constitute the basis for the exercise of property rights and are hence strongly linked with property rights. Those decisions are largely taken via public regulation. However, the emerging private regulation transfers 'power and authority from state to international and local level' and 'from public to private'.[47] The new forms of regulation, rather than acting as an alternative to public regulation and as 'the hollowing out of the state',[48] are increasingly recognized as closely linked with public regulation.[49]

The literature on institutions and environmental governance regimes has tried to categorize the types of interaction. According to interaction effects, Eberlein *et al.* categorize the relationship between public and private regulation as competition, coordination, cooptation and chaos.[50] In this research, the interaction between public and private regulation is examined with the goal of understanding their joint effect on various property rights.

Public and private regulation is often analysed in terms of the architecture of the regulatory regimes and the actors involved. Less attention, however, has been paid to the instruments used under different regulatory regimes. The literature on environmental instrument mixes is still piecemeal and mainly theoretical. However, instruments are an important component of rules and hence of institutions or regimes. They provide the linkage between the policy-makers and the targeted group, showing how policy-maker can influence the behaviour of the targeted group and further, to achieve the environmental goals. Public and private regulation is not possible 'without the use of a diverse set of policy instruments'.[51] Breaking down public and private regulation into specific instruments helps to understand how they interact, and further how such interaction influences property rights.[52] Therefore, this research focuses on instrument interaction. Instruments can either be relevant for the establishment of property rights (e.g. forest concessions) or aim at solving the limitations of property rights (e.g. performance standards for fishermen to protect

47 Scott, Cafaggi & Senden 2011, p. 4.
48 Jessop 1994; Peters 1994, p. 739; Rhodes 2005, p. 138.
49 Bartley 2011, p. 517; Cafaggi & Renda 2012; Scott, Cafaggi & Senden 2011.
50 Eberlein *et al.* 2014, pp. 11–12.
51 Zehavi 2012, p. 242.
52 For example, to understand whether private forest certification schemes complement, replace or contradict public regulation in a specific jurisdiction, one needs to first analyse the various instruments used under the domestic public regulation and certification schemes, such as forest tenure rights, zoning plan, biodiversity standards, logging permits and quota and so on.

benthic habitats). In case other interactions influence property rights, they will be incorporated as well. For example, governmental authorities and certifying bodies can coordinate their activities, share information, and reduce redundancies in auditing.

2.3.2 *The influence of public private interaction on the functioning of various property rights*

As discussed earlier, public and private regulation can have two important functions in relation to the different property rights.

First, public and private regulation can influence the preconditions of property rights' proper functioning. Hence we will address the question whether or not public and private regulation contribute to a clear definition of property rights, enhance enforcement capacity, promote coordination, induce the collection and distribution of information and address the right scale of governance. Second, public and private regulation can address the limitations of various property rights. For example, both public and private regulation can intervene in terms of market failures by commanding self-interested private actors to protect biodiversity and to establish riparian buffer zones. When governmental failures are present, private certification may also provide a complement or substitute for public regulation.

However, to a large extent the ability of a governance regime to remedy these externalities will also depend upon what can generally be referred to as 'good governance'. The importance of good governance in sustainable development has been largely stressed in the literature.[53] It is more particularly the importance of institutions that guarantee good governance that has been stressed.[54] The good governance concept has broadly been used also as a 'meta-concept' within the framework of globalizing administrative law in order to benchmark the effectiveness of particular institutions in (developing) countries.[55] Importantly for this topic is that one aspect of good governance is precisely the presence of well-defined property rights.[56] Therefore, the interaction between public and private regulation in relation to particular property rights can, within the context of a specific legal system, not be studied in isolation but will have to be related to particular criteria of good governance.

We will provide indications of good governance in the countries we have selected for case studies. Some general information with respect to the annual GDP of the country as well as a few quantitative indicators measuring the governance will also be provided. The World Bank has the

53 See, for example, Jordan 2008.

54 Chandra & Tisdell 1998 and Griendle 2007.

55 See more particularly Esty 2006 and Nanda 2006.

56 Prasad 2003.

worldwide governance indicators (WGI) project which has specific indicators of good governance. Countries have been ranked according to that WGI. For every country we discuss we will mention its scoring on the WGIs. Both in the WGI and Transparency International, an index on corruption is provided. The scoring on corruption will also be mentioned for every country. We do realize that these are just quantitative indicators which obviously have their limits. However, we believe that these already provide some context and background and some rough indication of the governance system in the particular country.

It is important to stress that the three crucial elements for our case studies (property rights, interaction between public and private governance and good governance) cannot be seen in isolation. It is precisely for that reason that in the case studies this is addressed as an indication of good governance on the basis of established criteria. That will better allow a holistic approach whereby the relationship between the three central elements of our study can be analysed.

2.4 Case selection

We have chosen case studies as our methodology in order to be able to accommodate the influences of different national contexts. A logical consequence of this choice, as far as the public regulatory system is concerned, is that this research will focus on domestic law. International law is only briefly sketched for the understanding of emerging private regimes (due to the lack of international law or because of the need to complement international law).

To understand how the interaction between public and private regulation as well as their influence on property rights play out in practice, nine countries have purposely been selected for case studies. The criteria for choosing these countries include: the prominence of the targeted environmental problem, the stages of development, the types of property rights adopted in these countries as well as the existence of comparatively well, institutionalized private regulatory regimes. This led us to the selection of some countries where private property played an important role (such as in the US) whereas in other countries public property was more prominent (for example in Canada). By also choosing some countries from the developing world (Indonesia and Bolivia) we could see how particularly in those countries the interaction between public and private regulation with the property rights regimes worked differently from that in developed countries like the US, Canada and Sweden. Moreover, by selecting both developed and developing countries which equally score differently as far as good governance indicators are concerned, we also had a broad sample of differing countries, thus allowing us to obtain some indication concerning the importance of good governance in relation to the establishment of property rights and the interaction with public and private regulation. The selection of the countries was hence determined by the relevant questions identified in the theoretical framework.

For each country study, the main forest or fishery problems are briefly introduced, followed by an overview of indicators for good governance and an introduction of the property rights and governance systems addressing such problems. Then the interaction between public and private regulation (as well as the instruments used in them) is examined. Individual case studies also examine whether in practice public and private regulation for forest and fishery interact in a way to fulfil the necessary preconditions for the proper functioning of property rights and to overcome their limitations in addressing common-pool resources problems. After individual case studies, the similarities and differences of the interacted systems across countries will be analysed through comparison.

The analysis may equally provide some indications on particular strengths and weaknesses in the property rights regimes in particular countries, more particularly in relation to the interaction with public and private regulation. This may allow for some recommendations. However, a warning is in place in that respect. Our case study approach does not claim any causal relationship between the structure of property rights, the interaction of public and private regulation and the level of common-pool resources problems. Our case study methodology does not allow the determination of such causal effects, though we will describe the interplay between different factors in national contexts. Moreover, in particular cases we may conclude that it is not the interaction between public and private regulation and the specific property rights per se that affect the effectiveness of addressing common-pool resources problems, but rather a lack of good governance. Given the importance of good governance we do provide indicators of it in the specific case studies, but a full analysis of the quality of governance in particular countries is clearly beyond the scope of our study. For the same reason we cannot engage in recommendations with respect to the improvement of good governance in particular countries.

3 Forestry

3.1 Introduction: forest problems and forest transition

Forest coverage is estimated to be over 4 billion hectares worldwide by 2010, composing 31 per cent of the total land area.[1] 53 per cent of the total forest area is located in five of the most forest-rich countries: Russia, Brazil, Canada, US and China.[2] By region, Russia and Northern Europe, North America, the Amazon, Congo basin and Southeast Asia have the richest forest coverage.[3] Of all the forest area, 36 per cent is primary forest. Planted forest has been increasing in recent years, which accounts for 7 per cent of the forest area. The rest is naturally regenerated forest.[4]

Forest plays vital and diversified roles in supporting ecosystems and human welfare, including providing supporting services (for species), provisioning services (the production of goods), regulating services (controlling climate, water and soil conservation) and cultural functions.[5] For example, forests had 289 gigatonnes of carbon stock in their biomass by 2010.[6] It also provides 'a habitat for nearly 90 per cent of the world's terrestrial biodiversity'.[7]

Forest problems have caused global concerns. It is estimated that forest conversion and loss continue at an alarmingly high rate: around 13 million hectares of forest disappear every year. Even when considering the large-scale planting in recent decades, the net forest loss is still around 5.2 million hectares per year.[8] South America and Africa mark the highest deforestation rates, at 4 and 3.4 million hectares annually from 2000 to 2010.[9] At national level, the largest annual net loss can be found in Brazil,

1 FAO 2010, p. 10.
2 Ibid.
3 Ibid., p. 12.
4 Ibid., p. xviii.
5 Millennium Ecosystem Assessment 2005.
6 FAO 2010, p. xviii.
7 Gupta, Van der Grijp & Kuik 2012, p. 7.
8 FAO 2010, p. xiii.
9 Ibid., p. xvi.

Australia, Indonesia, Nigeria, Tanzania, Zimbabwe, the Democratic Republic of the Congo, Myanmar, Bolivia and Venezuela.[10] The deforestation caused the decrease of around 0.5 Gt carbon stocks in forest biomass annually during 2005–2010.[11] Primary forests, which are crucial in maintaining biodiversity, are decreasing at a rate of 0.4 per cent annually.[12]

Although deforestation and forest degradation are global problems, different regions and countries face different types and extents of problems. Both problems are more prominent in developing countries, especially in countries with rich tropical forests. Forest coverage has been expanding in Europe and has been stable in North America.[13] One exception is Australia, which has experienced large losses of forests due to severe drought and forest fires between 2000–2010.[14] In addition, forest degradation is still a concern for developed countries. For example, outbreaks of forest pests mainly influence temperate and boreal zones of which countries such as Canada and the US are part.[15] The health of riparian forests is also a concern in North America.[16] Biodiversity is an important concern for Europe, where very few primary forests have remained except in Russia.[17]

The literature has identified a forest transition model which 'describes empirical patterns of change in forest cover resulting from the exploitation by human society of their forest resources'.[18] This stage model distinguishes four development patterns, using the percentage of forest cover and the forest change rate as the core criteria.[19] In the virgin forests stage (pre-transition), forest is relatively undisturbed by human intervention and is inaccessible for commercial uses. In this stage, countries have comparatively high forest cover and low deforestation rates. Economic development can lead to a frontier stage (early-transition) with an increasingly rapid forest change rate and shrinking forest cover. A continuing high deforestation rate leaves a small fraction of forests, hence slows down the deforestation and brings a forest and agriculture mosaics stage (later-transition). Finally, a forest stabilization stage (post-transition) can arrive in which forest coverage begins to increase again.[20] The literature has recorded these transition stages in North America, Europe and some tropical countries.[21] However, whether these transition stages materialize

10 Ibid., p. 21.
11 Ibid., p. xvii.
12 Ibid., p. xviii.
13 Ibid., p. xvi.
14 Ibid.
15 Ibid., p. xx.
16 McDermott, Cashore & Kanowski 2010, p. 95; Sweeney *et al.* 2004, p. 14132; Verry & Dolloff 2000.
17 Paillet *et al.* 2010, p. 101.
18 Gupta, Van der Grijp & Kuik 2012, p. 25; Angelsen & Rudel 2013, p. 91.
19 Hosonuma *et al.* 2012, p. 1.
20 Ibid., p. 3; Gupta, Van der Grijp & Kuik 2012, p. 26.
21 Rudel *et al.* 2005.

depends on 'the interplay of the contextual circumstances, the driving factors and the policy regime in specific localities, as well as on other transitions occurring in society'.[22] Therefore, in order to be able to predict the future forest status of a country, one needs to consider which drivers and governance regime are present in that particular country. After all, countries in different development stages of the forest transition model have different drivers for deforestation and forest degradation which call for different policy responses.[23]

The forest transition model suggests that countries in the frontier and mosaics stages mainly suffer from deforestation. When a country enters the stabilization stage, deforestation ceases to be a major concern. However, these countries may still suffer from forest degradation. Deforestation and forest degradation are hence two different types of problems, which also require different governance systems. Property rights can be used to limit the access to forests and to create incentives for preventing deforestation. Moreover, public and private regulation can create the necessary preconditions for the proper functioning of different types of property rights. However, even if property rights are well-established and deforestation is solved, forest degradation may still occur. Many forest services, such as supporting riparian functions and protecting biodiversity, have non-market values. Establishing property rights does not guarantee that these non-market values are internalized. Therefore, public and private regulation are called upon to overcome such limitations of property rights in addressing forest degradation. This chapter will examine the following two questions: first, in what respect do five different countries (Indonesia, Bolivia, Canada, the US, and Sweden) meet or not meet the necessary preconditions for the proper functioning of different types of property rights in countering deforestation? Second, how do public and private regulation interact in these countries in establishing these property rights as well as in overcoming their limitations in countering forest degradation?

The next section gives a historical overview of the international law, domestic law, and private regulatory regimes that have been developed to address deforestation and forest degradation. Section three describes how regulatory regimes addressing both forestry problems have evolved in the five countries. After accounting for the selection of these five countries for case studies, each case study starts with a description of the most important problems concerning forestry. Subsequently, the case studies describe how property rights have been established in a particular country and what additional public and private regulations have been implemented to counter deforestation and forest degradation. After describing the

22 Gupta, Van der Grijp & Kuik 2012, p. 25. In the other words, the forest transition model is a heuristic device rather than a 'deterministic prediction'. Angelsen & Rudel 2013, p. 91.

23 Gupta, Van der Grijp & Kuik 2012.

evolution and functioning of the regulatory mixes, the case studies evaluate the presence or absence of the necessary preconditions for property rights' effectiveness in the regulation of forestry – establishing property rights, enforcing regulations, coordinating actions, collecting and distributing information, scaling governance – and whether or not the limitations of property rights in handling forest degradation have been overcome by public and private regulation.

Although basically the same structure is followed in each case study, the focus of the analysis varies given the particular problems that arise with forests in the different countries. For instance, when discussing the interaction between public and private regulation, the focus is on deforestation in Indonesia and Bolivia. The same part in Canada, the US and Sweden, however, focuses on whether the rules concerning riparian zone protection and biodiversity protection are compatible under public and private regulation in order to prevent forest degradation. Moreover, since the public and private regulation of property rights is no longer a major issue in Canada, the US and Sweden, but is a major issue in Indonesia and Bolivia, the case studies of the former countries will focus on overcoming the limitations of property rights in protecting forests, while the case studies of Indonesia and Bolivia will also address the role of public regulation and certification in satisfying the necessary preconditions for the proper functioning of property rights in protecting forests.

Each case study ends with a summary of the main characteristics of the regulatory mixes in that particular country. The chapter ends with a comparison of the five case studies in terms of the satisfaction of the necessary preconditions for property rights' effective functioning in managing forests.

3.2 Forest governance: international law, domestic law and private/hybrid regimes

3.2.1 International law

Deforestation in rainforests started to trigger global attention in the 1960s. Since the 1970s, several international environmental conventions have started to address forest protection, such as the 1973 Convention on International Trade in Endangered Species of Wild Fauna and Flora (banning international trade in vulnerable forest species), the 1975 World Heritage Convention (establishing protected sites, including forest land) and the Ramsar Convention on Wetlands of International Importance (preventing the conversion of forested wetlands). The first forest-focused international agreement – International Tropical Timber Agreement (ITTA) – was reached, accompanied by its implementing organization, the International Tropical Timber Organization, ITTO in 1983. The ITTO was found to make too many compromises for economic reasons, and was not able to deal effectively with deforestation.[24] The ITTA was followed by a failure to reach a legally binding global forest convention in the 1992 UN Conference on Environment and Development in Rio. In the post-Rio era, efforts to negotiate a legally binding document continued under the auspices of the Intergovernmental Panel on Forests (IPF), the Intergovernmental Forum on Forests (IFF) and the United Nations Forum on Forests (UNFF) consecutively. However, a legally binding agreement has not been reached. Instead the UNFF adopted a non-legally binding instrument on all types of forest in 2007. The forest-related rules are still dispersed among various non-forest-focused international environmental conventions. In addition to the earlier conventions mentioned above, other forest-related conventions include the United Nations Convention to Combat Desertification, the Convention on Biological Diversity and the United Nations Framework Convention on Climate Change, which respectively protect forests to combat desertification, to protect biodiversity and to address climate change.[25]

3.2.2 Domestic law

Different property rights have been adopted under domestic law. Forest ownership is usually vested in the state, private actors or customary communities.[26] In some states, such as Bolivia, government claims ownership over all forests, but the land that forest grows on can be owned by private actors or customary communities.[27] When the forest or land is owned by the state, sometimes the rights to access, use and management of the resources are granted to other parties, for example, via logging concessions.[28] Public forests can also be set aside for local communities, including indigenous people.[29] As mentioned in the theoretical chapter, a clear cut boundary between public, private and community forest does not always exist. This research treats specific forest types as public, private or community property according to the party in whom the management authority is vested.

In addition to property rights, many other instruments have been adopted to regulate forest management and timber harvesting practices. Such instruments can be traditional command and control regulations, economic or suasive instruments. Command and control instruments include, for example, spatial planning (the establishment of protected areas), logging and transport permits, performance standards for forest management (e.g. annual allowable cut, clear-cutting standards,

24 Gulbrandsen 2010, pp. 44–45.
25 For an overview of such conventions, see Srivastava 2011.
26 White & Martin 2002, pp. 4–6.
27 Article 4, Bolivia Forest Law; Pacheco 2011.
28 White & Martin 2002, p. 8.
29 Ibid., p. 4.

reforestation requirements). Economic instruments include tax, subsidy, payment for ecosystem services, carbon offsetting projects and so on. Suasive instruments include principles, soft targets, research, information and education and so on.[30]

3.2.3 Private/hybrid regimes

3.2.3.1 Introduction

When the international community failed to reach a legally binding forest agreement, and existing international law and organizations provided unsatisfactory solutions for forestry problems, private and hybrid regimes started to develop in the 1990s. The prevalence of private regulation has been attributed to social campaigns targeting companies since the 1980s and to the rise of a neoliberal political climate.[31] The private and hybrid regimes cover not only timber products, but also a variety of forest-related commodities.

The earliest and most influential of these private regimes concerns forest certification, represented by the Forest Stewardship Council (FSC) and the Programme for the Endorsement of Forest Certification (PEFC). The forest certification schemes set standards for forest management and the supply chain of timber production. It requires an accredited, independent third-party certifier to evaluate and audit the production processes or methods according to predefined environmental and social sustainability standards.[32] The standards are defined by an independent governing body, and audits are conducted by private actors.

A more recent regime regulating forest management and the timber industry is the legality regime. In the last decade, developed countries are increasingly concerned about the legality of timber products imported from emerging economies and developing countries where the risk of illegal logging is high. Therefore, they require a verification of the legality of imported products, usually by third parties.[33] Responding to the requirements, many emerging economies and developing countries have started to develop such legality regimes. In a legality regime, standards are defined by domestic law, and timber producers usually can choose their own measures to verify and to guarantee the legality of their products, often by means of certification or other third-party verification schemes.[34] The

30 For an overview of the regulatory instruments, see Gupta, Van der Grijp & Kuik 2012, pp. 34–44; McDermott, Cashore & Kanowski 2010.

31 Bartley 2003.

32 Steering Committee of the State-of-Knowledge Assessment of Standards and Certification, 2012.

33 Overdevest & Zeitlin 2014.

34 Bartley 2014, pp. 97–98.

legality regimes are of a hybrid nature: the government plays an important role in deciding the standards and in regulating the process. Third parties are often used to conduct the audits.

In addition to these regimes regulating forest management and timber harvesting directly, similar initiatives have been developed to regulate other commodities, the production of which may be an important driver for deforestation and forest degradation. For example, commodity roundtables for agricultural products started to emerge in early 2000s, such as the Roundtable on Sustainable Palm Oil (RSPO), the Roundtable on Responsible Soy (RTRS) and the Roundtable on Sustainable Biofuels (RSB). These roundtables are modelled on the FSC, which involves multiple stakeholders to develop environmental and social standards and provides third-party certification programmes. Instead of the roundtables focusing on one type of agricultural product, some certification systems apply to a wide range of crops and livestock, such as the Sustainable Agricultural Network.[35]

In the certification regimes discussed above, civil society has voting rights in deciding the standards and third parties are used to audit the compliance by producers. Industry can also use self-imposed guidelines without or with lower levels of oversight. The Audit Protocol of the Leather Working Group[36] and the Consumer Goods Forum (CGF)[37] constitute examples. In addition, some buyers have also signed agreements to avoid purchasing products with deforestation risks, such as the Soy Moratorium[38] and the G4 Cattle Agreement.[39] Several voluntary information disclosure initiatives have started to develop. For example, the Carbon Disclosure Programme requires its members to disclose how they address forest risk commodities in their supply chain.[40]

Among the various private regimes, forest certification is the most important and earliest one targeting forest management directly. Though legality regimes also address forest management, they have only started to develop recently and their joint effectiveness with other instruments still needs to be tested in practice. Most other private initiatives target commodities which influence forests in a more indirect way and have a much shorter history. Therefore, the discussion of private instruments here focuses on forest certification.

35 Walker *et al.* 2013, pp. 15–16.

36 Leather Working Group, LWG Audit Protocol, 2012, www.leatherworkinggroup.com/about/protocol.htm.

37 Consumer Goods Forum, 2010, http://sustainability.mycgforum.com/images/sustainability-pic/Press_Release_-_2010-11-29-ClimateProtection.pdf.

38 Soy Moratorium, 2012, www.mvo.nl/Portals/0/publicaties/Magazine/2012/17/02%20press%20release%20ABIOVE.pdf.

39 www.greenpeace.org/international/en/news/Blogs/makingwaves/jbs-recommits-to-cattle-agreement-in-the-amaz/blog/43470/.

40 Carbon Disclosure Project, 2013; the Forest Footprint Disclosure Project, www.cdproject.net/en-US/Programmes/Pages/forests.aspx.

3.2.3.2 Forest certification

The FSC was initiated by the World Wildlife Fund (WWF) and other environmental NGOs, with the engagement of timber product retailers and other commercial interests in 1993.[41] Upon its creation, countries responded differently to this scheme. Some countries were favourable from the very beginning, such as Sweden.[42] Other countries were more hesitant, and established industry-dominated alternatives, such as the Sustainable Forest Initiative (SFI) in the US, the Canadian Standards Association (CSA), and the Norwegian Living Forests standard. Many such national certification schemes later became connected through endorsement by another influential international certification scheme, the Programme for the Endorsement of Forest Certification (PEFC).[43] Together, the FSC and PEFC covered 10 per cent of the global forests by 2014.[44]

Both the FSC and PEFC include standards for forest management and for the chain of custody, applying to forest management and to the supply chain respectively. The FSC has developed a global standard for well-managed forests, the FSC Principles and Criteria.[45] To tailor the general standards to the local situation in different jurisdictions, the FSC delegates the authority to elaborate them to its national affiliates.[46] Many countries have published their own FSC standards. Forest management is audited by certification bodies, which are accredited by an independent organization Accreditation Services International (ASI). Usually, forest managers or owners are the clients of forest management certification.[47] A typical certification process includes both a preliminary assessment, a formal assessment (field inspection, consultation, preparing preliminary report, peer review of the report, certification issuance) and annual audits following the issuance of certification.[48] The preliminary assessment indicates the changes applicants need to make to achieve certification, hence it helps them to move towards the formal assessment. After the formal assessment, a decision will be made on certification, which may '(1) approve the application unconditionally; (2) grant provisional approval on condition that specified "corrective actions" are taken to rectify specified "minor non-compliances" within a certain time (minor Corrective Action Requirements, minor CARs); (3) indicate that approval will be granted

41 Gulbrandsen 2010, p. 52.
42 Hysing 2009a.
43 Liu 2014.
44 Georgia-Pacific, Forest Certification Around the World, available at: www.gp.com/~/media/Corporate/GPCOM/Files/Sustainability/Sustainability-Document-List/Forest_Certification_Around_the_World.ashx?force=1.
45 https://ic.fsc.org/en/certification/principles-and-criteria.
46 Gulbrandsen 2010, p. 55.
47 https://ic.fsc.org/en/certification.
48 Meidinger 2006, pp. 70–71.

after certain "major non-compliances" are corrected (major CARs); or (4) deny the application'.[49] The FSC has established a procedure to handle complaints. Under the FSC system, complaints should first be addressed to the party against whom the complaint is made. When no satisfactory explanation is provided, the complainant can first file a complaint to the certification body (CB).[50] If a CB finds evidence for the deviation from FSC standards, it can take different decisions, varying from Observations, Minor CAR or Major CAR to the suspension or withdrawal of the certificate, with increasing levels of coercion.[51] If the complainant is not satisfied with the decision of the CB, he can file a complaint to the ASI and finally to FSC International.[52]

Unlike the FSC, which is a global programme, the PEFC is a mutual recognition framework (an umbrella organization), which endorses multiple independent national certification schemes according to common Sustainability Benchmarks.[53] Such benchmarks include both standards for forest management and chain of custody. How the certification process and complaint handling work depends on the individual national schemes.

The FSC and the PEFC have different governance structures. The FSC is organized independently of national governments. The FSC is accountable to its 'wide range of environmental, social and economic stakeholders'.[54] Its governance structure, stringent standards and transparency have attracted broad support among NGOs, market players and governments.[55] Most PEFC programmes are initiated and operated by industry or landowner associations and are often more closely linked with the government while the participation of social and environmental groups is more limited.[56] Therefore, many NGOs criticize the credibility of PEFC programmes. However, the competition between the FSC and PEFC systems, together with the efforts of governments and NGOs to compare and benchmark[57] certification schemes, have contributed to ratcheting up the standards under the two schemes and to convergence between them.

49 Ibid., p. 71.

50 Agblede 2010, available at: http://stud.epsilon.slu.se/1837/.

51 Ibid., pp. 38–39; Sahlin 2013, pp. 7–10, available at: www.naturskyddsforeningen.se/sites/default/files/dokument-media/rapporter/2013_engelsk_rapport_skog_credibility_ at_stake.pdf.

52 https://ic.fsc.org/preview.fsc-pro-01-008-v2-0-en-processing-complaints-in-the-fsc-certification-scheme-tracking-changes.a-3412.pdf.

53 www.pefc.org/standards/endorsement-mutual-recognition.

54 Gulbrandsen 2010, p. 70.

55 Ibid.

56 Ibid., p. 71.

57 Overdevest 2010.

3.3 The selection of countries

As discussed earlier, a number of factors need to be considered in choosing countries for further study: the prominence of forestry problems, the stage of development, the prevailing property rights and the presence of an institutionalized certification system. The countries we have selected are confronted with typical forest problems, either deforestation and/or forest degradation; they are in different stages of forest transition, have adopted different types of forest property rights and have comparatively wide certification coverage.

According to such criteria, five countries were chosen: Indonesia, Bolivia, Canada, the US, and Sweden. They all have substantial forest coverage: 51.4 per cent in Indonesia, 52.2 per cent in Bolivia, 34.1 per cent in Canada, 33.3 per cent in the US and 69.2 per cent in Sweden.[58]

Indonesia and Bolivia are two developing countries suffering from rapid deforestation. Bolivia is in the frontier stage[59] and Indonesia is in the transition from frontier to forest and agriculture mosaic stage.[60] They both have adopted diverse types of property rights: public, private and communal. Although fewer forests have been certified in Indonesia and Bolivia than in the three developed countries, given the low penetration rate of forest certification in developing countries, they represent examples with comparatively well-institutionalized certification schemes for developing countries.

Canada and the US will be analysed together, given their relatively similar structure and geographical proximity albeit with different property rights regimes. Canada, the US and Sweden are developed countries in the stabilization stage of forest transition. In these countries, forest degradation is a more prominent problem than deforestation. In the US, 42 per cent of forests are publicly owned while the rest are owned by individuals, families, corporations, and indigenous tribes. Over 90 per cent of forests in Canada are publicly owned. 81 per cent of forests in Sweden are privately owned.[61] Community-owned or managed forests do exist in these three countries (such as those owned or managed by Native American tribes in the US, First Nations in Canada and Sami people in Sweden) but only on a small scale.[62] Forest certification has started to develop in these countries and has a wide coverage there: around 36 per cent of forest areas in North America are certified[63] and more than half of the forests are certified in Sweden.[64]

58 http://data.worldbank.org/indicator/AG.LND.FRST.ZS.

59 Hosonuma *et al.* 2012, p. 9.

60 Gupta, Van der Grijp & Kuik 2012, p. 242.

61 McDermott, Cashore & Kanowski 2010, pp. 74, 80, 136.

62 Sunderlin, Hatcher & Liddle 2008.

63 www.unece.org/fileadmin/DAM/timber/publications/FPAMR-2014-final_01.pdf, at p. 17.

64 FSC Sweden 2013, p. 3, available at: https://se.fsc.org/preview.the-contribution-of-fsc-certification-to-biodiversity-in-swedish-forests.a-661.pdf.

Indonesia and Bolivia are confronted with rapid deforestation. For forest resources, if access is not restricted, different users tend to free-ride and to harvest the resources to the maximum possible, without inclination to conserve. Property rights can be used to limit access. In Indonesia and Bolivia, public, private, and community forests co-exist. The property rights are often less clear and less secure than in the three developed countries. This ambiguity has led to conflicts between forest users and has contributed to high deforestation rates. The case studies of these two countries will raise the question why public regulation and certification have failed to satisfy the necessary preconditions for the proper functioning of property rights and consequently have failed in addressing deforestation.

Besides, public and private regulation are called upon to overcome the limitations of property rights in addressing forest degradation. The case studies of Canada, the US, and Sweden will examine two prominent degradation problems: the loss of ecological function caused by insufficient protection of riparian forests and the loss of biodiversity. In these countries, a comparatively clear boundary between forests has been established. The deforestation problem has largely been solved but not the problem of forest degradation. To address the problem of forest degradation, Canada and the US have established riparian buffer zones under both public and private regulation. In Sweden, both public regulation and certification include forest management standards directly related to biodiversity protection. This study will focus on how public regulation and certification interact to address these forest degradation problems.[65]

3.4 Case study: Indonesia

3.4.1 Environmental problem: deforestation

Indonesia is a country with rich forest resources. Of its 187.67 million hectare extent, 98.559 million hectares (52.5 per cent) were covered by forest in 2010.[66] Around 240 million people, including over 300 ethnicities, live in Indonesia.[67] The importance of forestry in Indonesia's economy decreased to 0.80 per cent of the GDP in 2009, but its downstream industry possessed a higher proportion of the annual GDP (wood products industry: 1.43 per cent and paper and printing industry 1.09 per cent).[68] Forest products represented 6 per cent of the exports from Indonesia in 2010.[69] The EU

65 In the case studies the present tense is used to describe the most recent period. The reader may want to take into account that the situation referred to in the present tense may have already changed and will certainly change in the future.

66 Indrarto *et al.* 2012, p. 1.

67 Gupta, Van der Grijp & Kuik 2012, p. 120.

68 Royo & Wells 2012, p. 3.

69 ITS Global 2011, p. 6, available at: www.itsglobal.net/sites/default/files/itsglobal/ITS_Indoforest_Economic_Report.pdf.

and North America are important export markets for forest products of Indonesia; however the market has been increasingly directed to China and other Asian countries.[70]

Indonesia has experienced fast deforestation in the past decades. Although the data vary from source to source, most confirm the alarming forest loss rate. For example, the data from the Ministry of Forestry in Indonesia (MoF) show that the deforestation rate was 1.87 million ha per year from 1985–1997. After the fall of Suharto in 1998, Indonesia witnessed a sharp increase in the deforestation rate to 3.51 million hectares per year from 1997 to 2000. This was followed by a decreased rate to 1.08 million hectares per year from 2000–2005.[71] The FAO (Food and Agriculture Organization of the United Nations) statistics show a lower, but still alarming deforestation rate: 0.3 per cent per year during the period from 2000 to 2005 and 0.7 per cent per year during the period from 2005 to 2010.[72] It is estimated that the process has led to an accumulation of 24.1 per cent of Indonesia's forest cover lost from 1990 to 2005,[73] or 40 per cent from 1950 to 2000.[74] In spite of all the efforts to curb forest loss, the deforestation rate increased again in recent years, to more than 2 million hectares per year in 2011 and 2012.[75] Generally speaking, Indonesia is in a transition from the frontier to the forest and agriculture mosaic stage, with a long lasting and still alarming deforestation rate.[76] However, significant regional differences exist. For example, forest cover in Java has been reported as increasing in recent years, while Sumatra and Kalimantan still experience very fast deforestation.[77] 'Forest degradation is also at pace, with a continuous transformation of primary into secondary forest'.[78] Corresponding to this process, biodiversity is being lost at high speed in Indonesia.[79]

3.4.2 Good governance

Indonesia is a lower-middle income country with annual GDP of US$888.5 billion in 2014.[80] The World Bank has initiated the Worldwide Governance Indicators (WGI) project, which evaluates six dimensions of governance: voice and accountability, political stability and absence of violence, government effectiveness, regulatory quality, rule of law and control of

70 www.grida.no/graphicslib/detail/indonesian-exports-of-forest-products_f403#.
71 Indrarto *et al.* 2012, p. 3.
72 FAO 2010, available at: www.fao.org/forestry/fra/fra2010/en/.
73 http://rainforests.mongabay.com/deforestation/archive/Indonesia.htm.
74 Arnold 2008.
75 Hansen *et al.* 2013; Purnomo & Anand 2014, p. 28.
76 Gupta, Van der Grijp & Kuik 2012, p. 225.
77 Indrarto *et al.* 2012, p. 3.
78 Gupta, Van der Grijp & Kuik 2012, p. 120.
79 Indrarto *et al.* 2012, p. 3.
80 http://data.worldbank.org/country/indonesia.

corruption.[81] The WGI project ranks countries according to percentile range, with 0 as the lowest rank percentile and 100 as the highest percentile rank. Indonesia ranks 53.20 regarding voice and accountability; 31.07 regarding political stability, 54.81 regarding government effectiveness, 49.04 regarding regulatory quality, 41.83 regarding the rule of law and 34.13 regarding the control of corruption.[82] The scoring of the World Bank WGI on control of corruption is similar to the rank of corruption control by Transparency International. According to Transparency International, Indonesia ranks 88 out of 167 countries.[83] This means that overall Indonesia is ranked in the lower half in terms of good governance.

3.4.3 Property rights as well as public and private regulation

3.4.3.1 Property rights

The forest governance system in Indonesia dates back to the Dutch colonization, when a scientifically based, top-down forestry policy was formally adopted.[84] Although in practice, 'the Colonial Government only directly controlled the forests in Java, with forests on the outer islands being subject to adat (customary law)'.[85] The adat regulates the community-based land-use rights. Although the adat system varies among communities, it usually adopts a communal approach, where land rights are vested in the communities and the exercise of individuals' rights depends on the consent of the community.[86] This communal approach has been incompatible with the western categorization of land rights enshrined in written law. However, it was recognized 'to the extent that this did not interfere with Dutch commercial or state interests'.[87] After independence, the state took over the control of forests in Java from the Dutch, but the status on outer islands remained unclear until Suharto came into power in 1966, representing the start of the New Order regime.[88]

When Suharto came to power, Indonesia suffered from poor economic performance and external debts, forcing it to rely on its natural resources, including forest exploitation, to boost economic growth.[89] The 1945 Constitution authorized the state to 'control' the land, waters and natural

81 http://info.worldbank.org/governance/wgi/index.aspx#home.

82 http://info.worldbank.org/governance/wgi/index.aspx#reports.

83 www.transparency.org/cpi2015?gclid=CMK59ZPMwMsCFRBmGwodsSQPKg.

84 Arnold 2008, p. 78.

85 Adat usually refers to the customary law in Indonesia and Malaysia, which 'was the unwritten, traditional code, governing all aspects of personal conduct'. Instead of being a uniform system, the adat varies widely, even over short distances. See Arnold 2008, p. 79; Gold & Zuckerman 2014, p. 45.

86 Gold & Zuckerman 2014, p. 45.

87 Anold 2008, p. 79.

88 Ibid.

89 Palo & Vanhanen 2012, pp. 173–174; Lindayati 2002.

resources in Indonesia.[90] This was further confirmed in the Basic Agrarian Law of 1960 (BAL 1960)[91] and the Basic Forestry Law of 1967 (BFL 1967).[92] The BAL 1960 was intended to 'unify all the land law of Indonesia into a single system'.[93] The BAL recognized three kinds of land ownership: state land (untitled land, tanah negara), private land (titled land, tanah hak) and customary land (tanah ulayat).[94]

The BAL was intended to act as 'an umbrella law for all other laws and regulation on land and natural resources'.[95] However, it is not always referred to in relevant legislation. The passing of the BFL in 1967 led to a de facto division of the authority between the BAL and the BFL. The BFL 1967 differentiated Forested Areas and Non-Forested Areas and granted the MoF authority to regulate over 143 million Forested Areas (70 per cent of land area).[96] The Forest Areas were designated according to legal processes and often did not match actual forest cover.[97] In practice, only around 68 per cent of the designated Forest Areas actually was covered with forest and 15 per cent of Non-Forested Areas were covered by forests nonetheless.[98] As mentioned, the application of the BAL 1960 was restricted to Non-Forested Areas.[99] The BFL 1967 put forests under the control of the state and prohibited private or community ownership.[100] In addition to maintaining direct control on Java, the state also 'outsourced the management of forests on the outer islands to timber companies through the granting of timber concessions'.[101] In the highly centralized system, the central government had an exclusive right to grant timber concessions and by 1995, over 60 million hectares of forest had been granted as forest concessions, leading to a concentration of forest concessions in the hands of political elites.[102] Small-scale loggers and local communities were marginalized by the concession rules, since a minimum threshold of 50,000 hectares applied.[103] The status of local communities was further weakened by the creation of protected forests on their traditional lands,[104] a mass state-sponsored Transmigration Programme and the

90 Article 33(3) of the 1945 Constitution.
91 Article 2(2) of Basic Agrarian Law.
92 Article 5(1), Law No. 5 of 1967 concerning Basic Forestry Principles.
93 Szczepanski 2002, p. 239.
94 Indrarto *et al.* 2012, p. 36.
95 Safitri 2010, p. 71.
96 Arnold 2008, p. 79.
97 Gupta, Van der Grijp & Kuik 2012, p. 120.
98 Ibid.
99 Indrarto *et al.* 2012, p. 36.
100 Szczepanski 2002, p. 242.
101 Arnold 2008, pp. 19–80.
102 Barr *et al.* 2006; Resosudarmo 2005, p. 117.
103 Arnold 2008, p. 80; Singer 2008, pp. 528–529.
104 Campbell 2002, p. 83.

erosion of customary governance institutions via the Village Governance Law.[105] During this period, the status of 'indigenous Indonesians' was not recognized and even the term 'adat' (customary communities in Indonesia) was removed from the official vocabulary. The Village Governance Law tried to restructure the local villages according to the Javanese village and hence systematically disempowered other indigenous governance institutions and forest management systems.[106] In spite of this marginalization process, the close partnership of government, business interests and military power suppressed massive land conflicts. Most illegal logging was conducted by industrial permit holders, for example, by logging outside the permitted areas.[107]

The Asian Financial Crisis of 1997 triggered the end of the Suharto regime and a rapid transition to a democratic governance system, the Reformasi, followed. In the Reformasi period, the double track of BAL and BFL persisted. Under the BFL, in addition to state forests, the new law allowed private forests and started to recognize customary forests to a limited extent.[108] Closely mimicking the BAL 1960, Article 4 of the BFL 1999 stipulates that 'Forest control by the state shall respect customary law, as long as it exists and its existence is recognized and not contradicting national interests'.[109] The law requires governmental regulation of the process of recognizing or abolishing adat communities.[110] Such a governmental regulation, however, has never passed. This half-hearted recognition of customary forests makes the status of local communities weak. The MoF 'routinely allocates logging concessions with little regard for community claims and some regional governments deny the existence of adat communities in their territory'.[111] A landmark case was adjudicated by the Indonesia Constitutional Court in 2013, which invalidated the classification of forests under the BFL 1999.[112] According to the judgment, customary forest is to be excluded from state forest and is to be admitted

105 Arnold 2008, p. 80.

106 Ibid., pp. 80–81.

107 McCarthy 2000, p.105.

108 This is believed to relate to the demise of the military role in social affairs and the emergence of civil society. Under such conditions, communities began to claim their rights and the conflicts between them and transmigrants/concessions operators started to rise. See McCarthy, Barr, Resosudarmo & Dermawan 2006; Arnold 2008, p. 81.

109 English copy see http://theredddesk.org/sites/default/files/uu41_99_en.pdf; www.elaw.org/node/2644.

110 Article 67(2)(3).

111 Bartley 2011, p. 531.

112 PUTUSAN, Nomor 35/PUU-X/2012, www.forestpeoples.org/sites/fpp/files/news/2013/05/putusan_sidang_35%20PUU%202012-Kehutanan-telah%20ucap%2016%20Mei%202013.pdf; for an English introduction, see http://news.mongabay.com/2013/05/in-landmark-ruling-indonesias-indigenous-people-win-right-to-millions-of-hectares-of-forest/.

113 Sari 2013, p. 21.

as an independent type of forest ownership.[113] In spite of this progress, the exact meaning of 'customary forest' and the process of establishing it are still unclear. In practice, the existence of the adat community depends on the recognition by the government and the adat rights can be revoked or it can be determined that adat rights contradict national interests.[114] The lack of clarification in the process of recognizing adat rights and their exact content also contribute to the insecurity of the adat rights.[115] Consequently, although the adat system has alternately been suppressed and endorsed in the history of Indonesia, it has rarely enjoyed 'more than minimal force in practice'.[116]

In summary, under the current legal system, forests in Indonesia are subject to two regulatory systems. The Non-Forested Areas are subject to the BAL and the Forested Areas are subject to the BFL 1999, while three types of land ownership exist for both regulatory systems: public, private and customary. However, the property rights are not clearly established, do not internalize the externalities of forestry well, and are often contested because of ambiguous legislation and conflicts of interest between officials and stakeholders. Consequently, it is estimated that 52.6 per cent of the 31,864 villages in Indonesia were located in forest areas in 2007, of which only 23.6 per cent remained located in forest areas in 2009.[117] Many of these areas are claimed by the government as state forest, and allocated to concessionaires. It is reported that '22.5–24.4 million hectares were subject to conflicts as a result of unclear boundaries between villages and state forests'.[118] The unclear land tenure does not only influence the welfare of local communities, but also the performance of communities and concessionaries in terms of protecting the forests.[119]

3.4.3.2 Public regulation

In addition to domestic property rights, a number of other public policies also have a crucial impact on recognizing and protecting forest tenure, such as decentralization, spatial planning, forest permits and policies promoting community-based forest management.

Decentralization. The Reformasi period has witnessed an effort of decentralization, the emergence of civil society and increasing recognition of indigenous rights. A series of decentralization laws (known as otda laws) have been passed to transfer the political, administrative and fiscal author-

114 Arnold 2008, p. 86.
115 Ibid.
116 Bartley 2011, p. 530.
117 Indrarto *et al.* 2012, p. 13.
118 Ibid.
119 McCarthy 2000.
120 Arnold 2008, p. 77.

ity to regional governments.[120] Immediately following the end of the Suharto period, two otda laws[121] were passed in 1999, to transfer power in sectors not exclusive to central government but to districts and municipalities.[122] Forestry was not on the list of exclusive jurisdiction of central government and hence was to be decentralized. For example, the heads of districts and municipalities were authorized to issue small-scale timber extraction permits (for areas up to 100 hectares and 10,000 hectares respectively).[123] Since regional governments are allowed to share the revenues from issuing permits, they have strong incentives to abuse the small-scale permits: by allowing logging in broader areas, multiple permits are issued to the same area and no sustainable logging principles are required.[124] There are other disincentives for local politicians to prevent deforestation. For example, 'local politicians use forests as a means of gathering campaign funds for regional elections' and hence tend to be lenient towards the forest industry.[125] 'Regional electoral candidates being given shares in certain oil palm companies in their jurisdictions, so that when they are elected and take office, they will help facilitate permits for the conversion of forest areas'.[126]

In spite of the decentralization policy, the central government tried to regain power via the revised Basic Forestry Law of 1999 (BFL 1999) and new otda laws of 2004. The new otda laws 'provide the Central Government more opportunities to monitor and intervene in regional affairs', such as dismissing regional heads and scrutinizing regional budgets.[127] The BFL 1999 was drafted by the MoF, reflecting a systematic effort of the central government to reconsolidate their authority in the forestry sector.[128] It differentiates forests according to their functions: production, protection, and conservation. Timber concessions can only be granted to production forests and a sub-category of conservation forest. The BFL 1999 grants the authority to classify forests and to grant the status of 'Forest Areas' to the central government.[129] Decentralization issues require

121 Law No. 22 of 1999 concerning Regional Governance; Law No. 25 of 1999 concerning Fiscal Balance between the Centre and the Regions.

122 In Indonesia, there is another level of government between the local authorities and central government: provinces. Following the fall of New Order, resource-rich provinces started to demand greater autonomy and separatist movements even gained strength in some provinces. For fear of the possibility of encouraging separatism, the otda laws bypassed provinces. Arnold 2008, pp. 81–82.

123 Government Regulation No. 6/1999 on Forest Enterprises and Extraction of Forest Products from Production Forests. Indrarto *et al.* 2012, p. 27.

124 Indrarto *et al.* 2012, p. 27.

125 Ibid., p. 10.

126 Ibid., p. 11.

127 Arnold 2008, p. 84.

128 Barr *et al.* 2006, p. 14.

129 Article 1(14) and Articles 5, 6.

130 Article 66.

a governmental regulation,[130] which was passed in 2002 and later revised in 2007.[131] The Government Regulation only transferred minor powers to regional governments, such as issuing 'lesser permits', including timber extraction licences for non-commercial purposes and non-timber forest product extraction permits.[132]

Given the abuse of regional governments in issuing small-scale timber extraction permits mentioned above, the MoF has tried to revoke the authorization since 2000 through ministerial decrees. However, some regional governments tended to have different understandings of which laws should have priority in this case and continued their issuance of permits until 2004.[133] The continuing decentralization reforms and subsequent recentralization efforts led to a 'tug of war' between the central and regional governments.[134] One important contributor to the different interpretations by the MoF and the regional government lies in the ambiguous hierarchy between different types of legislation in Indonesia. According to Law 10.2004 concerning Law Making in Indonesia, the hierarchical order of legislation in Indonesia is: Constitution; Law; Governmental Regulation; Presidential Regulation and Regional Regulation.[135] Ministerial Regulations, however, have no clear position in this hierarchy.[136] The authority for regional governments to issue small permits was granted by a Governmental Regulation, but was revoked by Ministerial Regulation. It has hence been left undecided which rule should apply.

Spatial planning is another public regulatory instrument that has had an important influence on forest tenure besides rules on property rights. Spatial planning categorizes land and determines how forest is to be used. The first spatial planning law was issued in 1992 and revised in 2007.[137] It allows both the central and regional (provinces and districts) governments to conduct spatial planning at their respective levels. However, the regional plans regarding Forested Areas need to be approved by the MoF.[138] Many conflicts have arisen over plans developed by different levels of governments.[139] Many other activities than forestry take place in the forests, such as agriculture (especially estate crops), mining, establishing eco-regions and so on. Therefore, non-forestry agencies also play a role in deciding the destiny of forests, according to sector-based legislation.[140] For example, the

131 Governmental Regulation 34/2002 and Governmental Regulation 6/2007.
132 Arnold 2008, p. 87. Barr *et al.* 2006, pp. 47–48.
133 Indrarto *et al.* 2012, p. 28.
134 Agrawal & Ribot 1999; Purwanto 2005; Singer 2009, available at: http://b-singer.fr/pdf/Forest_policies_in_Indonesia.pdf.
135 The Law 10.2004 concerning Law Making in Indonesia, Article 7; Safitri 2010, p. 69.
136 Ibid.
137 Law 24/1992 and Law 26/2007.
138 Gupta, Van der Grijp & Kuik 2012, p. 128.
139 Ibid.
140 Arnold 2008, p. 95.

Ministry of Energy and Mineral Resources designates the mining areas while the Ministry of Environment designates eco-regions.[141] Although the 2007 Spatial Planning Law requires coordination among departments in planning land use, this is usually not what happens in practice.[142]

Permits. To conduct the above-mentioned activities in forests, a permit is needed. As discussed above, the MoF is responsible for issuing most permits for natural forest logging and regional government can only issue some 'smaller' permits. For the state forests to be used for non-forestry activities, the MoF needs to first release the forests. Then other authorities, such as regional heads, can issue permits for mining and estate crops plantations.[143]

Community-based forest management. In addition to the recognition of the customary forest, since 1999, Indonesia has also adopted several other policies to promote community-based forest management (CBFM) in state forests (such as Forest Area with Special Purpose, Village Forest, Social Forest, People's Plantation Forest, company–community partnership in forest management, Collaboration in Managing Conservation Forest)[144] and in private forests (such as Hutan Rakyat and Family Forest).[145] The CBFM models do not allocate ownership to local communities, but allow their participation in forest management and access to the benefits from it to different extents.[146] These models of CBFM can either be established by government licensing or agreements with companies (company–community partnership) and conservation offices (Collaborative Conservation Management).[147] In the licensing models, the rights granted to local communities are usually limited, sometimes prohibiting the transfer of rights, or the use of these rights as collateral for credit. They also allow limited participation of communities in the evaluation of licences, during which the duration of the rights is determined.[148] The rights local communities obtain from the agreements depend on the provisions of such agreements. Usually these rights are quite limited. The weak bargaining power of communities compared to companies and conservation offices, together with the lack of a fair judicial process also fail to guarantee good and fair agreements.[149]

3.4.3.3 Private regulation

Indonesia's interest in forest certification started in the early 1990s, when the government was stimulated by the International Timber Trade

141 Indrarto *et al.* 2012, p. 21.
142 Ibid.
143 Ibid., p. 31.
144 Safitri 2010, p. 52.
145 Sari 2013, p. 9.
146 Ibid., pp. 116–121.
147 Ibid., p. 121.
148 Ibid., pp. 293–295.
149 Ibid., pp. 295–296.

Organization meetings and wood products were facing boycott initiated by international NGOs.[150] At the same time when the international community was negotiating the start-up of the FSC, Indonesia was considering establishing its own certification system. Two voluntary schemes started to operate in Indonesia in 1999, the FSC and a national certification scheme: the Indonesian Eco-label Institute (LEI). The first FSC natural forest management certificate was granted in Indonesia in 2001.[151] The FSC has not developed a national standard in Indonesia. Instead, the accredited certification bodies have been conducting evaluations against their own standards which are locally adapted according to the FSC Principles and Criteria. Not until 2013 was a harmonized version of these locally adapted standards published by the FSC, as a first step towards developing a full-fledged Indonesian standard.[152]

The LEI standards have developed much faster. The Minister of Forestry announced that he had asked the former Minister of the Environment, Emil Salim to establish a national certification scheme by the end of 1993. Emil Salim established the LEI working group to negotiate the standards, involving a variety of stakeholders, such as government, academia, NGOs and the private sector.[153] The standards were drawn from international and national documents, such as FSC C&Is, ISO 14,000 standards, ITTO C&Is and Indonesian national C&Is. The standards drafted by this multi-stakeholder process were later approved by the government (the Indonesia National Standards Body).[154] The LEI institute was officially established as a foundation in 1998, and later became the accreditation body, which validated its own certifiers.[155] The LEI was transformed into a constituent-based organization in 2009, with its members consisting of indigenous people, a community chamber, a business chamber, an NGO chamber and an eminent person chamber. The government has no seat in the governing body.[156] In order to gain public trust, LEI collaborated with the FSC and established a Joint Certification Programme (JCP). According to the JCP, the forest management units are obliged to pass both LEI and FSC standards to obtain certification. The penetration rates of both FSC and LEI are still low. There were 21 units covering 1.52 million hectares of forest that held certification from the FSC by the end of 2013;[157] LEI certification covers 1.89 million hectares of forest.[158]

150 Muhtaman & Prasetyo 2006, p. 43.
151 Klassen, Romero & Putz 2014, p. 256.
152 FSC-STD-IDN-01-01-2013.
153 Tacconi 2007, p. 265.
154 Muhtaman & Prasetyo 2006, pp. 43–46.
155 Tacconi 2007, p. 265.
156 www.lei.or.id/anggota-lei.
157 www.tff-indonesia.org/index.php/programs/certification-support/list-of-fsc-certified-forest.
158 www.lei.or.id/iges-study-on-forest-certification.

Both FSC and LEI have made efforts to promote certification for community-managed forests to both enhance the communities' welfare and to strengthen forest protection. FSC has developed a group certification system for smaller actors, more specifically the Small and Low Intensity Management Forests Standards (SLIMF). LEI has also established a Sustainable Community-based Forest Management System (PHBML). The FSC SLIMF applies to small forests (areas not exceeding 1,000 hectares) and low-intensity management forests (harvest rate below 20 per cent of the mean annual increment within the production forest area, and annual harvest not exceeding 5,000 m^2).[159] By all means, the SLIMF coverage in Indonesia is still very limited, with around 2,000 hectares.[160] By June 2013, 32,683 hectares of community forests had obtained an LEI PHBML certificate.[161]

In addition to these voluntary schemes, a mandatory certification scheme (PHPL) was introduced in the early 2000s. All logging concessions were required to obtain the PHPL certificate. Community forests are exempted from this requirement.[162] Under this scheme, independent auditors (LP) accredited by the MoF assess the performance of the industries according to sustainable forest management standards, issued via several decrees in 2002 and 2003.[163] Although this certification scheme is mandatory, only 62 forest concessions have been certified before March 2013, composing only 35 per cent of the total number of forest concessions.[164] This certification system is also criticized for lacking independence and transparency, and for being paper-based with little meaningful impact.[165]

The effort to introduce certification originally yielded little uptake. Nonetheless, Indonesia continued to face strong pressure from consumer countries and civil society campaigns to tackle its rampant illegal logging, particularly when some developed importing countries started to impose legality standards in their domestic legislation. Such a legality regime was implemented by the EU, the US and Australia, requiring all the timber products entering into the market to comply with the law in the country of harvest.[166] Due to a lack of trust in the public administration of developing countries, a common approach in these jurisdictions is to require the establishment of a legality verification system, with third-party auditors evaluating the performance according to exporting countries' domestic legislation.[167] In this political climate, the legality verification scheme

159 Hinrichs, Muhtaman & Irianto 2008, p. 26.

160 www.tff-indonesia.org/index.php/en/map-of-tff-activity/list-of-fsc-certified-forest-in-indonesia-61198.

161 www.lei.or.id/program-kerja-2009-2013.

162 Nurrochmat *et al.* 2014.

163 Wells 2008.

164 Wijaya *et al.* 2014.

165 Wells 2008, p. 181.

166 Cashore & Stone 2014, p. 52.

167 For details, see Cashore & Stone 2012; Obidzinski *et al.* 2014.

(TLAS) started to develop in Indonesia.[168] The PHPL and TLAS were later incorporated in the voluntary partnership agreement (VPA) which Indonesia concluded with the EU in 2011.[169] Under the new system, forest operators can apply for both mandatory certification and legality verification. LPs are still responsible for conducting the assessment under the certification regime, while other independent verifiers (LVs) conduct the assessment under the TLAS. Both LPs and LVs are accredited by a National Accreditation Committee. If the assessment standards are satisfied, LPs can issue the SFM certificates and the LV can issue the legality certificates and export licences. The MoF does not issue certifications itself, but acts as a regulatory body of the LPs and LVs.[170]

Another voluntary forest certification scheme, the Indonesian Forestry Certification Cooperation (IFCC) was established in 2011. The IFCC published its SFM standard and chain of custody standard, which were endorsed by PEFC in October 2014. The Standardization Committee, the standards setting body of IFCC, is composed of businesses, NGOs, local communities, the scientific community, and governmental authorities and so on.[171] IFCC relies on the national accreditation body to accredit the certifiers.[172] Since the introduction of TLAS and IFCC is comparatively recent, empirical evidence of their impact is not yet available. PHPL has been functioning for more than one decade now, but it does not apply to community forests and is criticized as having little practical impact. Therefore the following discussion focuses on the FSC and the LEI.

3.4.4 The interaction between public and private regulation

3.4.4.1 The interaction on paper

As mentioned earlier, customary rights are only given reluctant recognition under domestic law.[173]

Clear land tenure and the protection of indigenous people or local communities are required under forest certification schemes. As discussed

168 The TLAS was introduced by the 2009 regulation on the 'Standards and Guidelines on the Assessment of Performance of Sustainable Forest Management and the Verification of Timber Legality in the State and Privately-Owned Forests' and was incorporated in its voluntary partnership agreements with the EU in 2011.

169 Voluntary Partnership Agreement between the European Union and the Republic of Indonesia on Forest Law Enforcement, Governance and Trade in Timber Products into the European Union, Official Journal of the European Union, L 150/252, L150/299, 2014.

170 VPA, Annex 5, sections 4.1; 4.2.

171 Form International, PEFC Conformity Assessment: Indonesia Forest Certification Cooperation Scheme 2014, p. 19.

172 Ibid., p. 146.

173 See *supra* section 3.4.3.1.

earlier, auditors evaluate the forest management performance against locally adapted FSC P&C standards in Indonesia. Principle 2 of the FSC P&C requires 'Long-term tenure and use rights to the land and forest resources shall be clearly defined, documented and legally established'. Principle 3 concerns indigenous peoples' rights: 'The legal and customary rights of indigenous peoples to own, use and manage their lands, territories, and resources shall be recognized and respected'. In addition, Principle 4 requires forest management to enhance the well-being of local communities. The FSC also has a scheme tailored to CBFM forests, the SLIM certificate. It allows small and community producers who are eligible as 'small or low-intensity forest management units'[174] to form a group and apply for a single FSC certificate. Such an arrangement can reduce the costs and workload for the group members; hence it can ease the burden for local communities who would like to apply for FSC certification.

LEI has three standards for forest management: Sustainable Production Natural Forest Management Certification (PHAPL), Sustainable Plantation Forest Management Certification (PHTL) and Sustainable Community-Based Forest Management (PHBML). Using PHAPL as an example, it establishes standards of production, ecological and social sustainability. For the sustainability of social functions, it requires a 'guaranteed community-based forest tenure system', 'guaranteed resilience and community and employees' economic development', 'guaranteed continuity of social and cultural integration of community and employees', 'realization of the responsibility to safeguard nutritional status and prevent negative impacts on community' and 'assurance of workers' rights'.[175] PHTL also includes similar requirements.[176] PHBML is a tailored standard for CBFM, including more robust standards concerning local communities, such as requiring an 'established symmetrical social relationship pattern in the production process' and 'fair benefit sharing in accordance with community interest'.[177]

The above analysis shows that three types of rules exist in Indonesia regulating land tenure and rights of indigenous people/local communities: the written law, customary law and private standards. Generally speaking, certification schemes respect customary law, while the latter is, at best, half-heartedly recognized under national law. Only until recently, a customary forest has been admitted as a category of ownership independent from state ownership. However, the criteria for the acknowledgement of such a forest and the process of establishing it remain obscure. The potential conflicts of these three types of rules fail to create a secure land tenure system, and create challenges for the protection of customary communities.

174 FSC-STD-01-003 V1-0.
175 LEI 5000-1.
176 LEI 5000-2.
177 LEI 5000-3.

3.4.4.2 The interaction in practice

The land tenure policies do not only fail to create a reliable system and to achieve satisfactory improvement in local communities' access to and benefits from the forest, but also have a negative influence on the forest management behaviour of both communities and concessionaires. Historically, agriculture and timber harvests in forests out of Java were regulated by the adat system. Based on traditional knowledge, such a system includes rules to ensure land fertility and to protect the forests. The adat systems had their own governing institutions under the leadership of Pasirah, who had authority to manage the forest, decided on individual use of the forest and protected the ecological functions.[178] The self-governing institutions under the adat system used to be the same as the village government. However, the implementation of the Village Government Law restructured local villages and their governing institutions. Villages were 'rearranged into administrative units containing a standard 100 families'.[179] The standardization sometimes joined geographically isolated units into one village, preventing effective governance. Moreover, the Pasirah lost authority over forests and the village head became the official village government, implementing policies from higher levels of government without considering local conditions and local rules.[180]

The decline of adat institutions weakened the traditional control of forest use. The rapid growth of logging companies and agriculture plantations, as well as the large increase of transmigrants, has further endangered the forests in outer islands. The concessionaires obtained certificates from the government, marginalizing the local communities living in and around the forests. Finding it difficult to identify agricultural land according to adat rules, local people have been forced to cultivate vulnerable land or protected forests and to shorten the fallow period of farmland. Such activities have increased land erosion and have reduced fertility. The conflicting interests of timber or plantation companies, transmigrants and local communities have also led to uncontrolled forest fires in the 1990s. The companies used fire to clear forests, to facilitate access and to reduce compensation to local communities. Local communities traditionally used fire under the control of the adat system to prepare land for cultivation, but now deliberately set fire to their land, which was occupied by companies, in retaliation. Migrants farming in forests have never been subject to traditional adat rules and hence have also used fire to gain access to farmland without controlling it.[181]

The weakening of adat institutions and the conflicting interests among various actors has led to rapid forest clearing and high deforestation rates

178 McCarthy 2000, p. 111.
179 Ibid.
180 Ibid.
181 Ibid.

in the 1980s and 1990s. Even after the Reformasi period, when customary rights of local communities started to be recognized, strong incentives for deforestation remained. With the decrease of government control, land conflicts began to increase. Both companies and local communities have continued to clear forests for more profitable use to support their land claim. Moreover, the marginalization of local communities has incentivized them 'to directly engage in illegal logging or to support deforestation by commercial operators in exchange for employment and/or other forms of payment'.[182]

As discussed above, in addition to establishing customary forest, the government has also introduced CBFM policies to enhance local communities' engagement in and benefits from forest management. However, the adoption of these CBFM models is still limited. For example, although adat communities can apply for permits for a Forest Area with Special Purpose, in practice, it has only been used once. Less than 0.6 million hectares of the People's plantation forest have been awarded, far less than the target set by the government.[183] The rights allowed under these CBFM models have often been limited, unable to empower local communities and incentivize their sustainable management performance. Many CBFM policies take little account of customary practice. 'Imposed regulatory instruments have proved cumbersome and inappropriate in customary contexts, even weakening pre-existing forms of community organization and excluding the poorest (often landless) households and the informal forest use on which they most depend'.[184] As a result, some communities have abused licences and have conducted destructive logging.[185] However, other CBFM models have led to more positive results, such as the Village Forest model. The stakeholders in licensed areas have reported empowered communities and better conservation of forests.[186] Still, the application of the Village Forest model has also been limited: only six villages had obtained a licence by 2011.[187]

The forest certification schemes have not provided satisfactory solutions for the public regulatory system in Indonesia that fails to create secure forest tenure and to incentivize sustainable forest management. Although both the FSC and the LEI have already started to operate in Indonesia since the late 1990s, their coverage is still limited.[188] Both schemes require clear and secure land tenure as a requisite to grant certification.[189] However, the above analysis shows that to achieve such a requisite has been

182 Arnold 2008, p. 97.
183 Gupta, Van der Grijp & Kuik 2012, p. 132.
184 Royo & Wells 2012, p. 11.
185 E.g. Social Forestry according to Ministerial Decree 677/1988. Safitri 2010, p. 298.
186 Gupta, Van der Grijp & Kuik 2012, p. 129.
187 Ibid.
188 *Supra* section 3.6.2.3.
189 *Supra* section 3.6.3.1.3.

a great challenge in Indonesia so far. The FSC standards require companies to obtain 'free and prior informed consent' (FPIC) from the communities, which seems a crucial mechanism to empower the systematically marginalized local communities. However, to achieve FPIC is also wrought with difficulty. The logging permits are often issued before the start of consultation with local communities. The power differentials lead to the question whether the 'consent' comprises a fair outcome.[190]

To promote the CBFM, both the FSC and the LEI have established tailored schemes: the SLIMF and PHBML. Even with the vision to reduce the burden to communities caused by certification, these standards have only engaged very limited participation.[191] Numerous challenges prevent their uptake. Institutional capacity is an important one of them. To get certified and remain competitive in the market, an institution is required to 'help members secure markets and supplies, to achieve suitable economies of scale, and gain market power through joint bargaining, processing and purchasing of supplies and service'.[192] Neither an adat institution nor local government is suitable to conduct such tasks under market-based certification schemes. The communities need to develop specific organizations, usually cooperatives to fulfil the requirement for certification.[193] Certification costs remain high for communities who can seldom afford those costs without donor support. What is more, the expected price premium is not guaranteed.[194]

Field studies have been conducted on the sites which have received their certificate between 2004 and 2007.[195] All such sites were private land covered with small-scale teak plantations (rather than natural forests). With the aim of 'land rehabilitation, water conservation and the utilization of barren land', local communities had already started tree planting on these sites under government support (a forestation and reforestation programmes) long before certification.[196] The environmental quality had already improved significantly by then. Forest certification was later introduced to the local communities by NGOs and the private sector, who supported their application. The earlier government programmes had already motivated local communities to conduct environment-friendly activities and had prepared them for certification. A recent study also finds this positive synergy between public and private regulation in an FSC certified community forest. The certified villages had already joined the Social Forestry Programme (one CBFM model) before getting certified. Positive effects have been noted, including economic benefits to community

190 Bartley 2011, p. 533.
191 *Supra* section 3.6.3.1.3.
192 Hinrichs, Muhtaman & Irianto 2008, p. 50.
193 Ibid.
194 Ibid., pp. 58–61.
195 Ibid.
196 Ibid., p. 43.

members and increasing forest conservation behaviour.[197] On the one hand, such case studies show positive interaction effects between public regulation and forest certification have been achieved. On the other hand, they question the added value of certification schemes to governmental support. Moreover, the above analysis shows that local communities are still marginalized in the current land tenure system, and the progress of CBFM has been slow. This may suggest that most forest communities are still not prepared for the current certification schemes.

3.4.5 How does the interacted system address deforestation?

In what respect has Indonesia satisfied or not the necessary preconditions for the proper functioning of different types of property rights in countering deforestation? How do public and private regulations interact in Indonesia in establishing these property rights as well as in overcoming their limitations in countering forest degradation?

3.4.5.1 Public forests (state forests)

As discussed earlier, the BAL differentiates between public, private and customary land. The BFL 1999 defines public (state) forests as forests on public land. However, in practice, Forested Areas are often misunderstood as state forests. Due to conflicts within the BFL and between BFL and other regulations, the definition of Forested Areas is unclear, leading to different applications by the MoF and other government agencies/regional governments.[198] The different interpretations by various agencies contribute to conflicting permits and rapid deforestation.

70 per cent of land areas have been claimed as Forested Areas, creating challenges for forest management agencies to patrol the vast areas and to ensure the enforcement of forest regulation. Weakness of forest management institutions has been identified as a key cause of the rapid deforestation in Indonesia.[199] 'Limited resources and poor management of forestry data and information create difficulties for institutions responsible for forestry in demarcating state forest boundaries' and for monitoring forest-related activities.[200] In addition, collusion and corruption have also been important issues hampering forest management.[201]

Moreover, the capacity of various government agencies to regulate state forests in the public interest has been weakened by the decentralization process. Decentralization was supposed to increase local political

197 Harada & Wiyono 2014.
198 *Supra* section 3.6.3.2.2.
199 Indrarto *et al.* 2012, p. 12.
200 Ibid., pp. 12, 20–24.
201 Ibid., pp. 11–12, 24.

participation and to allow decision-making by those having a good understanding of local conditions.[202] However, it has been riddled with conflicting interests and attempts by the central government to regain power and reap the increasing regional revenues and local benefits from forest management.[203] This 'tug of war' between different levels of government has been aggravated by the passing of inconsistent legislation. The conflicts between the MoF and regional governments have been escalated by the lack of clear hierarchy between different legislations. Hence, whether the decentralization really has steered local participation in forest management and addresses deforestation is seriously questioned.

There has also been lack of coordination between governance agencies responsible for different sectors, such as forestry, agriculture and mining. These various agencies tend to adopt compartmentalized programmes and focus narrowly on sectoral objectives, 'partly because bureaucratic performance is assessed against sector-based targets'.[204] The lack of coordination has led to conflicting planning and permitting as well as to weak monitoring.

Certification has limited capacity to address the above-mentioned limits of public regulation. Clear forest tenure is a precondition for forests to get certified. Therefore, the lack of clear forest property rights acts as a barrier to the wide coverage of forest certification. Certification provides an additional layer of monitoring and enforcement. However, the coverage is still very limited. Moreover, certification has often been limited to the relatively well-performing sites and has limited capacity to address the large number of bad performers.

3.4.5.2 Private forests

As discussed above, a land certificate and the designation of local or regional governments are needed to establish private forests. However, the land registration process is slow and costly, leaving many forests unregistered, adding difficulties in formally establishing private forests. The regulation to allow the conversion of private land in protected and conservation forests to state forests further adds to the uncertain status of private forests.

Private enforcement requires the capacity of private rights holders to monitor the behaviour in and the status of their forests, and to resort to the judicial systems when disputes arise. The corrupt judicial system, however, has made the costs of resorting to private enforcement in the courts very high.[205]

202 Ibid., p. 28.
203 Ibid.
204 Indrarto *et al.* 2012, p. xi.
205 Ibid., p. 24.

As in the example of public forests, certification has only limited possibilities in addressing the above-mentioned problems.

3.4.5.3 Community forests

Property rights. Communal property rights lack clear boundaries of resources and users as well as suitable appropriation and provision rules, which define the content of the property rights. In Indonesia, the law only provides minimum recognition of customary rights over forests. Only from 2013 onwards, was it admitted as one type of forests distinct from state forests. However, the content and the procedure of establishing customary forests remain unclear. Moreover, because of the 'tug of war' among different levels of government and the lack of coordination between agencies of different sectors, conflicting decisions are common, leading to unclear boundaries for forests. The CBFM policies try to grant more rights concerning forest management to local communities. However, the rights granted to communities are often limited, failing to create sufficient protection incentives.

Certification recognizes customary rights and requires clear forest tenure as a precondition to certify forests. However, challenges to get certified exist for many community-managed forests, due to the high costs, demanding expertise and institutional needs associated with it. Though communities can also apply for certification by collaborating with forest companies, their bargaining power is weak, especially when their rights are not formally recognized by law. Once certified (for the already better performers with clearer boundaries), however, it can consolidate the rights of clients and strengthen clear boundaries.

Enforcement. Traditionally, forests were governed by the forest-dwelling communities, who developed their own governing rules 'that were woven around the socio-cultural features as well as the economic activities of local communities' and had their own self-governing institutions.[206] Under the communal property rights system (the community-based tenure system), 'a complex bundle of group and individual property rights' exists across communities.[207] The different systems have various rules concerning the conditions under which community members can clear new plots of land for agriculture and harvest forest products.[208] For example, based on local experiences, such rules control the use of fire to clear land, include conditions to prevent erosion and maintain fertility. 'In this way, the adat system aims to maximize livelihoods by ensuring fertility and protecting the ecological processes upon which agriculture depends'.[209] As mentioned

206 McCarthy 2000, p. 102.
207 Ibid., p. 109.
208 Ibid., p. 102.
209 Ibid., p. 108.

earlier, the communities used to have their own governing institutions under the leadership of Pasirah, who manages the forests and decides the scope of individual rights. In these governing institutions, community members play a key role in determining management rules, with clear rights and responsibilities. Hence they have incentives to be involved in the daily monitoring of the forestry-related activities.

Since the colonial period, the state has tried to gain control over the forests, replacing the traditional adat governing systems with a public regulatory one. During the New Order period, the government tended to grant concessions to well-positioned commercial interests in state forests, where traditional communities used to live. Conflicting rights claims, however, are prevalent. Moreover, since 1979, the government has reorganized the local governing institutions according to the Village Government Law. The traditional adat institutions lost their role in governing forests and were replaced by village heads, who were responsible for implementing policies top down, rather than designing and implementing rules that fitted the adat systems. In the reorganization, sometimes geographically isolated communities were thrown together, making it difficult for the local government officials to effectively manage forests. Consequently, community members, who used to benefit from the protection of forests and took part in the self-governing institutions, were excluded from decision-making. Many forests became subject to conflicting claims from local communities and commercial interests who received concessions from the government. Therefore, the traditional way of relying on local communities to monitor forest-related behaviours has been seriously undermined. As mentioned before, the capacity of public regulators to protect vast state forests is limited. The lack of monitoring is followed by weak sanctions and a costly judicial system. Therefore, the replacement of customary self-governing institutions with governmental control has led to the dismantling of the traditional enforcement system. The limited capacity of the public regulatory system has often left a vacuum in forest governance.

Certification schemes can provide an additional layer of monitoring. However, current certification coverage in Indonesia is still limited, let alone for community-managed forests. As discussed earlier, special organizations such as cooperatives have been established to conduct the daily management. However, such cooperatives have often been criticized for 'poor governance and obscure management practices allowing for collusion and corruption'.[210] Therefore, unlike under the traditional governing institutions, reluctance of present day community members to engage in the cooperatives has been noted in some cases.[211] In addition, under the

210 Hinrichs, Muhtaman & Irianto 2008, p. 51.
211 Ibid.

certification schemes, it is the third-party certifiers who conduct monitoring rather than the resources users under the customary governing system. Compared to resource users, the certifiers are at a greater distance from the daily management of the forests. Moreover, certifiers may compete with each other in order to attract clients (forest managers). Sometimes, they 'offered to conduct minimum services (auditing), lowering standards to win contracts and satisfy funders' expectations'.[212] Therefore, whether certification can always provide an effective alternative to customary self-governing institutions in monitoring is questioned. In addition, unlike self-governing institutions and public regulation which can, in principle, rely on a variety of sanctions, the potential 'sanctions' that certification schemes can provide are limited. Private certification schemes are voluntary. Under the FSC scheme, as mentioned above, if the CB finds a violation of certification standards, it can issue minor CARs, major CARs and the suspension or revocation of a certificate. The strongest form is the latter option. Therefore, the deterrent effect certification schemes place on forest managers is limited.

Coordination. For community-managed forests, many individuals have stakes in them and coordination between these individuals is crucial. Under the adat system, self-governing institutions involving the broad participation of community members exists. This system, however, has been largely dismantled. Replacing traditional customary forests by state forests and allowing commercial interests to manage them makes the use of forests more complicated. More actors are involved, with conflicting claims, significantly increasing coordination costs. The public regulatory system itself does not offer a platform to solve the conflicts, but often simply requires the companies to negotiate with local communities. The bargaining power difference is further enlarged by legal and administrative practices that exclude customary rights. Therefore, a fair negotiation is not guaranteed.[213] Certification requires local participation in decision-making, by for example requiring companies to obtain 'free and prior informed consent' from communities before starting forestry. However, as discussed above, such requirements are often violated.

Information. The replacement of traditional communal governing institutions with public forests has also had a devastating influence on the availability of information. Traditional communities have lived in the forests for a long time and have depended on forests for their subsistence. Traditionally knowledge has been accumulated by local members which has helped them to maintain sustainability for many generations. Under the current system, it is up to the professionals, public/private regulators,

212 McCarthy 2012, p. 1876.
213 Ibid., pp. 1875–1878.

Table 3.1 Summary of the scores for the presence or absence of the necessary preconditions for property rights' effectiveness

Necessary preconditions for the proper functioning of property rights					
	Definition of property rights	*Enforcement*	*Coordination*	*Information*	*Scale of governance*
Public forests	Public forests are conflated with Forested Areas; the procedure for establishing Forested Areas is not clear, leading to conflicting permits. Clear forest tenure is a condition for forest certification. Therefore the unclear status of a public forest is a barrier for forests to obtain certification.	Limited government capacity to patrol the vast forests; corruption. Certification provides an additional layer of monitoring. However, the coverage is limited and certification usually applies to already well-performing forests.	Decentralization and compartmentalized legislation lead to conflicts between different departments and levels of government. Neither of them incentivizes the protection of forests. Certification has limited capacity to incentivize forest protection.	–	–
Private forests	Many private forests are not registered, therefore they have an uncertain status. Regulation allows changing private land in conservation forests into public land. Certification is limited by unclear forest tenure.	Enforcement relies on the capacity of rights holders to enforce their rights and when necessary, resorting to justice. However, the judicial system is slow and costly. Certification does not address this issue.	–	–	–

Table 3.1 continued

Necessary preconditions for the proper functioning of property rights

	Definition of property rights	*Enforcement*	*Coordination*	*Information*	*Scale of governance*
Community forests	Legislation only provides minimum recognition of customary rights. CBFM policies provide limited rights to communities. Certification recognizes customary rights. However, land conflicts, high costs and expertise requirements make certification difficult. Once certified, it can consolidate the rights and strengthen clear boundaries.	Public regulation replaced traditional self-governing institutions. However, the public enforcement capacity is insufficient. Certification adds an additional layer of monitoring. Certification relies on newly established cooperatives and third-party auditors for monitoring and enforcement. This is more costly than under traditional governing institutions, wherein local members enforce.	Traditional coordination mechanisms have dissolved to a large extent. Current public and private regulation implies more parties are involved: local community members, commercial enterprises, public and private regulators, adding to the costs of coordination.	Traditional knowledge is no longer appreciated when the regulatory system replaced traditional governing institutions.	–

to collect information concerning forestry, which is much more costly than relying on local members under the customary institutions.

How Indonesia performs regarding the preconditions for the proper functioning of property rights is summarized in Table 3.1.

3.4.6 Conclusion

Three layers of rules regulate forest tenure in Indonesia: customary law (adat), national law and private certification schemes. The interaction between the multi-layered systems has not been available to establish clear forest tenure, let alone to overcome the limitations of forest tenure in controlling deforestation (for summary, see Table 3.1).

First, inconsistencies exist in defining property rights over forests in the multi-layered system. Traditionally, most forests are managed by local communities under the adat system. The communal rights, however, have been replaced by state ownership in modern Indonesia. Forest certification schemes require secure land tenure and respect of customary rights. However, the recognition and protection for customary rights under national law have been tortuous. Customary rights were excluded during the New Order regime and got limited acknowledgement in the BFL 1999. The independence of the customary forest from the state forest was only confirmed recently. Since the requirement of clear forest tenure is not fulfilled, certification schemes have limited capacity to address the chaotic definitions in law in Indonesia. Even state forests suffer from unclear boundaries, due to the lack of coordination between levels of government and departments. Private property rights over forests also face uncertainties due to the large amount of unregistered forests.

Second, the enforcement of property rights also presents challenges. Traditional self-governing institutions have been dismantled and replaced with public regulation. Due to limited enforcement capacity and corruption, government has been unable to fill the enforcement gaps left by adat systems. The costly and slow judicial system also prevents effective private enforcement. Certification schemes provide an additional layer of enforcement, but have limited impact because of limited coverage.

Third, a coordination mechanism exists under the traditional self-governing institutions. Excluding customary communities and promoting commercial interests in the forests traditionally controlled by communities has led to many conflicts between different actors, and to increasing coordination costs as well as to a loss of information gathering and sharing.

Altogether, a multi-layered forest governance system has developed in Indonesia, which suffers from contradictions and lacks coordination both within public regulation, and between public and private regulation.

3.5 Case study: Bolivia

3.5.1 Environmental problem: deforestation

Bolivia has 57.2 million hectares of forest, composing 53 per cent of its total land area.[214] 80 per cent of its forest coverage is located in the lowlands while the rest is in the highlands.[215] Bolivia is a landlocked and poverty stricken country.[216] Around 62 per cent of the total population self-identify as members of an indigenous group.[217] 80 per cent of the indigenous population is categorized as poor.[218]

Although being a forest-rich country, Bolivia has suffered significant deforestation and forest degradation. Since the mid-1980s, the deforestation rate has increased steadily from 0.33 per cent (during 1987–1991) to 0.53 per cent in 2006 and 2010.[219] Generally speaking, Bolivia is in the second stage of forest transition (frontier stage) – with still large forest coverage and an increasing deforestation rate.[220] However, regional differences exist, including some areas still in the virgin forest stage, others in the frontier stage, and yet others in the forest and agriculture mosaic stage.[221] The deforestation is most prominent in lowland forests.[222]

The Bolivian lowlands were traditionally occupied by indigenous groups. Historically, three waves of 'land occupation' (by investors and labourers for rubber tapping since the 1890s, by colonists from the west for food production since the mid-1950s, and by medium and large commercial agriculture since the 1980s) have created conflicts between indigenous people and other groups, and furthered pressure on the forests.[223] Migration from the west, population growth in the colonist settlement areas and increased demand for food associated with urban population growth are still important drivers of deforestation today.[224] The problems associated with demographic factors and conflicts between different groups have further been amplified by institutional changes, such as policies concerning access to land and land tenure, strong pro-agriculture policies, and weak forest regulations.[225]

214 FAO 2010, pp. 223, 228.
215 Müller, Pacheco & Montero 2014, p. 3.
216 Nittler & Nash 1999, p. 32.
217 Barrera 2011, p. 8, available at: http://papers.ssrn.com/sol3/papers.cfm?abstract_id=1920297.
218 Uberhuaga, Larsen & Treue 2011, p. 81.
219 Müller, Pacheco & Montero 2014.
220 Hosonuma *et al.* 2012, p. 9.
221 Müller, Pacheco & Montero 2014, p. 52.
222 Ibid., p. 8.
223 Ibid., p. 3.
224 Ibid., pp. 17–18.
225 Ibid., pp. 17–19; Pellegrini & Dasgupta 2011.

3.5.2 Good governance

Bolivia is a lower-middle income country with annual GDP of US$33.00 billion in 2014.[226] According to the World Bank WGI project, Bolivia ranks 48.28 regarding voice and accountability, 31.55 regarding political stability, 29.81 regarding government effectiveness, 20.19 regarding regulatory quality, 12.50 regarding the rule of law and 29.81 regarding control of corruption.[227] The score of the World Bank WGI indicator on control of corruption is similar to the rank of corruption control from Transparency International. According to Transparency International, Bolivia is ranked 99 of 167 countries.[228] Hence, in terms of good governance, Bolivia scores average or below average in all respects.

3.5.3 Property rights as well as public and private regulation

3.5.3.1 Property rights

The first Forestry Law of Bolivia was enacted in 1974, which declared the state owner of all forests and required users of forests to obtain permits from the state.[229] These permits were issued in the form of logging contracts, which 'could only be obtained by registered enterprises' and hence excluded local and indigenous communities.[230] The exclusion resulted in fierce social conflicts among stakeholders and culminated in a demonstration by indigenous people in 1990.[231] During this period, forest companies dominated forest extraction, which operated over 21 million hectares of forest.[232] This period was characterized by corruption and weak enforcement capacity, which made the logging virtually unregulated and led to deforestation.[233]

The indigenous movement, and deteriorating economic conditions, combined with growing concerns about deforestation and forest degradation led to the land and forest law reform in the mid-1990s.[234] The reform was also influenced by the rise of neoliberalism in the 1980s and 1990s.[235]

226 http://data.worldbank.org/country/bolivia.

227 http://info.worldbank.org/governance/wgi/index.aspx#reports.

228 www.transparency.org/cpi2015?gclid=CMK59ZPMwMsCFRBmGwodsSQPKg.

229 Benneker 2008.

230 Peña-Claros & Dockry 2010, p. 202, available at: www.tropenbos.org/file.php/645/11-m.pe%C3%B1a-claros-r.guzman-m.dockry.pdf.

231 Ibid.

232 Fourteen million of them, however, located in private land or traditional indigenous areas. Pacheco, De Jong & Johnson 2010, p. 272.

233 Peña-Claros & Dockry 2010, pp. 202–203.

234 Ibid.; Müller, Pacheco & Montero 2014, pp. 24–25.

235 Redo, Millington & Hindery 2011, pp. 231–233. Since the 1980s, Bolivia started a deep structural reform with neoliberal orientation. See Müller, Pacheco, & Montero, 2014, p. 24.

Typical measures included, 'deregulation of domestic market, trade liberalization, privatization of public companies, public subsidy reduction and state role reduction'. The neoliberal reform on the one hand created pressure on forests by promoting agricultural development and on the other hand shaped the forest rights and regulation policies.

The documentation of land ownership had long been a problem in Bolivia and conflicting land claims were very common.[236] In order to solve such conflicts and to establish a national, reliable property register, the National Institute for Agrarian Reform Law (INRA Law) was enacted in 1996.[237] It clarified the process for land entitlement and granted indigenous communities an exclusive right to communal lands via establishing Tierras Communitarias de Origen (Native Community Lands) (TCOs).[238] Since then, three types of land ownership have existed in Bolivia: public, private, and communal ownership via TCOs. The INRA Law also created new institutions such as the INRA ('monitoring the implementation of the rural cadastre, the delimitation of public lands and the expropriation of private lands') and the Agrarian Superintendence (hereinafter SA, classifying lands and supervising land-use plans).[239]

Also in 1996, the new Forestry Law was approved (Law No. 1700). Although it still claimed all forests are state-owned, it broadened the rights of indigenous people and local communities to utilize forests.[240] It also regulated forestry practices to promote good forest management.[241] An independent public agency established as the competent authority under the Forestry Law, the Forestry Superintendence (hereinafter SF, responsible for approving forest management plans, supervising and controlling forest operations).[242]

A new round of reforms started in 2006, when President Evo Morales came to office, marking a shift towards post-neoliberalism.[243] In this period, 'the role of government in promoting economic and social development' was re-emphasized and many hydrocarbon and mining companies were renationalized.[244] The reform also changed the regulation of land rights and forest management. The INRA Law was revised by the 'Law of Community Redirection of the Agrarian Reform' (Law No. 3545).[245] The land

236 Boscolo & Vargas Rios 2007, p. 192.
237 Ibid., pp. 192–193.
238 Ibid., p. 193; Redo, Millington & Hindery 2011, p. 232.
239 Pacheco 2004a, p. 11, available at: www.fao.org/forestry/19583-06647b501b992f13fa80edbbb8942712c.pdf; Boscolo & Vargas Rios 2007, p. 193; Müller, Pacheco & Montero 2014, pp. 25–26.
240 Article 4, 28 Forestry Law 1700; Real 2002, p. 51.
241 Peña-Claros & Dockry 2010, p. 203.
242 Müller, Pacheco & Montero 2014, p. 25.
243 Redo, Millington & Hindery 2011, p. 233.
244 Pacheco, De Jong & Johnson 2010, p. 275.
245 Müller, Pacheco & Montero 2014, p. 30.

regularization and titling process under the 1996 INRA Law had long been criticized as bureaucratic and costly and has been modified under the new law. An important institutional change had also been implemented during this period: the SF and SA have been replaced by the Forest and Land Inspection and Social Authority (ABT). It is no longer an independent agency but under the aegis of the Ministry of Rural Development.[246] Efforts have been made to revise the 1996 Forestry Law, but consensus has not been achieved yet.[247]

3.5.3.2 Public regulation

Several public regulatory instruments have had significant influence on the forest tenures, including decentralization, concessions, and performance standards for forest managers.

Decentralization. Bolivia has traditionally been a centralized country, where local government had quite limited authority and resources to decide and implement policies.[248] However, in the 1990s, a decentralization reform started, marked by the Popular Participation Law of 1994 (No. 1551) and the Administrative Decentralization Law of 1994 (No. 1654).[249] The decentralization process started in response to the need to institutionalize local participation and forest regulation reforms.[250] The reform tried to shift authority from central government to prefectures (via the Administrative Decentralization Law) and municipalities (via the Popular Participation Law), with special focus on the latter.[251] The Prefectures are responsible for formulating forest development plans at departmental level, public forest education, forest research and developing a programme to strengthen municipal institutional capacity.[252] The municipalities are responsible for monitoring logging activities, raw material supply and processing, deciding municipal forest reserves to be assigned to local communities (ASLs) and formulating land-use plans at municipal level.[253] To carry out these responsibilities, municipalities are required to create municipal forestry units (Unidades Forestales Municipales, Municipal Forestry Unit UFMs).[254] In the new round of reforms since 2006, the earlier decentralization which shifted power to municipalities has lost support and prefectures have started to regain more importance.[255]

246 Ibid.
247 Ibid.
248 Ibid., p. 8.
249 Pacheco 2005, available at: www.cifor.org/publications/pdf_files/interlaken/pablo_pacheco.pdf.
250 Ibid., p. 3.
251 Ibid., pp. 9–10.
252 Ibid., p. 15. Boscolo & Vargas Rios 2007, p. 196; Pacheco 2004b, p. 91.
253 Pacheco 2004b, p. 91; Pacheco 2004a, p. 12.
254 Pacheco 2005, pp. 15–16.
255 Müller, Pacheco & Montero 2014, p. 30.

Concessions. Since the 1996 reforms, concessions have been used to grant rights to access and harvest forests on public land to private actors and indigenous people. Private enterprises can choose to maintain their long-term logging contracts awarded according to the previous forestry law (20 years) or convert them to concessions (40 years with automatic renewal every five years).[256] Local communities can obtain concessions for forests on public land by organizing themselves as associations (ASLs).[257] The municipalities are required to delimit the boundary of municipal forest reserves (20 per cent of public forests within municipalities) and the reserves are to be allocated to traditional forest users and indigenous people organized in as ASLs under concession contracts.[258] Authorization is also needed for actors to use forests on private land and community land. The owners of private land can access the forests on their land when a few legal requirements are satisfied.[259] The same applies for TCOs on communal lands.[260] In summary, nowadays, five types of forest tenure exist: industrial concessions, former long-term contracts; local communities associations (ASLs), indigenous territories (TCOs) and private properties.[261]

The 2006 reform tried to further strengthen the position of indigenous people. Since then, public forests are no longer allocated to concessions but only to communities.[262]

Performance standards regarding forest management. Different rules apply according to the types of forest tenure. Forest concessionaires and private landowners (for forests of more than 200 hectares) are required to develop and implement a forest management plan and an operational annual plan, for which detailed, scientifically based technical standards are required.[263] The regulation for small private properties and forest community operations (TCOs) has been modelled on this regulation and has created barriers preventing participation in commercial forest exploitation.[264] Another major change introduced by the revised Forestry Law is the replacement of the harvest volume-based tax system with an area-based tax system, with the aim of restricting corruption and encouraging more intensive (rather than selective) logging.[265] The regulatory regime for

256 Article 29.

257 Article 31.

258 Müller, Pacheco & Montero 2014, p. 29.

259 Article 32, Forestry Act. Boscolo & Vargas Rios 2007, p. 194; for details, see following section 3.7.3.1.

260 Ibid.

261 Boscolo & Vargas Rios 2007, pp. 193–194.

262 Noting this does not influence the existing forest concessions. Pacheco, De Jong & Johnson 2010, p. 275.

263 The technical standards include for example, a minimum logging cycle of 20 years, a minimum cut diameter by species, mandatory reserves in harvested areas, infrastructure development, stream protection and so on. See Pacheco 2004a, p. 9.

264 Müller, Pacheco & Montero 2014, p. 29.

265 Boscolo & Vargas Rios 2007, p. 197.

non-timber forest products mimics the regime for timber products.[266] The competent authority under the Forestry Law, the SF, was designed as an independent agency deriving its direct finance from forest taxes.[267] To promote the transition to the new regime, Fonabosque, a national forestry development fund, was created.[268]

3.5.3.3 Private regulation

The 1996 Forest Law was developed following a multi-stakeholder dialogue involving the private sector, environmental NGOs, indigenous groups, governments, and international assistance agencies.[269] A parallel approach at that moment was to develop a certification regime with the goal of encouraging sustainable forest management and to boost its timber exports.[270] A non-governmental organization, the Bolivian Council for Voluntary Certified Forest Management (Boliviana Certificación Forestal Voluntaria, Bolivian Council for Voluntary Forest Certification CFV) was established in 1995 to develop national certification standards. The CFV standards were approved in 1998 and endorsed by the FSC in 1999, making it the first FSC national initiative in the developing world.[271] The certification standards are 'virtually the same as' the standards under the revised Forestry Law.[272] The certification presents more stringent standards only in social issues, such as labour rights, community organization and consensus building.[273] After the first certification was awarded to an indigenous community, the certification coverage increased quickly in Bolivia and reached 2.21 million hectares in 2005.[274] This was the country with the largest certified tropical forest area,[275] which was very notable in this poor country with widespread corruption and limited access to export routes.[276]

In recent years, the Bolivian forest market experienced a significant change. Exports to the US and America fell due to the economic downturn. At the same time, the regional (South America), Asian and domestic market increased significantly, which diluted the incentives to seek forest certification.[277] The land reform decreased the forest areas under concessions and made communities the most important forestry players. The latter, however,

266 Pacheco 2004a, p. 9.
267 Boscolo & Vargas Rios 2007, p. 196.
268 Ibid.
269 Contreras-Hermosilla & Vargas Rios 2002, pp. 4–5.
270 Nittler & Nash 1999, p. 32.
271 Ibid.; Quevedo 2006, p. 318.
272 Nittler & Nash 1999, p. 34; Jack 1998.
273 Nittler & Nash 1999, p. 34; Quevedo 2006, p. 319.
274 Espinoza & Dockry 2014, p. 82.
275 Duery & Vlosky 2005; Espinoza & Dockry 2014, p. 82.
276 Ebeling & Yasué 2009, p. 1145.
277 Espinoza & Dockry 2014, p. 83.

faced more challenges in obtaining certification.[278] The government also shifted away from its early support of the FSC certification scheme, fearing that foreign NGOs would dominate the forest governance narrative.[279] Instead, a national certification scheme was created in November 2014 under the ABT.[280] A national system is believed to be less costly and more accessible since the NGOs lack financial capacity to seek FSC certification.[281] Against this background the area of FSC certified forests started to decrease sharply since 2009. Only 0.87 million hectares of forests remained certified in 2013.[282]

3.5.4 The interaction between public and private regulation

3.5.4.1 Interaction on paper

To tackle the chaos created by land conflicts and to promote a more equitable distribution of benefits between various stakeholders, a land reform started in the mid-1990s. The reform merged the land rights and forest use rights.[283] Under the new regime, five categories of access rights exist. On the public land, long-term logging contracts, commercial concessions, and ASLs can be established. In addition, the reform recognized private property rights and 'the collective property of indigenous people over the land that these groups occupied traditionally' as TCOs.[284] Both the TCOs and private landowners are allowed to commercially use the forest on their own landholdings.[285] However, the indigenous people and private owners need to justify their economic and social function (FES) to acquire and maintain ownership.[286] 'The FES was relatively broadly defined, including not only agricultural but also forest and conservation activities'.[287]

The land regularization and titling process was slower and more expensive than originally expected. To accelerate the process, especially the titling of TCOs, a new round of land reform started in 2006. The forest concessions were cancelled and all public lands were to be allocated to communities and landless people.[288] A more specific definition of FES was provided: the actors satisfy the FES if 'they intend to meet the "best

278 Ibid., p. 84.
279 Ibid., p. 86.
280 http://abt.gob.bo/index.php?option=com_content&view=article&id=653:lanzamiento-certificacion-forestal&catid=8&Itemid=211. Since this is a totally new certification scheme unfortunately no further information is available as to its practical implementation, nor on its effectiveness.
281 Espinoza & Dockry 2014, p. 86.
282 Ibid., p. 82.
283 Pacheco 2011, p. 3.
284 Ibid., p. 4.
285 Boscolo & Vargas Rios 2007, p. 194; Pacheco 2011, p. 3.
286 Pacheco 2011, p. 3.
287 Pacheco & Benatti 2015, p. 471.
288 Ibid., p. 473.

interests" or welfare for all residents or achieve economic development through the "best use" of the land'. The best use may take the form of agricultural activities, forestry, biodiversity conservation or ecotourism.[289] If the FES is not complied with, the state can expropriate the land for redistribution.[290] Therefore, through the two rounds of land reform, the law formally recognized the land tenure of indigenous people as well as local communities and attempted to remove overlapping land claims and conflicts, at least on paper. In spite of the clearer land tenure, its security is not guaranteed due to the frequently changing policies.

The CFV (FSC Bolivia) approved the 'Standards for Certification of Forest Management of Timber-Yielding Products in the Low Lands of Bolivia' (FSC Bolivia standards) in 1998.[291] The standards adhere to the Principles and Criteria (P&C) of the FSC and require the forest management to be environmentally appropriate, socially beneficial, and economically feasible.[292] The FSC Bolivia standards contain several principles related to forest tenure and rights of indigenous people or local communities. Principle 2 (Tenure and Use Rights and Responsibilities) requires that 'long-term tenure and use rights to the land and forest resources shall be clearly defined, documented and legally established'. Principle 3 (Indigenous People's Rights) stipulates 'the legal and customary rights of indigenous people to own, use and manage their lands, territories and resources shall be recognized and respected'. These general principles are further illustrated by specific criteria and indicators. In Principle 4 (Community Relations and Worker's Rights), there are also some indicators requiring the enhancement of local communities' well-being.[293]

As discussed earlier, the standards established under the FSC Bolivia certification and the domestic forest regulation are very similar. Regarding land tenure and the rights of indigenous people/local communities, their standards are also highly compatible. The rights of indigenous people and local communities are formally recognized in the two waves of land reform. Institutions have been created to regularize and register the landholdings and relatively clear standards and processes are established on paper. These are in line with the requirements of the FSC standards.

3.5.4.2 Interaction in practice

3.5.4.2.1 THE INTERACTION ACROSS SECTORS AND GOVERNMENT AGENCIES

In addition to forest agencies and forest policies, many other government

289 Redo, Millington & Hindery 2011, p. 233.
290 Ibid.
291 Standards FSC-STD-BOL-04-2000.
292 Ibid., 2.
293 Criterion 4.1. The communities within, or adjacent to the forest management area should be given opportunities for employment, training, and other services.

agencies and policies also have an important influence on forest management. The lack of coordination among the agencies and policies however, causes many conflicts in forest tenure.

For example, the INRA is responsible for titling TCOs according to the INRA Law of 1996. According to the Forestry Law of 1996, municipalities are provided with 20 per cent of public forests in their jurisdictions as municipal forest reserves to be allocated to ASLs. For fear that the claims for TCOs may endanger their portions of municipal forests reserves, the municipality may oppose the TCO adjudication. Therefore, the lack of coordination between different authorities may aggravate the contested, long, land regularization process.[294]

Lack of coordination also exists between INRA and SF, which are responsible for awarding TCOs in indigenous land (according to INRA Law) and concessions on public land (according to Forestry Law) respectively. The INRA Law and Forestry Law were issued in 1996, when indigenous rights had just begun to be recognized and logging contracts were established on much indigenous populated land. The INRA is required to respect indigenous rights. However, according to the Forestry Law, these logging contracts can automatically be converted to concessions. This has resulted in the fact that 27 out of 86 concessions awarded by 1997 were overlapping with indigenous territories.[295]

In addition, 'the nesting of land reform in the agrarian reform paradigm' and the neglect of forestry issues have also contributed to the failure in promoting sustainable forest management.[296] In Bolivia, most land identified for redistribution is covered by forest.[297] However, both the conditions for land titling and accompanying policies provide preferable treatment of agricultural over forest uses. As mentioned earlier, to be titled with private ownership or TCOs, the private actors and local communities need to prove their use satisfies an economic and social function (FES). In theory, both forest management and agricultural use can be used to justify FES. Conducting forest management according to the legal standards is costly. Many actors choose to clear forests and cultivate crops or livestock to prove FES and secure their land tenure.[298] In addition, many supportive policies are in place to support agricultural development, such as zero interest loans, subsidies and other instruments aimed at achieving food sovereignty.[299] However, few such accompanying policies are available for forest management activities.[300] Even in the new land reform since 2006, the

294 Real 2002, pp. 79–81; Pacheco 2004a, p. 19.
295 Real 2002, pp. 50–52.
296 Pellegrini & Dasgupta 2011, p. 278.
297 Ibid., p. 279.
298 Redo, Millington & Hindery 2011, p. 232; Pacheco 2015, p. 471.
299 Pellegrini & Dasgupta 2011, p. 281.
300 Ibid.

neglect of forest issues is hardly addressed. Even the Forestry Directorate is excluded from the revision process of the INRA Law.[301]

3.5.4.2.2 THE INTERACTION BETWEEN FORMAL AND CUSTOMARY INSTITUTIONS

The above analysis shows that the lack of coordination between governments in different levels and sectors makes the establishment of TCOs difficult. Even when TCOs are established, they are often criticized for insufficient consideration of the existing informal institutions.[302] The TCOs grant collective rights to indigenous people at the territorial level and a new governing body has also been established at this level.[303] How to allocate the collective land and forests at the individual level, however, is usually decided by the traditional local governing organizations, usually at the village level.[304] A territory often includes diverse communities with different cultures and governing traditions. The competition between authorities has created undesirable effects in governing the forest resources. The territorial governing institution usually acts as a collective movement to advocate indigenous interests and to secure their land tenure, and is poorly equipped to allocate and administrate the resources internally.[305] Traditional local institutions are also weakened by the introduction of new rules and the long titling process. Based on different traditions, the interaction of territorial and local institutions has created very diverse modalities of land distribution and forest resources access modalities.[306] The land legalization process has been longer and more costly than expected. According to the 1996 INRA Law, 106.7 million hectares of land needed to be regularized by 2006. However, only 9.2 million had been regularized by 2006 and 5.7 million hectares of them were titled as TCOs.[307] The average costs of land regularization are estimated as US$2.75 per hectare.[308] Only after the 2006 reform, the process has been accelerated. Sixty-four million hectares of land had been titled by 2012 and 22.5 million of them as TCOs.[309] Besides, the conflicting claims concerning the settlement areas have often been settled by allocating titled lands to TCOs far away from the communities' traditional settlement, rendering it difficult for members to effectively manage them. Moreover,

301 Ibid.
302 Pacheco *et al.* 2008.
303 Ibid., p. 31.
304 Ibid.
305 Ibid., pp. 31–32.
306 The modalities vary from 'free access forest resource management, common-access and individual natural resources management' to 'individual tenure and natural resources management'. See Cardona *et al.* 2014.
307 Müller, Pacheco & Montero 2014, p. 36.
308 Ibid.
309 Ibid.

the delay in awarding property rights to disputed lands has induced illegal transactions between indigenous people and farmers who claim the same lands, leading to further forest clearing.[310]

3.5.4.2.3 THE ILL-SUITED PERFORMANCE STANDARDS UNDER PUBLIC REGULATION AND CERTIFICATION

The forest regulation under the Forest Law and forest certification scheme in Bolivia has also failed to incorporate and adapt to the local forest practice and governing institutions. As mentioned earlier, both public regulation and the CFV certification scheme adopt a scientifically based approach to forest regulation. They require the development of forest management plans and detailed technical requirements. Such requirements apply both to commercial concessions and to small private operators and local communities. Such an approach is, however, better suited to the bigger operators who are integrated in the market economy than to smaller producers.[311] Local communities often have norms concerning subsistence use or small-scale commercial use of forest resources, but are not prepared for a large-scale commercial use in terms of management organization, techniques, knowledge and finance.[312] Therefore, local communities have to create new organizations, 'community enterprises', or to seek collaboration with companies (CCP) to operate formally under the Forest Law and certification scheme. However, both modalities are not easily achievable.

With the support of NGOs, national government agencies or foreign donors, some communities have managed to establish community enterprises. A few communities have also got their forests certified, which was expected to bring them a securer international market and a price premium. However, the cultural differences between local communities and the market economy have provided the local members little incentive to engage actively in forest management and to learn the management, accounting and other skills needed to maintain a self-sufficient project.[313] Once the external support stopped, few of them could maintain their own forest management and certification.[314] Moreover, to secure market access through certification, the communities are required to provide a secure supply of timber (products) and to achieve high quality standards, which have proved difficult to satisfy.[315] Therefore it is not surprising that among the land titled to local communities and indigenous people, few had a management plan. Recalling that 5.3 million hectares had been titled as

310 Pacheco *et al.* 2008, pp. 32–33.
311 Ibid., pp. 40–44.
312 Ibid., De Pourcq, Thomas & Van Damme 2009.
313 De Pourcq, Thomas & Van Damme 2009; Markopoulos 2002; Becker & León 2000.
314 Ibid., p. 19.
315 Ibid., p. 15.

TCOs by 2006, only 0.723 million had a forest management plan by 2007.[316] Even after the 2006 reform, the development and approval of forest management plans in TCOs was still slow, varying from 0.045 million hectares to 0.417 million hectares per year from 2006 until 2013.[317] Only 0.9 million hectares obtained concessions via ASLs by 2007.[318] After that, only 0.074 million hectares of additional land were awarded concessions.[319] As for forest certification, there has even been a decrease. Although some communities had once obtained certification, all of them have lost their certification in the following inspections, due to failing to meet the requirements of the certifiers.[320]

Another opportunity for communities to conduct commercial use and get certification is to resort to CCP, relying on companies for their technology and organizational skills. However, this alternative also has obvious negative effects, 'including bad working conditions, limited development of communities' bargaining power and high transaction costs'.[321]

For many forest operations, the CFV standards include respecting land tenure, indigenous people's rights and ameliorating the relationship with local communities. However, an overview of the FSC certification in Bolivia shows that auditors primarily focus on documentation, monitoring, environmental issues, and compliance with laws on workers rights which can be immediately complied with. Social issues, such as respecting the rights of local communities only form an insignificant part in the required conditions.[322]

In summary, the difficulties for local communities to conduct formal forest management under public regulation and forest certification pushed them to conduct illegal logging and to engage in informal markets. Almost two decades after the land and forest law reform, illegal logging in Bolivia is still prevalent and the deforestation rate remains high.[323]

3.5.5 How does the interacted system address deforestation?

In what respect has Bolivia satisfied or not the necessary preconditions for the proper functioning of different types of property rights in countering deforestation? How do public and private regulation interact in Bolivia in establishing these property rights as well as in overcoming their limitations in countering forest degradation?

316 Peña-Claros & Dockry 2010, p. 202.
317 www.ine.gob.bo/indice/EstadisticaSocial.aspx?codigo=80301.
318 Peña-Claros & Dockry 2010, p. 202.
319 www.ine.gob.bo/indice/EstadisticaSocial.aspx?codigo=80301.
320 De Pourcq, Thomas & Van Damme 2009, pp. 14–15.
321 Medina, Pokorny & Campbell 2009, p. 410.
322 Nebel *et al.* 2005, p. 183.
323 Pacheco, De Jong & Johnson 2010, p. 274; Redo, Millington & Hindery 2010, p. 236.

3.5.5.1 Public forests

Although the state owns all forests in Bolivia, the land on which forests grow can be owned by public or private parties, or indigenous people/communities. Public forests here refer to the forests that occupy public land. As discussed earlier, three types of forest tenure can be established in public forests: long-term contracts, concessions with private actors, and ASLs for indigenous people.

Property Rights. Before 1996, only companies were allowed to obtain logging contracts to access public forests. The 1996 reform converted the logging contracts to concessions and allowed indigenous use of public lands via ASLs. This reform tried to increase the certainty of property rights and provided incentives for sustainable management. As mentioned earlier, during the same period, certification schemes started to develop. The Forestry Law and certification scheme include similar standards. The similarity in standards, together with external support and the expectation of enhanced access to the international market, promoted the fast spread of certification in the initial stage. Because the standards were more suitable for large-scale commercial use than indigenous forest use, it is mainly the concessioners that got certified.

Since 2006, however, the concession system has been cancelled and no more concessions have been awarded. The ensuing uncertainty of property rights has provided little incentive for concessioners to continue seeking certification.[324] Moreover, clients expected certification not to add much to the burden of public regulation given the similarities in standards. However, as discussed later, the lack of enforcement of public regulation has left substantial gaps between the two regulatory schemes. Together with the shift to non-environmentally sensitive markets and the loss of support from the government, certified areas have reduced sharply.

Enforcement. Under public property rights, the enforcement capacity of governmental agencies is crucial. From 1996 until 2006 SF has been the responsible agency for the implementation of forest regulation. It was designed to be financed by timber tax, part of which was also expected to be transferred to municipalities for their establishment of UFMs and local regulation. However, SF was poorly staffed. Tax revenues shrank, generating insufficient finances to SF.[325] Even less funding was transferred to municipalities and UFMs have never even been established in some of them.[326] The lack of finance, staff and technology has left forest regulation largely unimplemented.[327] The introduction of forest certification has provided an additional layer of monitoring. However, only a small part of the vast public forest area has been covered by certification.

324 Espinoza & Dockry 2014, pp. 85–86.
325 Pellegrini 2009, pp. 14, 34 note 7.
326 Pacheco 2004b, p. 95.
327 Ibid.; Pellegrini 2009, p. 14; Peña-Claros & Dockry 2010.

3.5.5.2 Private forests

After the 1996 reform, the rights for private owners to use forests on private land have been acknowledged. However, in many cases land ownership continued to be contested by different groups, due to the waves of land occupation mentioned above.[328] The land regularization process started under the INRA Law, trying to sort out the chaotic land tenure picture and granting titles to private parties and indigenous people. The process, however, has been protracted and costly. Moreover, to obtain the titles, private parties and indigenous groups needed to prove that their land use satisfied FES requirements. The application of such requirements, however, is more favourable to agricultural use than to forestry use.

Relying solely on private parties to enforce forest tenure is difficult, due to frequent invasions of forests and illegal logging.[329] Hence, the support from public regulators is important; however, it is criticized as slow and unreliable.[330] The SF is sometimes also reluctant to enforce the law strictly and to evict outsiders.[331]

3.5.5.3 Community forests

Property Rights. The 1996 reform has allowed indigenous groups to access forest resources, via either establishing indigenous ownership over the land (TCOs) or by establishing concessions on public land (ASLs). The new government has further prioritized the needs of indigenous communities in the land regularization process. This is in line with the certification standards, which require the protection of customary rights and a clear forest tenure.

However, recognition has been insufficient to guarantee the establishment of clear property rights. The land regularization and titling process has been criticized for being bureaucratic, costly and slow. For the TCOs, an additional governing level has been introduced (territorial level) on top of traditional governing institutions (usually villages). Diverse and sometimes geographically remote villages shared the same TCO, weakening the traditional governing institutions. Formal territorial level boundaries co-exist with traditional boundaries, adding confusion and difficulties in daily management. In addition, disputes concerning settlement areas have often been settled by allocating titled land to TCOs far away from the communities' traditional settlement. This has rendered it difficult for members to effectively manage them.

Moreover, lack of coordination between different agencies and levels of government has extended the land regularization process. Delay in the

328 Ebeling & Yasué 2009, pp. 72–73.
329 Ibid., p. 73.
330 Ibid.
331 Ibid.

adjudication of disputed land claims has promoted illegal transactions between indigenous people and ranchers/farmers, leading to further forest clearing.

Enforcement. In theory, self-governing institutions of indigenous groups can monitor behaviour in the forests and ensure the enforcement of the communities' property rights. Current Bolivian law recognizes indigenous ownership over land via TCOs. However, the introduction of formal institutions and norms has jeopardized the traditional governing institutions and monitoring which used to be in place. For example, a study of forest governance in Yuracare communities before and after the introduction of community forest management required by law has exposed such dangers.[332] The Yuracare people used to have their indigenous rules of timber harvesting, which, although not scientifically founded, were congruent with local ecologies. The resource users who move around the forests were to conduct monitoring and report violations. The traditional monitoring by each individual of the communities has been replaced by a forestry professional '[who] the government requires the forest management plan holders to utilize for periodic assessment studies'.[333] As a result, the indigenous people no longer participate actively in the monitoring of the territory, leading to the loss of many potential observers. Therefore the neglect of local context has created challenges in controlling illegal logging and promoting sustainable forest management.[334]

Certification has provided an additional layer of monitoring. However, due to cultural differences, expertise and budget limits as well as the unfavourable standards for community-managed forests, all the certified community forests have lost their certificates.

Coordination. In the traditional governing institutions community members have been involved in decision-making and enforcement. However, competition between the newly introduced TCOs and traditional institutions has weakened the traditional collective-choice arrangement. To get certified, communities need to either establish a community enterprise or resort to collaboration with companies. Either way, a more complicated governance system has been established, rendering the coordination among various actors more demanding.

Information. In designing the specific forest management standards, both public law and certification have adopted a western scientifically based approach and have ignored local knowledge. Such standards are based on specialized knowledge increasing the demand of forestry professionals (for designing and implementing management plans).[335] They are hence more favourable to commercial concessions than to a community-managed forest.

332 León *et al.* 2012.
333 Ibid., pp. 202–203.
334 Pacheco *et al.* 2008.
335 Pellegrini 2009, pp. 13–14.

Table 3.2 Summary of the scores for the presence or absence of the necessary preconditions for property rights' effectiveness

Necessary preconditions for property rights' proper functioning					
	Definition of property rights	*Enforcement*	*Coordination*	*Information*	*Scale of governance*
Public forests	The 1996 reform allowed long-term contracts and concessions, adding to tenure security. Certification spread fast. The subsequent reform of 2006 cancelled concessions, hence generating uncertainty for concessionaires. Certification shrank consequently.	Weak public regulatory capacity; lack of financial resources and staff. Limited additional monitoring by certifying bodies due to limited coverage.	–	–	–
Private forests	Protracted and costly land regularization process. The requirement (FES) to establish private land ownership is more favourable to agricultural than to forestry use.	Private enforcement is difficult due to conflicts and illegal logging. Governmental support of proprietors is limited.	–	–	–
Community forests	The law recognizes customary rights. It established a new level of governing institutions on top of the existing one, sometimes joining remote villages, causing conflicts about entitlements. Traditional and formal boundaries co-exist. Lack of coordination between governmental agencies leads to protracted land regularization process.	Formal institutions replaced traditional ones, which are less suitable for the local culture, making monitoring more difficult and costly. All community forests which were once certified have lost their certification.	A more complicated governance system came into existence: TCO under the law; traditional institutions; cooperatives or partnership under certification. The complicated system adds to coordination costs.	Both public and private regulation focus on western scientifically based knowledge rather than on local experiential knowledge.	–

This has impeded many communities from getting certified. A few communities have managed to get certified with the support of NGOs and the government. However they were not able to become self-sufficient and to maintain their certificates due to cultural, financial and expertise limits. Hence, the role of certification in securing forestry management by communities has been limited.

How Bolivia performs regarding the preconditions for the proper functioning of property rights is summarized in Table 3.2.

3.5.6 Conclusion

The state owns all forests in Bolivia, but the land on which forests grow can be managed by the public, private parties and indigenous groups (via TCOs). Three types of forest tenure can be established on public land: long-term contracting, concessions, and ASLs. None of them meet the preconditions of property rights' proper functioning, nor do public and private regulation overcome the limitations of the different types of forest tenure (for a summary, see Table 3.2).

Large-scale 'land occupation' movements have led to many conflicting claims. Since 1996, efforts have been made to regularize land tenure and grant titles to private and indigenous people. The new law also prioritizes indigenous rights. Such efforts are in line with the requirements of certification schemes. Therefore, comparatively clear forest tenure only exists on paper. The weakening of traditional institutions, an expensive and slow land titling process, uncoordinated behaviour from different government agencies and policies prioritizing agriculture over forestry have failed to put the formally clear forest tenure into practice.

Even after the reform, public enforcement remained weak, due to budget and staff constraints. In many respects, the Forest Law has not been implemented. Private parties have difficulties in enforcing their property rights by themselves, as indicated by widespread land conflicts and rampant illegal logging. For community forests, although they are formally recognized under law, and traditional governing institutions are not formally abandoned, the introduction of an additional level of institutions has weakened traditional communal enforcement. In theory, certification could provide an additional layer of monitoring. However its impact is limited due to a low coverage rate.

3.6 Case study: North America

3.6.1 The environmental problem: the health of riparian forests

Canada and the US are quite similar in terms of forest coverage. They are countries with the third and fourth largest forest areas in the world: respectively 310 million and 304 million hectares.[336] 34.1 per cent of land in

Canada and 33.3 per cent in the US, is covered by forests.[337] The land clearance for agriculture caused by the invasion of European immigrants during the nineteenth century has led to the loss of one third of the original forests in the US and less than 10 per cent in Canada.[338] Forest cover has been stabilized in both countries in the past 100 years, though the percentage of primary forest continues to decrease.[339]

Though both countries have completed forest transition and stabilized forest cover, forest fragmentation and losses still occur in some regions.[340] For example, forest areas are still decreasing in the west.[341] In addition, many other issues still influence forest health and the service forests provide, (for example, in terms of supporting ecosystems, enhancing the health of streams) in both countries, such as outbreaks of forest insect pest damage and forest fires.[342] As mentioned before, in Indonesia and Bolivia unclear property rights have led to high deforestation rates. Canada and the US face challenges from a different type of problem: forest degradation, which is associated with the loss of services provided by the forests. The establishment of rights has not sufficed to address degradation problems. Therefore, the focus is on how public and private regulations interact in order to overcome the limitations of property rights in addressing forest degradation.

The case studies of Canada and the US focus on one important service provided by the forests: maintaining healthy riparian areas. Poor management of riparian areas can deteriorate water quality, influence the terrestrial and aquatic ecosystems, increase flood and erosion risks and so on.[343] Some of these negative impacts have indeed taken place in North America. For example, the dramatic decline of wild salmon stocks on the Pacific Coasts of the US and Canada have been associated with forest harvesting practices.[344] Deforested riparian areas are also recorded which have led to narrowing stream channels and increasing water pollution.[345] This research focuses on how different governance instruments regulate forest management activities in order to protect riparian areas.

336 FAO 2010, p. 13.
337 http://data.worldbank.org/indicator/AG.LND.FRST.ZS.
338 Bryant, Nielsen & Tangley 1997; Smith & Darr 2004.
339 McDermott *et al.* 2010, p. 71.
340 Wulder, White & Coops 2011.
341 McDermott *et al.*, 2010, p. 71.
342 Ibid., p. 6; FAO 2010, p. xx; United States Department of Agriculture 2011, p. vii.
343 Harris 2010, p. 8; Phillips, Swift & Blinn 2000, p. 274.
344 McDermott *et al.* 2010, p. 95.
345 Sweeney *et al.* 2004; Verry & Dolloff 2000.

3.6.2 Good governance

Canada is a high income country with an annual GDP of US$1.785 trillion in 2014.[346] According to the World Bank WGI project, Canada ranks 95.57 regarding voice and accountability (out of 100), 91.26 regarding political stability, 95.19 regarding government effectiveness, 97.60 regarding regulatory quality, 94.71 regarding the rule of law and 93.75 regarding the control of corruption.[347] The World Bank WGI indicator on control of corruption is similar to the rank of corruption control from Transparency International. According to Transparency International, Canada ranked ninth out of 167 countries.[348]

The US is also a high income country with an annual GDP of US$17.42 trillion in 2014.[349] According to the World Bank WGI project, the US ranks 79.80 regarding voice and accountability, 66.99 regarding political stability, 89.90 regarding government effectiveness, 88.46 regarding regulatory quality, 89.90 regarding the rule of law and 89.42 regarding the control of corruption.[350] The World Bank WGI indicator on control of corruption is also similar to the rank of corruption control from Transparency International. According to Transparency International, the US ranked sixteenth out of 167 countries.[351]

3.6.3 Property rights as well as public and private regulation

3.6.3.1 Property rights

In Canada, over 90 per cent of forests 'are publicly owned and most of the authority for managing these forests rests with provincial governments'.[352] Such forests are usually referred to as crown forests.[353] The rest are mainly owned by private actors (individuals and firms). By 2008, only around 1.46 million hectares of forest were owned by local communities.[354]

Private parties can gain rights to harvest public forests via area or volume-based licences.[355] In addition to gaining rights to access and harvest, the licensees are obliged to prepare management plans according to regional land-use plans.[356]

346 http://data.worldbank.org/country/canada.
347 http://info.worldbank.org/governance/wgi/index.aspx#reports.
348 www.transparency.org/cpi2015?gclid=CMK59ZPMwMsCFRBmGwodsSQPKg.
349 http://data.worldbank.org/country/united-states.
350 http://info.worldbank.org/governance/wgi/index.aspx#reports.
351 www.transparency.org/cpi2015?gclid=CMK59ZPMwMsCFRBmGwodsSQPKg.
352 McDermott *et al.* 2010, p. 74.
353 Ibid.
354 Sunderlin, Hatcher & Liddle 2008, p. 8.
355 McDermott *et al.* 2010, p. 76.
356 Ibid.

In the US, 58 per cent of the forests are privately owned. Of the public forests, around a half (20 per cent of total forests) are owned by the federal government, administered by the US Forest Service (USFS).[357] The rest are owned by the states (8 per cent) and others (14 per cent). Only a small portion of the public forests are designated for use by local communities.[358]

The USFS is responsible for the management of federal forests and it is required to prepare forest management plans. With declining staff since the 1990s, harvesting is increasingly contracted out to private parties. However, the management is kept under government control.[359] How state-owned forests are managed and how the forest tenure is established varies between states.[360]

This research focuses on the dominant types of forests in North America: the public forests in Canada and the private forests in the US.

3.6.3.2 Public regulation

In Canada, the legislative competence concerning 'the development, conservation and management of forestry resources in the province' rests exclusively at provincial level.[361] Therefore, most provinces have adopted their own Forest Act, including basic rules concerning sustainable forest management.[362] The Forest Act usually applies to public forest and is implemented by forest ministers. In some provinces, private forests are subject to separate provincial or local legislation, implemented by separated authorities.[363] Generally speaking, the regulation of private forests is less intensive than that of public forests.[364]

Shared competence between federal, territorial and provincial governments does exist with some forest-related issues, such as 'science and technology, industrial and regional development and the environment'.[365] For example, the Federal Fisheries Act and some provincial laws regulate the forest activities which have a potential influence on both fish and fish habitats.[366] Some courts have defined riparian forests as fish habitats.[367] The Federal Fishery Act hence also provides the federal basis to protect ripar-

357 Ibid., p. 80.
358 Sunderlin, Hatcher & Liddle 2008, p. 8.
359 McDermott *et al.* 2010, p. 83; Berry 2006, p. 4.
360 McDermott *et al.* 2010, p. 83.
361 The Canadian Constitution Act of 1867, section 92A.
362 For a list of such Acts, see C&I Canada 2000, Table 4.4b.
363 E.g. New Brunswick, Quebec, see McDermott, Cashore & Kanowski 2012, p. 76; O'Carroll 2004, p. 19.
364 O'Carroll 2004, p. 13.
365 Canadian Council of Forest Ministers 2000, p. 60.
366 Ibid., p. 61.
367 E.g. R. v. British Columbia Hydro and Power Authority [1997] B.C.J. No. 1744 at par. 20 (B.C.S.C.) available at: www.canlii.org/bc/cas/bcsc/1997/1997bcsc11119.html.

ian forests.[368] It, however, does not require the establishment of riparian forest buffer zones.[369] At provincial level, riparian issues are regulated by provincial laws, such as Forest Acts and specific riparian forest regulations/guidebooks for public forests and other legislation for private forests.[370] The riparian zone protection can be mandatory in some provinces while voluntary in others.[371]

The US also has no single federal Forest Act, but relies on 'a complex array of interlocking state and federal laws'.[372] The environmental effects of forestry activities are addressed in several federal laws, such as the Clean Air Act, the National Environmental Policy Act, the Clean Water Act (the CWA), the Federal Insecticide, Fungicide and Rodenticide Act, the Coastal Zone Management Act and the Endangered Species Act.[373] States also regulate forestry activities, either by specific forest laws or general environmental statutes.[374] State and private forests can be subject to the same or different regulation.[375]

As far as riparian protection is concerned, the CWA plays an important role through regulating non-point source water pollution (NPS). The CWA requires each state to adopt programmes to achieve the NPS reduction and water quality goals.[376] To fulfil such obligations, states have developed best management practices (BMPs) for forest operations. The BMPs are 'a practice or usually a combination of practices that are determined by a state or designated planning agency to be the most efficient and practicable means of controlling point and non-point source pollutants at levels compatible with environmental quality goals'.[377] The BMPs are mandatory in some states, while voluntary in others. The states that opted for voluntary BMPs rely on education and technical assistance to promote their application and states that adopted regulatory BMPs use mandatory controls and enforcement strategies.[378] The CWA and its NPS clauses apply to all types of forests, including federal, state and private forests.[379]

One important instrument used by the provincial regulations/guidebooks in Canada and state BMPs in the US is the establishment of riparian buffer zones. In the riparian buffer zones, forest management activities and other potentially harmful activities are restricted or prohibited. It usually

368 O'Carroll 2004, p. 13.
369 Ibid., p. 17.
370 For an overview of the provincial laws, see ibid., pp. 19–65.
371 Ibid.
372 Lucier & Shepard 1997.
373 Ibid., pp. 194–196.
374 Ibid., p. 193.
375 www.defenders.org/publications/state_forestry_laws.pdf.
376 33 USC 1329.
377 Helms 1998.
378 Phillips, Swift & Blinn 2000, pp. 284–285.
379 Brown, Brown & Binkley 1993, p. 10.

consists of two elements: the widths of buffer zones for a certain stream type and the extent of management restrictions. The restrictions or prohibition may concern activities such as road construction and maintenance, timber harvest, the use of silvicultural chemicals, and various other practices.[380]

3.6.3.3 Private regulation

In North America, three certification regimes are actively operating: the regional standards of the FSC in both the US and Canada, the CSA which applies to all Canadian forests and the SFI which applies to forests in both countries. By July 2014, around 14 million hectares of forest in the US and 55 million hectares of forest in Canada were certified by the FSC, comprising 6 per cent and 22 per cent of the total forest area respectively.[381] CSA covered 40 million hectares of forest in Canada by November 2014.[382] SFI has certified more than 100 million hectares of forest in Canada and the US by 2014.[383] Together, 36 per cent of forest areas in North America are certified.[384] These certification schemes have standards covering various environmental, economic and social issues related to forest management. The adoption of riparian buffer zones is an important one of them. How the riparian buffer-zone rules under public regulation and forest certification interact in order to provide protection are further analysed in the following sections.

3.6.4 The interaction between public and private regulation

As discussed earlier, the goal of the case studies of Canada and the US is not so much to examine how public and private regulation interact in influencing the establishment and functioning of property rights to address deforestation. Rather the goal is to examine whether public and private property rights help to overcome the limitations of property rights in addressing forest degradation (riparian zone protection).

3.6.4.1 Interaction on paper

Riparian buffer zones are a good candidate for the comparison of the standards under certification and public regulation. This is because two quantifiable indicators exist for the comparison of their prescriptiveness and compatibility: the widths of buffer zones for a certain stream type and the extent of management restrictions. McDermott *et al.* have compared

380 Ice *et al.* 2004; O'Carroll 2004, pp. 19–65.
381 https://ic.fsc.org/preview.2014-fsc-market-info-pack.a-3730.pdf.
382 www.csasfmforests.ca/docs/fact_sheet_2014.pdf.
383 www.sfiprogram.org/files/pdf/2014-sfi-progress-report-spreads/, p. 14.
384 www.unece.org/fileadmin/DAM/timber/publications/FPAMR-2014-final_01.pdf, p. 17.

the standards of a few commonly used certification regimes (CSA, SFI and some regional FSC standards) and public regulation in some states/provinces in the US and Canada, and find that certification standards 'closely mimic government policy approaches'.[385]

As discussed earlier, riparian buffer zones are regulated under the BMPs in the US and under provincial laws in Canada. Three certification schemes, FSC, SFI, and CSA, apply broadly in both countries. McDermott *et al.* analysed the riparian protection requirements under the CSA, SFI standards and three regional FSC standards (the FSC-SE which covers ten US states, the FSC-PC which covers three US states and the FSC-BC which covers British Columbia). The CSA prescribes the procedures for developing substantive management thresholds regarding riparian buffer zones, but does not provide on-the-ground performance requirements. SFI requires forest operators to follow the BMPs for water quality protection and establishes procedural requirements for riparian protection measures where BMPs do not exist. BMPs hence became a mandatory requirement under the SFI scheme. FSC-SE also provides procedural requirements and requires the management to meet or exceed state BMPs. Both FSC-PC and FSC-BC have their own substantive requirements on the size of the riparian buffer zones and the extent of restriction.[386]

For the schemes with substantive requirements themselves (FSC-PC and FSC-BC), McDermott *et al.* compare the prescriptiveness of their standards with public regulation according to the widths of riparian buffer zones and the extent of restriction. They finds the FSC-BC standards closely mimic provincial policies. The FSC-PC standard resembles an average of the legal requirements in the three states it covers.[387] The FSC-SE and SFI refer to the governmental BMPs regarding riparian protection and the CSA also requires legal compliance.

The above analysis shows that forest certification usually adopts standards closely mimicking public policies regarding riparian protection in North America. In other words, a high degree of coherence exists between public regulation and certification regarding riparian zone protection.

3.6.4.2 Interaction in practice

In addition to the coherence in the rules regulating riparian buffer zones, public and private regulators also coordinate with each other in implementing the rules. British Columbia and Ontario are such examples.

385 McDermott, Noah & Cashore 2008, p. 47.

386 Ibid., pp. 62–64.

387 Ibid., p. 67. Though the substantive requirements under the FSC-PC are less restrictive than some state laws, all the discussed certification standards require forest operators to comply with applicable law. The legal compliance requirements hence ensure the certification standards regarding riparian protection are not more lenient than public regulation.

As mentioned before, the competence to regulate forest issues generally, and riparian buffer zones specifically is mainly vested in the provinces in Canada. In British Columbia, most forests (97 per cent) are publicly owned.[388] The Forest Practices Board is the regulatory body for public forests, which audits 'forest management performance by government and industry and to investigate public complaints'.[389] In addition to the organizations responsible to enforce public regulation, independent auditors are active in auditing the performance of forestry operations against certification standards. Most forests in British Columbia are certified according to one of the certification schemes.[390] The above analysis shows that certification standards are very similar to public regulation concerning riparian protection issues and certification regimes require legal compliance. Hence many synergies can be found in the monitoring and enforcement of the Forest Practices Board and auditors. With the aim of reducing duplication of effort by the auditees and of saving audit costs, the Forest Practices Board has made an assessment of land audited by the SFI, and established a process to rely on the work of auditors. The auditors and licensees agreed to open the certification audit files and share them with the Forest Practices Board. The SFI places high management, monitoring and documenting requirements on the forestry operations. Though such requirements may not directly lead to on-the-ground performance improvement, they make performance data more accessible and save time and costs for the Forest Practices Board in conducting their inspection. In addition, the Forest Practices Board can raise questions in their inspection, which help the industry and certifiers to re-examine the audit process and to improve their future audit.[391] Similar arrangements can also be found between the regulatory agency in Ontario (Ontario's Minister of Natural Resources, the OMNR) and two certification schemes, the CSA and FSC.[392]

In the US, before the introduction of private certification schemes, the states' BMPs had already been in place to set the specific riparian buffer-zone requirements. To evaluate whether the BMPs were actually applied and complied with and whether they were effective in achieving environmental goals, many states established monitoring and assessment programmes.[393] Such assessment programmes show that during the early

388 http://cfs.nrcan.gc.ca/pubwarehouse/pdfs/31835.pdf, p. 10.

389 Kamieniecki 2000, p. 181; noting that the Forest Practices Board is not one level of government, but an independent organization responsible for the enforcement of forest law in public lands in British Columbia.

390 The certified areas were estimated as 51.9 million hectares in British Columbia by 2009, http://cfs.nrcan.gc.ca/pubwarehouse/pdfs/35713.pdf, p. 46.

391 Cafferata *et al.* 2003.

392 Wood 2009, p. 94.

393 Ice *et al.* 2004.

days when BMPs were issued, the lower bound compliance rate was about 80 per cent, for example 80 per cent in Washington in 1980s, 84 per cent in Florida in 1985 and 78 per cent in Montana in 1990.[394] Some assessments show that the BMPs are helpful in reducing water quality impacts, preventing erosion and achieving other environmental benefits. Deficient effectiveness is usually associated with the non-compliance with the BMPs.[395] Lots of efforts have been made to increase the compliance rate of the BMPs.[396] The compliance rate of the BMPs has been increasing since the 1990s, and was estimated to be 89 per cent on average by 2009.[397]

Many factors, such as federal and state legislation and regulation, public pressure as well as forest certification, are believed to have influenced the compliance rate of the BMPs.[398] Public assessment programmes show some weaknesses in public monitoring, such as insufficient funding, a lack of (qualified) staff or a high turnover for staff, inconsistency in monitoring among states and so on.[399] As discussed earlier, forest certification schemes in North America have highly compatible standards with public regulation regarding riparian issues. Private auditors are auditing the performance of forest operators against these private standards. They hence can provide a beneficial complement to public regulation which suffered from insufficient funding and staff. Forest certification, moreover, can provide an additional incentive to comply with public regulation.[400] It can also create 'influences on the expansion of state BMPs to include additional subject areas'.[401] Some empirical research confirms the positive role of certification. For example, in Texas, the compliance rate of forest operations which transport timber to SFI mills is 94.9 per cent, and the compliance rate of those which transport timber to non-certified mills is 85.7 per cent.[402] This example, however, does not exclude the possibility of 'self-selection' problems. It is possible that these operations which seek SFI coverage already have a better BMP compliance rate than others. In addition, though the compliance rate of BMPs has been increasing since the 1990s, it is not clear whether this is because of the introduction of certification or other contributing factors such as governmental efforts.[403]

394 Ice, Schilling & Vowell 2010.

395 Ibid., p. 268; Ice *et al.* 2004; Brown, Brown & Binkley 1993.

396 Including technical assistance, education programmes, cost-share programmes and so on. For details, see Kilgore & Blinn 2004.

397 In some states, the increase in compliance rate is obvious, for example, the compliance rate in Montana increased from 78 per cent to 96 per cent, see Ice *et al.* 2004, p. 161; Ice, Schilling & Vowell 2010; Carraway *et al.* 2002, p. 31.

398 Ice, Schilling and Vowell 2010; Shepard 2006.

399 Ellefson *et al.* 2001, 2006; Kilgore, Ellefson & Phillips 2003; Husak *et al.* 2005.

400 Ice, Schilling & Vowell 2010, p. 268.

401 Kilgore, Ellefson & Phillips 2003, p. 120.

402 Carraway *et al.* 2002, p. 26.

403 Ice, Schilling & Vowell 2010. Carraway *et al.* 2002, p. 31.

In spite of these limitations, research shows that it is common for independent auditors to require forest operators to address riparian protection issues. An examination of 80 forestry operations certified by SmartWood (a major FSC certifier) according to FSC standards in the US shows that the percentage of operations given one condition concerning the improvement of aquatic and riparian areas ranges from 38 per cent to 100 per cent, depending on the type of FSC standards.[404]

In summary, the certification schemes adopt similar standards to public regulation regarding riparian buffer-zone protection; therefore they have limited capacity in promoting beyond-compliance behaviours. However, certification does have added values. By adding an additional layer of monitoring and inspection, certification has the potential to strengthen the enforcement and compliance of public regulation (the BMPs in the US). In addition, the converged standards can enhance the efficiency and effectiveness of enforcement. Private auditors can complement public regulation, given the limited resources to monitor the BMPs implementation. The coordination between public and private regulators can also enhance the efficiency of enforcement. To what extent certification's potential to strengthen public regulation has indeed been realized in practice, has not yet been firmly established.

3.6.5 How does the interacted system influence riparian zone protection?

In Canada and in the US, comparatively clear forest tenure has been established. Private property rights bring the costs and benefits of the right holders closer to the societal ones and create incentives to internalize the externalities caused by forestry activities. However, non-market values of the forests, such as the ecological function of the riparian forests are not materialized in the market and hence may not be considered by the private rights holders. The remaining externalities can act as a justification for external interventions, including both public and private regulation. The above analysis has shown the interaction between public regulation and certification concerning riparian buffer zones. This section examines how the interaction of such rules contributes to internalizing the externalities.

Different aspects concerning the transaction costs of property rights may also influence the capacity of public and private regulation to incentivize internalization, such as information creation, enforcement strengthening and scale setting. Providing information on forests, status and forestry activities, enables forest owners, managers and public and private regulators to make informed decisions about the protection of

404 Newsom, Bahn & Cashore 2006, p. 204. When the certification body finds the performance of its client does not satisfy the indicators required in the certification standards, it issues a condition requiring the improvement of performance within a specified time frame.

riparian forests. Monitoring can be helpful to ensure that public and private regulation is actually complied with and hence that the externalities are internalized as designed. Many externalities, such as riparian forest protection, concern both local characteristics and landscape considerations.[405] Therefore, the capacity of public and private regulation to incorporate considerations at different scales is crucial.

3.6.5.1 Information

The above analysis shows that certification standards regarding riparian buffer zones closely mimic public regulation. Nevertheless, certification schemes do have added values. In addition to the substantive rules regarding riparian buffer zones, many certification schemes also have rules concerning the documentation processes related to forest management.[406] Such requirements make more information about the environment and behaviour of the forest owners/managers available and transparent, complementing the information made available by public regulation. Moreover, as discussed above, public regulators sometimes coordinate with private regulators to share information. For example, in BC, the auditors and licensees agreed to open the certification audit files and share them with the Forest Practices Board.

3.6.5.2 Enforcement

Information provides the basis for forest owners/managers and regulators to decide how to internalize the externalities of the riparian forests. Monitoring is also essential to ensure that regulation is actually complied with. In the US, many states have developed monitoring programmes to track compliance with BMPs. Such programmes collect information on the implementation of BMPs and create the basis to enhance compliance.[407] The introduction of certification schemes creates another layer of monitoring. The anecdotal evidence discussed above shows that certification may be helpful to enhance compliance of riparian buffer-zone rules. The coordination between public and private regulators also helps to reduce the redundancies in monitoring and hence control the monitoring costs.

3.6.5.3 Scale of governance

As for the scale of governance, riparian zone protection needs consideration over both local characteristics and landscape (watershed) coordination. Although certification is well institutionalized in North

405 Lee, Smyth & Boutin 2004, p. 175.
406 Gulbrandsen 2004, pp. 86–88.
407 Kilgore, Ellefson & Phillips 2004.

America, certification covers only 36 per cent of the forests. Moreover, certification schemes focus on individual management units.[408] Public regulation is hence more suitable to solve landscape issues. However, research shows that even public regulation has not incorporated landscape and watershed-scale consideration sufficiently.[409]

3.6.6 Conclusion

This case study analysed how public and private regulation interacts to address riparian buffer-zone protection in North America. The discussion focused on public and private forests in Canada as well as private forests in the US. In Canada, it are the concessioners of public forests who conduct the daily management and therefore, such forests are treated as private property in the analysis. Comparatively clear forest tenure has been established in North America, but the non-market values of the riparian forests are typically not considered by private owners or managers. Therefore, external intervention is used to internalize the externalities linked to the degradation of riparian zones. This research shows many positive synergies between public regulation and certification in addressing riparian zone protection. The interaction between public and private regulation means information is shared and monitoring strengthened. The standards under two types of regulation converge and public regulators and certification schemes coordinate their activities. To incorporate considerations at the watershed-scale, though, is still a challenge under both public and private regulation.

3.7 Case study: Sweden

3.7.1 Environmental problem: the loss of forest biodiversity

Sweden has high forest coverage. Of its 45 million hectares of land, more than 60 per cent (28.2 million) is covered by forests.[410] Sweden has a long history exploiting its forest resources and forest industry plays an important role in its economy, accounting for 11–12 per cent of industrial employment and 11 per cent of exports.[411]

In a developed country which has completed forest transition like Sweden, deforestation is not a major concern.[412] However, after a long history of intensive forest management, Sweden has experienced degradation since the nineteenth century. The intensive management has caused

408 Visseren-Hamakers & Pattberg 2013.
409 Richardson, Naiman & Bisson 2012, p. 236.
410 FAO 2010, p. 9.
411 McDermott *et al.* 2010, p. 138.
412 Boström 2003; Agblede 2010, p. 8.

'the widespread removal of standing dead trees, coarse woody debris and large old-growth pine and spruce', leading to dramatic biodiversity reduction in Sweden's forests.[413] The natural old-growth forests with high conservation values have been largely depleted, with only 5 per cent remaining below the montane region.[414] More than 1800 forest species are listed as endangered or threatened and the number is still increasing.[415] Therefore this section focuses on how public and private regulation addresses this pressing problem – loss of biodiversity – in Sweden's forests.

3.7.2 Good governance

Sweden is a high income country with an annual GDP of US$571.1 billion in 2014.[416] According to the World Bank WGI project, Sweden ranks 99.51 regarding voice and accountability, 86.89 regarding political stability, 95.67 regarding government effectiveness, 96.15 regarding regulatory quality, 97.60 regarding the rule of law and 97.60 regarding the control of corruption.[417] The World Bank WGI indicator on control of corruption is similar to the rank of corruption control from Transparency International. According to Transparency International, Sweden ranks third out of 167 countries.[418]

3.7.3 Property rights as well as public and private regulation

3.7.3.1 Property rights

The majority of forested areas are classified as productive forest land (23.2 million hectares).[419] Most of the productive forests are privately owned: 50 per cent by individual owners, 25 per cent by private sector companies, 6 per cent by other private owners.[420] The rest are owned by the state (3 per cent), state-owned companies (14 per cent) and other public owners (2 per cent).[421] There are indigenous communities – Sami people living in the forests – and they sometimes have conflicts with forest owners.[422] However,

413 McDermott *et al.* 2010, p. 137.

414 Sahlin 2011, p. 2.

415 Sahlin & Svenska Naturskyddsföreningen 2009, p. 2.

416 http://data.worldbank.org/country/sweden.

417 http://info.worldbank.org/governance/wgi/index.aspx#reports.

418 www.transparency.org/cpi2015?gclid=CMK59ZPMwMsCFRBmGwodsSQPKg.

419 Swedish Forest Agency 2014, *Swedish Statistical Yearbooks of Forestry*, p. 46, available at: www.skogsstyrelsen.se/Global/myndigheten/Statistik/Skogsstatistisk%20%C3%A5rsbok/01.%20Hela%202014%20-%20Entire%202014/Skogsstatistiska%20%C3%A5rsboken %202014%20(hela).pdf.

420 www.skogsstyrelsen.se/en/AUTHORITY/Statistics/Subject-Areas/Estate-and-Ownership-Structure/Estate-and-Ownership-Structure/.

421 Ibid.

422 McDermott *et al.* 2010, p. 139; Lantto & Mörkenstam 2008.

the group is not large.[423] Small-scale private forests are predominant in southern Sweden and industrial forest owners play a major role in the north.[424] It is estimated that 355,000 family forest owners existed in 2009, each having an average of 46 hectares of forest.[425] Around one third of the family forest owners, possessing around 50 per cent of the individual forests, have joined the regional forest owners' organizations, which 'take care of operational forestry and safeguard family forestry on a smaller scale'.[426] Forest industry is concentrated, with six industrial companies owning 39 per cent of the forests, including Sveaskog, a state-owned company and SCA, the largest private forest company in Europe.[427] Owing to the favourable conditions of their land, non-industrial forestry provides more than 60 per cent of the harvested timber volume.[428]

3.7.3.2 Public regulation

After a 'decade of confrontation' between timber production supporting policies and environmental campaigns, the current Forestry Act in Sweden was passed in 1993.[429] It was drafted by a parliamentary commission, involving multiple stakeholders, such as 'politicians, representatives from the labour unions, forest industry, forest owners and the leading environmental NGO'.[430] It represented a shift to the participatory and soft mode of governance. The 1993 Forestry Act sets the goals of timber production and environmental protection on an equal footing.[431] Instead of setting detailed criteria to achieve these two goals, the 1993 Forestry Act introduces only minimum performance standards and leaves forest owners substantial discretion to choose the proper measures. The principle of 'freedom with responsibility' has hence been enshrined in the 1993 Forestry Act.[432] It authorizes the competent authority to set standards for nature conservation and cultural heritage preservation, timber felling and regeneration, and maintenance of forest health via regulations.[433] A few regulations have been introduced to set the standards by the competent authority, the National Forestry Board, which was reorganized as the State Forestry Agency (SFA) in 2006.[434]

423 www.fao.org/docrep/w1033e/w1033e05.htm.
424 McDermott *et al.* 2010, p. 136.
425 Ibid.
426 www.nordicforestry.org/facts/Sweden.asp.
427 Gulbrandsen 2010, p. 91.
428 McDermott *et al.* 2010, p. 138; Schlyter, Stjernquist & Bäckstrand 2009, p. 377.
429 Swedish Gov. Bill 1992/93: 226; Nylund 2010, pp. 8–9.
430 Schlyter, Stjernquist & Bäckstrand 2009, p. 380.
431 Johansson & Keskitalo 2014, p. 121.
432 Ibid., p. 121.
433 The 1993 Forestry Act, section 30.
434 Johansson & Keskitalo 2014, pp. 122–123.

In addition to specific forestry legislation, other environmental laws can also apply to forests. As an EU Member State, Sweden has an obligation to implement EU legislation, such as the Habitat Directive.[435] At domestic level the Environmental Code 1998 is also crucial for forest protection, which 'sets out the requirements relating to the conservation of ecological values, including the protection of habitat for endangered species'.[436] In 2008, the 'Species Protection Ordinance (SFS, 2007:845) was adopted under the Environmental Code' to implement the EU Directives.[437]

In addition to the binding legislation, since the late 1990s, the Swedish parliament has introduced sixteen environmental objectives, with the ambition to 'hand over a society to the next generation in which the major environmental problems have been solved'.[438] The Environmental Objectives define the desired environmental status and the time scale to achieve them (usually by 2020).[439] 'Sustainable Forest' is one of the Environmental Objectives, with the target that 'the value of forests and forest land for biological production must be protected, at the same time as biological diversity and cultural heritage and recreational assets are safeguarded' within one generation (by 2020).[440] The State Forestry Agency is responsible for following up and evaluating the 'Sustainable Forest Objective'.[441] To implement this goal, the SFA published 'Quantitative Targets of Swedish Forest Policy' in 2005.[442] These quantitative targets include setting aside forestland with high conservation value, enhancing biological diversity (using dead wood, mature forest and old forest as indicators), the protection of cultural heritage and developing action programmes for threatened species.[443]

The above analysis shows that different layers of environmental goals exist in the Swedish public regulatory system for forestry: the Forestry Act requires a balance between timber production and environmental protection; while the 'Sustainable Forest' objective and Environmental Code have the single goal of promoting sustainable development.[444] Different instruments are used to achieve these goals, including command and control

435 Council Directive 92/43/EEC on the Conservation of Natural Habitats and of Wild Fauna and Flora.

436 Lister 2012, p. 177.

437 www.riksdagen.se/sv/Dokument-Lagar/Lagar/Svenskforfattningssamling/Artskydds forordning-2007845_sfs-2007-845/.

438 Edvardsson 2004, p. 170.

439 Ibid.

440 www.miljomal.se/sv/Environmental-Objectives-Portal/Undre-meny/About-the-Environ mental-Objectives/12-Sustainable-Forests/.

441 Ibid.

442 Swedish Forest Agency 2005.

443 Ministry of the Environment, The Swedish Environmental Objective – Interim Targets and Action Strategies, Summary of Gov. Bill 2000/01:130, p. 48 (2001).

444 Johansson & Keskitalo 2014, pp. 117–119; Uggla, Forsberg & Larsson 2016, p. 37.

instruments, suasive instruments and incentive-based instruments. The former two types of instruments are the most important instruments used for forestry regulation in Sweden.

The command and control instruments are prescribed under the Forestry Act, especially section 30 which requires environmental consideration during forest management. According to this provision, the SFA has adopted a number of regulations. These include some binding requirements during forest management, such as standards for clear cuts, leaving non-productive forest land untouched, avoiding damaging sensitive habitats, leaving protective buffer zones, leaving older trees on felling sites and so on.[445] The SFA monitors compliance and has an inventory system, Polytax, tracking progress.[446] However, the coercion of these instruments is not strong, and non-compliance by forest owners and operators is usually left unpenalized.[447]

Suasive instruments include education, information and counselling provided by the SFA, inventories and practical assistance.[448] These instruments are not legally binding, but are supposed to support the achievement of the 'Sustainable Forest' objective.[449] The advice for forest owners to set aside additional conservation areas voluntarily and the development of 'green forest management plans' guidelines are such examples.[450] The size and budget of the SFA (by then the National Forestry Board) have been reduced since the 1990s, as well as the use of suasive instruments has also been reduced.[451]

Incentive-based instruments include taxes and subsidies, and economic compensation to forest owners when their forests are listed formally as protected forests.[452] However, the taxes and subsidies were abolished under the new Forestry Act. The funding for protected areas, on the other hand, has experienced an increase since the 1990s.[453]

3.7.3.3 Private regulation

Due to environmental campaigns and demands of the European market for sustainable harvested timber, Sweden started to respond quickly to the creation of the FSC. WWF and the Swedish Society for Nature Conservation (SSNC, the largest Swedish environmental NGO) initiated an FSC working

445 Johansson & Keskitalo 2014, p. 123.
446 Ibid., pp. 122–124.
447 Ibid., p. 124; Hysing 2009b, p. 656.
448 Hysing & Olsson 2005, p. 512.
449 Johansson & Keskitalo 2014, p. 125.
450 Hysing & Olsson 2005, p. 512; Johansson & Keskitalo 2014, p. 125.
451 Hysing 2009b, p. 656.
452 Ibid. Johansson & Keskitalo 2014, p. 121.
453 Hysing 2009b, p. 656.

group to develop a national standard. This workshop group was later joined by other NGOs, forest companies, forest owners, labour unions, consumer industry and Sami organizations (indigenous peoples' organizations).[454] The group was divided during the negotiations. Forest owners' associations (representing the interests of individual owners) left the process due to conflicts with Sami people who demanded broad recognition of their customary rights over reindeer herding and concerns that proposed standards were geared to industrial forestry.[455] Greenpeace also withdrew its support due to the concern over intensive forest harvesting, which was allowed under the standards.[456] The other parties continued the process and adopted the first national FSC standard in 1997, which was endorsed by FSC International in 1998.[457] Immediately after having withdrawn from the FSC process, forest owners' associations joined forces to develop their own forest certification standard, which was endorsed by the PEFC in 2000.[458]

In the early 2000s, the Swedish FSC and PEFC started the Stock Dove process, with the aim of creating 'a bridging document to mutually recognize the two standards'.[459] Recommendations were developed in 2001 for the standards with a view to harmonize them.[460] In addition, several NGOs also published reports comparing the two standards criterion by criterion.[461] These reports concluded that the Swedish FSC was more stringent and prescriptive than the PEFC standard on several key issues.[462] With the exception of issues concerning the Sami, the Swedish PEFC Board revised their standards according to the recommendations reported in the comparisons. For example, the revised PEFC standards imposed more stringent requirements on set-aside areas, retention trees and landscape planning, transparency, labour rights and monitoring.[463] The revised standard was endorsed in 2006.[464] On the other hand, the FSC did not follow the recommendations to lower their standards due to the concern to 'maintain the rigour, independence and distinction of its standard'.[465] In spite of this, with the competition of PEFC, forest companies managed to lobby the FSC to

454 Gulbrandsen 2010, pp. 92–93; Schlyter, Stjernquist & Bäckstrand 2009, p. 380.
455 Gulbrandsen 2010, pp. 93–94; Schlyter, Stjernquist & Bäckstrand 2009, p. 380.
456 Gulbrandsen 2010, p. 94.
457 Ibid.
458 Ibid., www.pefc.org/component/pefcnationalmembers/?view=pefcnationalmembers&Itemid=48/10-Sweden.
459 Lister 2012, p. 182.
460 Ibid.
461 FERN 2001; see also Gulbrandsen 2010, pp. 97–98.
462 Gulbrandsen 2010, pp. 97–98.
463 Ibid., p. 99.
464 Ibid.
465 Lister 2012, p. 182.
466 Gulbrandsen 2005, pp. 350–351.

introduce more flexibility in the revised chain of custody standard.[466] Therefore, the competition between two standards led to the 'cross-fertilization and convergence of the two schemes': the PEFC standard was ratcheted up to boost credibility and the FSC allowed greater flexibility to accommodate business interests.[467] Therefore, later studies have only found minor differences between the two standards, such as 'the slightly higher harvesting restrictions in FSC certification and stricter commitment by contractors in PEFC certification'.[468]

The certification schemes have diffused rapidly since their inception in Sweden. By December 2013, 12 million hectare forests were certified under the FSC, with twelve companies covered by normal certificates and thirteen organizations covered by group certificates (certificates targeting smaller scale forest operations).[469] Eleven million hectares of forest obtained PEFC certificates by December 2012.[470] All major companies had already been certified by the FSC by 2000, and individual forest owners tended to be PEFC certified due to the lower financial costs.[471] By December 2012, 4.9 million smaller owners were PEFC certified via the group certification, and one forest owners' association was certified directly, covering 0.9 million hectares of forest.[472] This made around half of the individual owned forests certified by PEFC. In addition, some actors are dual-certified 'in order to facilitate chain-of-custody certification of their fiber supply and have enough production flexibility to meet customer demands'.[473]

3.7.4 The interaction between public and private regulation

This section examines biodiversity protection as a particular type of externality which has not been internalized by establishing property rights. It discusses how public and private regulations interact to protect biodiversity.

3.7.4.1 Interaction on paper

Different goals have been established under the forest regulatory systems in Sweden: the goals under the Forestry Act, Environmental Code, 'Sustainable Forest' Objective and interim targets as well as the goals under private certification schemes. Under the guidance of different goals, different standards have also been established.

The Forestry Act aims at achieving a balance between timber production and environmental protection (including biodiversity protection).

467 Ibid., p. 352.
468 Savcor Indufor 2005; Schlyter, Stjernquist & Bäckstrand 2009, p. 381.
469 https://se.fsc.org/statistik-och-fakta.242.htm.
470 http://pefc.se/statistik/.
471 Lister 2012, pp. 182–183.
472 http://pefc.se/wp-content/uploads/2013/05/PEFC-Statistik-121231Summering.pdf.
473 Lister 2012, p. 183.

However, this suggests both interests may conflict sometimes. For example, according to the law, 'conservation requirements must not be so far-reaching that they make on-going land use significantly more difficult'.[474] Moreover, neither the goals nor standards established under the legislation are specific. As a framework law, the Forestry Act only established minimum performance standards and left much room for forest owners to decide suitable measures to achieve the goals.[475]

The goal of the Environmental Code is to 'promote sustainable development which will assure a healthy and sound environment for present and future generations'.[476] The code is applicable to 'activities that may cause damage or detriment to human health, the environment or other interests that are protected by this Code'.[477] Therefore, the Environmental Code applies to forestry as well.

The 'Sustainable Forest' Objective is to protect the forests and their biodiversity, cultural and recreational values. Several interim targets have been established, which need to be achieved in 2005 or 2010. The interim targets are largely quantifiable and more specific than the goals under legislation.

Another layer of biodiversity protection standards is set by private certification schemes, which are developed to achieve the triple goals of ecological, economic and social sustainability.[478] The FSC Sweden standard has two principles directly related to biodiversity protection. Principle 6 requires forest management activities to conserve biodiversity[479] and Principle 9 concerns specifically the management activities in High Conservation Value Forests.[480]

These principles are further elaborated with concrete criteria on both performance and procedural issues. The criteria include some key issues such as: setting aside 5 per cent of productive forest land in the management unit from harvesting; protecting threatened species and their habitats; creating new habitats by burning regeneration areas, requiring retention trees and keeping dead wood.[481] In spite of some marginal difference, the PEFC scheme also has similar standards for protecting biodiversity.

474 The requirement is defined as 2 to 10 per cent of the net income from forestry. See Johansson & Keskitalo 2014, p. 122.

475 Lister 2012, p. 179.

476 Chapter 1, section 1, www.government.se/contentassets/be5e4d4ebdb4499f8d6365720ae68724/the-swedish-environmental-code-ds-200061.

477 Ibid., section 3.

478 FSC-STD-SWE-02-02-2010, Introduction. PEFC-SWE-002-V2, Introduction.

479 Principle 6 of FSC Sweden: 'Forest management shall conserve biological diversity and its associated values, water resources, soils, and unique and fragile ecosystems and landscapes, and, by so doing, maintain the ecological functions and the integrity of the forest'.

480 Principle 9: 'Management activities in High Conservation Value Forests shall maintain or enhance the attributes which define such forests. Decisions regarding High Conservation Value Forests shall always be considered in the context of a precautionary approach'.

481 Savcor Indufor 2005, pp. 56–59.

The literature has compared the legislative standards with the FSC and PEFC standards, and concludes that both certification standards are more demanding than the legislative requirements in a number of key issues, including biodiversity protection.[482] For example, the 5 per cent set-aside requirement exceeds the legislative requirement, where there is no quantifiable requirement; the certification standards require the creation of new habitats where legislation only requires the protection of existing ones; the quantifiable requirements for mixed stands and dead wood also go beyond the legal requirement;[483] in addition, the certification schemes provide protection to broader types of forests than the legally protected ones, such as woodland key habitats.[484]

The public regulatory systems and certification schemes closely interact with each other. First, public regulation often contains only open standards.[485] The forest owners, especially individual owners, sometimes do not know how to fulfil their obligation and to achieve the public goals. Therefore, forest certification provides a valuable approach for them to translate 'the government's broad environment/production goals into specific operational targets and plans that landowners understood and could implement'.[486] Second, certification standards can be incorporated into public regulation and hence make the legal standards more specified and rigorous. For example, leaving retention trees in harvested forests is an important measure to maintain biodiversity. In the early 1990s, the legislation had no clear standards for retention trees. The government's attempt to regulate it was not successful due to the opposition of forest owners. Ten years later, the FSC negotiation was able to create a consensus: ten trees per hectare should be retained. This certification criterion was incorporated into government requirements in 2003.[487]

482 Ibid., pp. 53–64.

483 Ibid., pp. 56–59.

484 FSC criterion 6.2.1; 91.1.

485 E.g. the 1979 Forestry Act mandates private forest owners to prepare a forest management plan with the support of the county board. However, the 1993 Forestry Act abolished such a requirement and reduced the governmental technical service. Instead, the SFA started to issue green management plan guidelines, with the view of balancing sustainable production and nature conservation goals. Therefore, the public regulation leaves the development of the management plan to the forest owners, who may find it difficult to figure out exactly what is expected from them. See Lister 2012, p. 190.

486 Hysing & Olsson 2005, p. 512. For example, both the FSC and PEFC require the forest owners to have a forest management plan with more detailed instructions than the guidelines provided by government. Therefore, through seeking certification coverage, forest owners could develop green management plans encouraged by government as well. See Lister 2012, p. 189. Another example concerns the 'sustainable forest' objective. Its interim targets include issues regarding set-aside areas, dead wood, deciduous trees and the conversion of spruce standards. Certification schemes also include concrete criteria on these issues and can hence contribute to the achievement of such targets.

487 Lister 2012, pp. 193–194.

The positive interaction between public regulation and forest certification, however, does not mean that their goals and standards are fully compatible. For example, the interim targets for the 'Sustainable Forest' objective requires setting aside 900,000 hectares of Forest with High Conservation Value (HCVF). The certification schemes also make it mandatory to set aside 'a minimum of 5% of the productive forest land area'.[488] The required types of set-aside areas and the restrictions on these lands, however, are not fully aligned.[489] Therefore, certification schemes could only be, at best, a partial mechanism to achieve this interim target.[490]

Another potential conflict brought by the different goals under public and private regulation is the requirement under the Forestry Act, which prescribes that the conservation measures should not be too burdensome (not beyond 2 to 10 per cent of net income from forestry). Such requirement provides a legal constraint in the choice of conservation measures by forest owners. Forest certification schemes do not have such requirements and may therefore allow more far-reaching conservation behaviour. In addition, the legal standards only require the protection of existing environmental values. Certification schemes, however, require the creation of new environmental values via, for example, retention trees.[491] However, 'these seemingly complementary requirements might actually be difficult to achieve together, especially on the small percentage of forest typically retained during clear-cutting'.[492]

3.7.4.2 Interaction in practice

As mentioned earlier, both the Environmental Code and the Forestry Act apply to forestry in Sweden. However, in practice, the application of the Environmental Code in the forestry sector remains limited.[493] The Species Protection Ordinance adopted later has precedence over the Forestry Act, and it also includes more stringent standards than the Forestry Act regarding the protection of habitats.[494] The co-existence of different legislation and sometimes conflicting standards lead to confusion on which rules should apply. Given the more friendly attitude to forestry under the Forestry Act, forestry issues are mainly handled under the Forestry Act. This 'resulted in a situation in which species (protection) has largely been overlooked'.[495]

Even compliance with the Forestry Act has not been satisfactory. The SFA monitors the compliance of forestry law through a monitoring system,

488 E.g. FSC, criterion 6.4.1.
489 Lister 2012, pp. 191–193.
490 Ibid.
491 Johansson & Keskitalo 2014, p. 128.
492 Ibid.
493 Ibid.
494 Ibid.
495 Ibid.

Polytax, which assesses the performance of forest owners and environmental considerations required by law. The inventory system consistently shows that the non-compliance of logging operations with legal requirements remains high (remaining in the range of 20–30 per cent since the 2000s), especially concerning regeneration felling and key biotope protection.[496] However, the SFA has not imposed sanctions for non-compliance due to the soft steering governance mode and constitutional constraints concerning property rights.[497] Whether the deregulated forest policy based on 'freedom with responsibility' needs to be revised is debatable. Governmental agencies find revisions necessary and intend to issue more direct orders and prohibitions to regulate activities concerning key biotopes. The forest sector, on the other hand, argues that the current regulatory style should be maintained.[498]

As the level coercion of command and control instruments is low, public suasive instruments and private certification schemes might be important complements to ensure sustainable behaviour. Rather than issuing orders and prohibitions directly, governmental agencies can also provide education, information and technical assistance to forest owners. However, research shows that these suasive instruments help to raise awareness of biodiversity protection, but usually fail to change the underlying values and preferences of the forest owners.[499] A case study shows that forest owners are conducting conservation measures, but largely based on 'their own terms rather than as an unconditional implementation of the forest policy'.[500] They implement the measures they find reasonable and acceptable, but ignore the more challenging and far-reaching ones such as leaving dead wood.[501] Moreover, the use of suasive instruments has been shrinking, due to the reduced size of the SFA[502] and the cutting back of funding.[503]

3.7.4.3 Added value of certification schemes?

Private certification schemes provide a means to monitor the behaviour of forest owners by third-party auditors. They are found to be an important complement to public regulation. However, this by no means suggests that the two systems are substitutes. First, the monitoring systems under public regulation and certification schemes operate differently. The governmental Polytax system records performance and environmental conditions in

496 Johansson & Keskitalo 2014, p. 122.
497 Ibid., p. 124; Hysing & Olsson 2005, p. 520.
498 Johansson & Keskitalo 2014, p. 124.
499 Hysing & Olsson 2005, p. 521.
500 Ibid.
501 Ibid.
502 Hysing 2009a, p. 656.
503 State funding for informational instruments has decreased by 48 per cent since 1990. Hysing & Olsson 2005, p. 520.

the whole country. Certification schemes only monitor the certified forests and not all forests.[504]

Second, as mentioned above, the goals and standards under the public regulatory system and the certification schemes are not fully compatible. Not all environmental values in certification schemes are included in the public monitoring system. Certification schemes may also not guarantee the full achievement of public goals. Using again, set-aside targets as an example, the interim targets required an additional 500,000 hectares of forests with high conservation values to be set aside by 2010. The size of voluntary set-asides has been exceeded. However, because there is no consistent accounting and reporting system for the set-aside areas, the quality and maintenance of these voluntary set-asides under certification schemes are unknown.[505] Some NGOs report that forest owners can shift the set-aside geographically and log a previous set-aside; they can also sell the set-asides and leave them logged by the purchasers.[506] Therefore the duration of set-asides is not guaranteed.[507]

Third, the credibility of certification schemes in steering sustainable forest management has been questioned in recent years. A wide coalition of environmental NGOs, the forest sector, government and the public was mobilized in the early stages to support the rapid emergence and spread of forest certification schemes. However, this coalition weakened in recent years. On the one hand, with the motivation to secure market and reputational accountability, the forest sector has increasingly been involved in forest certification and the coverage of certification schemes in Sweden has been rapidly expanding.[508] Many environmental NGOs, on the other hand, have withdrawn their support for the certification schemes. NGOs have long criticized PEFC standards as not credible.[509] In recent years, criticism against the FSC has also intensified. SSNC, one of the initial founders of FSC Sweden, withdrew from the Swedish FSC board in 2008 and further withdrew its membership in 2010. Some other NGOs have also left the FSC, such as the Youth Environmental Organization and Friends of the Earth Sweden.[510] They argued that on the one hand the governance system and accountability mechanism were not rigorous enough and that on the other hand there was wide non-compliance of certified forests with certification standards.[511]

504 Keskitalo *et al.* 2009; Johansson 2014.
505 Lister 2012, pp. 192–193.
506 Sahlin 2013.
507 Ibid., p. 20.
508 Johansson 2012, p. 433.
509 Ibid., p. 430.
510 Ibid., p. 433.
511 Ibid.; Sahlin & Svenska Naturskyddsforeningen 2009.

During 2007 and 2009, SSNC conducted field surveys on nearly 300 certified company-owned old-growth forests, including forests operated by the largest forestry companies such as SCA, Bergvik Skog, Holmen Skog, Korsnas and Sveaskog. It concluded that most of these forests were slated for logging, potentially violating FSC standards, according to which such forests should be set aside.[512] SSNC and Greenpeace filed a number of formal complaints against these forest companies.[513] In addition, other disputed issues include 'logging in areas of high biodiversity value', 'destruction of dead wood ... and soil compaction or destruction'.[514] Several issues were found to fuel disputes: first, though more specified than legal standards, the criteria in the FSC standards still do not give clear guidance on the way to achieve them regarding many issues, such as leaving dead wood and protecting red-listed species outside key habitats. The unspecified standard gives rise to different interpretations between forest companies and NGOs.[515] Second, the slow progress in public regulation also impedes the implementation of the FSC standard. This is especially true concerning the woodland key habitat, which is referred to as 'an area where there is occurrence of one or more red listed species or where the nature of the forest itself indicates a strong likelihood of finding a red listed species'.[516] The woodland key habitats are not formally protected areas under the legal system.[517] Nevertheless, the SFA has authority to classify forests as woodland key habitats and has started to carry out an inventory.[518] However, the inventory progress is very slow, with only 20 per cent of the key habitats identified.[519] Forest operators are committed not to log in woodland key habitats under the certification schemes. However, without formally recognizing them, companies sometimes have difficulty in identifying woodland key habitats, or adopt different priorities and mechanisms of identification from NGOs. This ambiguity in identifying woodland key habitats leads to logging in these areas and potential infringement of the FSC standard.[520]

Confusion concerning the proper behaviour expected from forest owners occurs more often due to the existence of multiple regulatory systems (mandatory legal standards, voluntary informational instruments and certification schemes). The SFA has acknowledged the difficulties in communicating the difference between legal requirements and recommendations to forest owners and the insufficient awareness of the forest

512 Sahlin & Svenska Naturskyddsforeningen 2009, p. 12.
513 Sahlin 2013, p. 9.
514 Agblede 2010, p. 9.
515 Ibid., pp. 28–29.
516 Ibid., p. 30.
517 Sahlin & Svenska Naturskyddsforeningen 2009, p. 11.
518 Agblede 2010, p. 30.
519 Sahlin & Svenska Naturskyddsforeningen 2009, p. 11.
520 Ibid., pp. 30–32.

policy goals and requirements.[521] Lack of coordination is a cause of ambiguity in the multi-layered governance system.

Another issue criticized by NGOs is the accountability mechanism of FSC. As discussed earlier, a system of complaints handling has been established by the FSC. If a certification body finds the deviation from the FSC standard by responding to the complaints, it can take different actions, varying from observations, a Minor Corrective Action Request (Minor CAR), a Major CAR, to the suspension or withdrawal of the certificate, with increasing levels of coercion.[522] SSNC claims that the CBs sometimes delay in responding to formal complaints and are reluctant to issue major CARs and the suspension of certificates.[523] Even though NGOs had already reported widespread non-compliance for many years, the first partial suspension of a certificate was only made in 2014.[524] The CBs are also criticized for inconsistency in handling complaints and providing instructions to complainants.[525] Without a strong and credible accountability mechanism, it is unclear whether the performance of certified actors will meet certification standards.

In summary, both public and private regulations have their own limitations. The coercive level of the law is low: non-compliance with command and control instruments is often noted without being followed by sanctions. Suasive instruments help to raise awareness of biodiversity protection, but usually fail to change the underlying values and preferences of the forest owners. Private certification complements the first two instruments and has a very high coverage rate in Sweden. However, monitoring differs between public and private regulation; goals and standards under public and private regulation are not fully coherent and the credibility of certification schemes has been criticized in recent years for their incapacity to enforce their own rules and for the lack of legitimacy of accountability mechanisms.

3.7.5 How does the interacted system influence biodiversity protection?

Biodiversity protection is regulated by public and private regulators because it is a non-market value of forests which private owners/managers do not automatically take into account. As was done in the case of North America, the capacity for public regulation and private certification in correcting market failures is analysed in terms of creating information, strengthening monitoring and setting the right scale.

521 Johansson & Keskitalo 2014, pp. 128–129.

522 Ibid., pp. 38–39; Sahlin 2013, pp. 7–10.

523 Sahlin 2013, p. 8.

524 Stora Enso 2014, p. 38, available at: http://assets.storaenso.com/se/com/DownloadCenterDocuments/Global_Responsibility_Performance_2014.pdf.

525 Sahlin 2013, p. 9.

3.7.5.1 Information

Private certification requirements and private auditing help to generate information concerning the environment and forest management.

Biodiversity protection in Sweden also provides an example of how multiple layers of regulatory systems accommodate uncertainties. Although biodiversity protection is a goal of the multi-layered systems, there is no scientifically based answer to the optimal extent of protection and how it should be done.[526] Such uncertainties may cause confusion for forest owners and managers in determining how to protect the forests. The capacity of public and private regulation to accommodate scientific uncertainties is hence crucial for the protection of biodiversity. As mentioned earlier, different goals have been adopted under the Forestry Act, the Environmental Code and forest certification. Some inconsistencies exist among the multiple layers of standards. The legal standards are sometimes vague, such as the rules concerning red list and key habitats.[527] The inconsistencies and ambiguities aggravate the differences of opinion between forest owners/managers and environmental groups, creating disputes in protecting biodiversity. Although certification standards are more specific than the legal standards in Sweden, they sometimes are still not specific enough and lead to confusion in implementation. Moreover, a recent study reviewing scientific literature on environmental standards under the SFA and FSC shows that the legislative and certification standards regarding many environmental issues do not reach the threshold recommended in scientific literature.[528] Therefore, scientific uncertainty has not been well accommodated under the public or the private regulatory system.

3.7.5.2 Enforcement

Forest law is 'de-regulated' with low levels of coercion. Public monitoring shows that non-compliance with legal requirements remains high. However, the non-compliance is usually not followed by sanctions. Suasive instruments have enhanced the environmental awareness among forest owners, but have failed to change their underlying values and preferences. Reduced budgets and personnel also have limited the use of suasive instruments. Certification schemes complement oversight to steer sustainable management.

However, due to different monitoring methods and standards, certification has not been able to replace public regulation. Moreover, NGO criticisms against certification schemes, especially concerning the more

526 Uggla, Forsberg & Larsson 2016, p. 4.
527 Ibid., pp. 3–4.
528 Johansson *et al.* 2013, pp. 107–108.

appreciated FSC scheme (relating to its ability to enforce and to accountability) have intensified in recent years.

3.7.5.3 Scale

The protection of forest biodiversity needs consideration at both local and landscape level. The capability of public and private regulation of addressing biodiversity at both scales is different. Certification applies at the level of individual management units and can hence address issues at a local level, but to a lesser extent at landscape level. For example, certification requires forest owners to set aside some areas and leads to large amounts of forest being set aside from harvesting.[529] However, as discussed earlier, the quality and duration of the set-aside areas are not guaranteed. Sustainability is hence doubtful. Moreover, the set-aside areas are fragmented. The poor connectivity impedes the 'habitat network functionality for conservation of viable populations'.[530] Public regulation has the potential to address these pitfalls. Establishing protected areas and setting them aside from harvesting can create a comparatively stable status of larger areas. Public regulation can reduce fragmentation by applying across individual managed units. However, public regulation is subject to legal and financial limits as well. For example, to set aside private forests from harvesting, public regulators need to compensate the private owners for lost production. Lack of funding is a major barrier to reaching formal protection targets.[531] Therefore, the quantitative goal to formally set aside forests has not been achieved.[532]

A comparison of the FSC and legal standards on biodiversity conservation in Sweden and north-west Russia shows the importance of scale for protecting biodiversity. It shows that the FSC standard in north-west Russia includes 'indicators for all spatial scales of biodiversity conservation, from tree and stand to landscape and ecoregion', while the Swedish standard only focuses on lower scales such as tree and stand scales.[533] In the case studies chosen for the two countries (the similar areas of forested lands), Russia has three times more formally protected forests than Sweden and a similar size of voluntary set-asides under the FSC. A comparison of the habitat network functionality shows that Russia performs better, both in terms of structural and functional connectivity.[534] This research suggests the

529 Lister 2012, pp. 192–193.

530 Angelstam *et al.* 2011, p. 1126.

531 Lister 2012, p. 192. To fill the gap, the state-owned company made a pool of land available to complement the funding. However, due to the lower than average economic value of these forests, whether the added funding would be sufficient to purchase sufficient private land is unknown. See Angelstam *et al.* 2011, p. 1126.

532 Swedish Forest Agency 2012, p. 106.

533 Elbakidze *et al.*, 1983.

534 Ibid.

importance of governance scale in biodiversity conservation.[535] However, the causal link between scale and protection of biodiversity needs further underpinning since the baseline levels of biodiversity before the introduction of the current governance system are not provided. It could be possible that, due to the much longer history of intensive forest management in Sweden, the remnants of natural forests are more abundant in Russia, regardless of the current governance systems.

3.7.6 Conclusion

This case study of Sweden shows similarity to the North America case study: public regulation and certification contribute to the generation and sharing of information and to strengthening of monitoring. Certification plays an even more important role in Sweden, not only because of its wider coverage but also because of its role in complementing and specifying the general and minimum legal requirements under the legal framework. In Canada and the US, the law is more detailed. However, the goals and standards under the Swedish Forestry Law, the Environmental Code and certification are not fully coherent, which creates confusion among forest owners and amplifies conflicting views associated with scientific uncertainty. This is a lesser problem in Canada and the US where public and private standards are more alike. Moreover, although certification is an important complement to public regulation in Sweden, the different standards and ways of monitoring do not make it a substitute for public regulation. In addition, this case study also confirms the difficulty for certification schemes to achieve a level of scale that takes landscape issues sufficiently into consideration.

3.8 Comparison and conclusion

Forests are common-pool resources, which may be subject to overharvesting when access is insufficiently controlled. Three types of property rights can be used to limit access: public, private and communal property rights. Clear definition of property rights, effective enforcement, coordination, information creating and sharing as well as setting the right level of governance are crucial preconditions for effectively controlling access. Moreover, public and private regulation may be invoked to overcome the limits of property rights in internalizing externalities in forest management.

3.8.1 Indonesia and Bolivia

Bolivia is at the frontier stage on the forest transition curve. Indonesia is in a stage transitioning to the mosaic stage. Both suffer from fast deforestation.

535 Ibid., 1993.

Unclear forest tenure and widespread land conflicts have contributed to deforestation in both countries. Three types of forests exist in Indonesia and Bolivia and community forests play a particularly important role.

Community forests have traditionally been governed by self-governing institutions. Such institutions are nested in an institutional background increasingly shaped by public and private regulation.

In Indonesia, customary forests have largely been replaced by public forests. Simply replacing community rights with state ownership, without considering the local ecology and communities who still live in and on the forests has had devastating effects, especially since government implementation is weak and enforcement capacity low. After the Reformasi period, the government started to recognize customary rights minimally. However, the content of such rights and the procedure of formalizing the rights remain unclear. In Bolivia, indigenous rights have been recognized under the legal framework. However, a clear and secure property rights regime which can address deforestation has yet to be established. These two cases show that the recognition of community rights and self-governing institutions by public and private regulation is a necessary but insufficient condition for effective protection against deforestation.

The content of the rights granted to communities as well as the design of governing institutions can also influence the effect of property rights in constraining deforestation. For example, in Indonesia, in addition to establishing customary forests under the Forestry Act, a variety of CBFM policies have been adopted, aiming to enhance the role of communities in forest management. Such policies often allow communities to participate in forest management and to benefit from it to some extent. However, in many cases, the transfer of rights, and the use of these rights as collateral for credit are either limited or prohibited and coverage is very low.

Certification schemes in both Indonesia and Bolivia require clear forest tenure and the protection of indigenous rights over forests. However, the very fact of conflicting forest tenure denies many forest operations the opportunity to get certified.

Indonesia and Bolivia present different ways of designing and reforming self-governing institutions. Indonesia replaced the adat governing institutions with formal government agencies – the villages. Bolivia has not formally abandoned the traditional governing institutions. However, the creation of TCOs has introduced additional governing institutions at the territorial level. Certification is a market-based instrument, requiring a managing institution which can fit into the market economy. Neither customary governing institutions nor formally established public agencies fit such a role. In both Indonesia and Bolivia, communities need to establish their own enterprises or collaborate with commercial forestry companies to get certified, which has completely failed in the latter case. The co-existence of multiple governing institutions has further weakened the traditional forest governing systems. Community members no longer participate actively in decision-making and patrolling the forests. Like in many other developing countries, public regulators in both Indonesia and Bolivia are subject to severe budget, staff and expertise constraints. Consequently, their resources are insufficient in patrolling the vast forests. Under a certification scheme, professionals rather than community members are relied on to monitor forest-related behaviour, which implies a huge reservoir for oversight in communal forests is no longer being used to the full extent. Moreover, the coverage of certification remains limited, especially for community forests. The co-existence of multiple governing institutions also increases coordination costs.

The decentralization process in Indonesia and Bolivia itself is a driver of deforestation. It is supposed to empower local people and their decision-making opportunities and hence to contribute to reducing deforestation. However, the systems are captured and corrupt, lack coordination among government agencies and conflicts between levels of government are aggravated by obscure and conflicting legislation. For example, in Indonesia, conflicting legislation concerning decentralization and recentralization reflects a tug of war between levels of government. Multiple agencies often conduct conflicting spatial planning and issue overlapping permits for the same area. The self-interest of bureaucrats issuing these conflicting permits cannot be easily overcome. However, legislation could address one factor which aggravated such conflicts – the lack of hierarchy between different levels of legislation and the conflicting clauses in the legislation. In Bolivia, conflicting interests among different levels of government and different departments also prevail, such as the reluctance of municipalities in supporting TCOs' applications and the nesting of the land regularization process in the agricultural model. Such conflicts are also reflected in the legislation. For example, the requirement of an FES in titling private land and TCOs being more favourable to agriculture than to forestry use. In sum, replacing community property with state ownership, disregarding the local ecology and communities, and decentralization of public regulation have contributed to deforestation in Indonesia and Bolivia.

3.8.2 North America and Sweden

In North America and Sweden, community forests remain important in certain regions. However, the long industrialization and modernization history has transformed most of the forests into state or private property. A clear and secure tenure system has been established. Deforestation is no longer a major problem, but different forms of degradation exist. Many of such problems concern the non-market value of the forests, which are not considered sufficiently by property rights holders, such as the protection of riparian zones and biodiversity. Therefore, external interventions, such as public and private regulation have been adopted with the goal of having proprietors internalize these non-market values of forests. The case studies show how public and private regulation have tried to contribute to the internalization of externalities by information creation and sharing; enhancing monitoring and setting the right scale of governance.

The interaction between public and private regulation in the two case studies shows some positive synergies in generating and sharing information and in oversight. Certification schemes' requirements concerning documentation, process, and transparency generate information and demand oversight. Although public and private regulation have the potential to improve information and enforcement, the actual effect is context-dependent. In North America, standards under certification schemes concerning riparian buffer zones closely mimic public regulation, reducing the risk of conflicting standards. However, this also raises the question of the added value of certification schemes in terms of going beyond compliance. In Sweden, public law provides only general and minimum requirements, which are specified and strengthened by private certification. Even inside the public regulatory system, there are also different goals and standards. Multiple layers of goals and standards therefore create some incoherence and confusion among resource users. Scientific uncertainties exist in how to manage the forests sustainably. How public and private regulation accommodate such uncertainties is crucial for the prevention of forest degradation. The case study of biodiversity protection in Sweden shows that the inconsistencies and ambiguities in multiple layers of governing systems have fuelled conflicts of opinion between forest owners/managers and environmental groups. Such conflicts have been aggravated by scientific uncertainties, creating difficulties in choosing an approach in protecting biodiversity.

As far as enforcement is concerned, the introduction of certification schemes adds an additional layer of inspection. In North America, the coordination between public regulators and certification schemes helps to reduce the redundancies in inspection. To rely on certification can save the resources of public regulators which can be allocated for other purposes. In Sweden, public regulation relies on a soft steering model and the coercion level of public regulation is low. In this case, the widely adopted certification provides a very useful complement to oversight. However, certification standards are not fully consistent with legal requirements and certification relies on a different mode of monitoring from public regulation. Therefore certification cannot replace public regulation.

Concerns over the scale of governance also present a challenge for public and private regulation. Both riparian forests and biodiversity protection require consideration at both local and landscape levels. Certification schemes are based on assessment on the Forest Management Unit (FMU) level. To incorporate higher level considerations needs coordination between different FMUs. Public regulation has the potential to address issues at the landscape level, but often lacks the political power to do so.

3.8.3 Putting the pieces together

The above analysis compares two groups of countries which are in different stages of forest transition and experience different types of forest problems: deforestation and forest degradation. The different problems addressed in these two groups of countries do not mean that there is no cross-learning between them. Before the limitations of property rights can be overcome, the preconditions for their proper functioning have to be satisfied.

A clear definition of property rights is crucial for their establishment and functioning. The examples of Indonesia and Bolivia show that the pure recognition of a certain type of property rights is not sufficient; establishing a suitable content of the rights and how they should be executed are also essential. In all the case studies, forest law which defines the content of property rights is based on modern management techniques. Such techniques are developed for modern, large-scale commercial forestry, and try to provide guidance on how to manage the forests by maintaining their productivity and sustainability. The Swedish law is criticized by some as not sufficiently reflecting recent scientific knowledge and hence not providing sufficient protection of the environment. On the other hand, some developing countries, such as Bolivia, fail to take into account local knowledge. Indigenous people still live in the forests and manage them in a traditional way, mainly for subsistence use. Modern forest regulation based on western science relies heavily on the specialized knowledge of forest professionals and the capacity of public regulation to monitor the whole process of forest management. Such a basis, however, often does not exist in many developing countries. A copy and paste of the western model without reflecting local knowledge fails to accommodate the indigenous interests and their communal property rights.

As for enforcement, the above analysis shows that public and private regulation can be complementary. North America and Sweden exemplified different models of complementarities. Arrangements can be made to reduce the costs of double inspection. However, the ways public and private regulation conduct inspections differ in Sweden and the standards they evaluate against are also different. This means private certification cannot substitute public regulation. The relation between public regulation and private certification is even more complicated when it comes to communal property rights, where traditional governing institutions and traditional enforcement exist abreast of public and private regulation. In Indonesia and Bolivia, the introduction of new public governing institutions has dismantled or weakened traditional governing institutions. Traditional enforcement relying on members of communities is replaced by professionals, public regulators or private auditors. The replacement has actually weakened the enforcement of property rights.

What the case studies have also shown is that the different elements of governing forests are closely linked. For example, information can influence both the definition and enforcement of property rights. As discussed above, incomplete information and scientific uncertainties exist over how to protect forests. On the one hand, some flexibility in regulation is needed to reflect incomplete information and uncertainties. However, the Swedish example shows that ambiguous and sometimes conflicting rules in public and private regulation may amplify the effect of scientific uncertainties, and lead to debates among forest stakeholders. Information availability also influences the enforcement capacity of different actors. The examples of Indonesia and Bolivia show that replacing communal property rights with public or private property rights has led to a substantial increase in enforcement costs. The community members who have easy access to experiential knowledge of forest management are no longer enforcing the traditional rules. The alternative agents policing the forests – professionals, governmental officials – are further away from daily forest management and have higher information costs.

The level of governance is also related to coordination and enforcement. In Indonesia and Bolivia, many actors still live or operate in forests, such as indigenous people, farmers and concessioners. The interests of these actors often conflict. This makes coordination costs very high. Bolivia has tried to solve the conflicting claims by establishing TCOs on indigenous land. The TCOs are established at a higher level than the traditional governing institutions. With higher levels of governance, the TCOs can lobby for indigenous interests and can promote the establishment of property rights, hence reducing external coordination costs. However, the determination of the content of property rights and the allocation among individuals still takes place at a lower level, that is, the village level. The coexistence of several governing institutions therefore increases the internal coordination costs.

As discussed above, the capacity of certification and public regulation to address issues at landscape level differs. Certification schemes operate at the level of individual forest management units. To solve things at a higher level, coordination among the individual units is necessary. Public regulation may address issues at a higher level. However the Swedish example shows that the possibility for public regulation to address issues at landscape level, for example ensuring ecological connectivity in set-aside forests, is subject to the limited enforcement capacity of regulators. Deforestation is also an issue which needs to be addressed at the landscape level. Spatial planning is an instrument which is normally used by the public regulator to address it. However, the Indonesian experience shows that without coordination between different departments and levels of government, the self-interested regulators may issue conflicting spatial plans, aggravating deforestation.

4 Fishery governance

4.1 Introduction: the status of fish resources

Capture fisheries and aquaculture provide valuable sources of protein for human health and play an important role in the local economy and employment of coastal areas. Fish production has increased steadily in the past few decades[1] and reached 91.3 million tons for capture fish and 66.6 million tons for aquaculture in 2012.[2] World fish consumption increased from 9.9 kg per capita in the 1960s to 19.2 kg per capita in 2012. Together with the increase in fish production, the health of fish stocks has been a global concern for decades. It is reported that in 2011, 57.4 per cent of fish stocks were fully exploited and 28.8 per cent were overexploited.[3] Overexploitation 'has led to the collapse of target species', to large quantities of by-catch and to 'alteration of entire marine ecosystems and the "fishing-down" of marine food webs to lower trophic levels'.[4] For example, the last century has witnessed the collapse of numerous important stocks, such as Atlantic cod and northern blue fin tuna, causing serious socio-economic and ecological implications.[5] The environmental impact has important consequences for human life and the economy, since fish is an important source of animal protein in the human diet and fishing and its ancillary activities[6] provide

1 In the last five decades, the fish food supply increased at an average rate of 3.2 per cent per year. FAO 2014, p. 3.

2 Ibid., p. 4.

3 Ibid., p. 37. By definition, fully exploited stocks mean that catches are 'at or very close to their maximum sustainable production' so that there is no room to further expand catch or some stocks 'may even be at some risk of decline unless properly managed'. 'Overexploited stocks produce lower yields than their biological and ecological potential', hence requiring 'strict management plans to rebuild stock abundance and restore full and sustainable productivity'. See FAO 2012a, p. 53.

4 Chu 2009, p. 219; see also Pauly *et al.* 1998; Hutchings 2000; Jackson *et al.* 2001.

5 Hutchings 2000; Schindler *et al.* 2002; Worm *et al.* 2006.

6 This is especially true in developing countries. Fish provides 19.2 per cent of animal protein in developing countries, and 24 per cent of low-income food-deficit countries. FAO 2012a, p. 5.

employment and livelihood support.[7] Many different actors are involved in these sectors, such as industrial actors and small-scale fishermen, including the traditional users, usually competing for limited fish resources. Small-scale fishing is especially important, providing employment for 90 per cent of the capture fishers[8] and producing 70 per cent of the total world catch.[9] Therefore, the depletion of fish stocks is not purely an environmental issue, but also a social and economic one.

Although finding causes of fluctuations in marine fish stocks has been difficult, given the frequent co-existence of long-term environmental trends and human behaviour, it is accepted that overexploitation is a main driver for declining fish stocks.[10] The underlying drivers of overexploitation include the general overcapacity of the fishing industry,[11] ways of fishing, processing, storing and transporting,[12] market demand and price,[13] the lack of regulatory capacities among governments, and their willingness to enforce them and so on.[14] Though the specific drivers may be context-dependent, it is widely accepted that the fundamental issue is that fish stocks are common-pool resources.[15] Without proper governance regimes to limit the access to fishery resources, 'too many people chase too few fish'.

For inland fisheries, the main drivers influencing fish stocks are no longer the exploitation rates, but other factors, such as 'habitat quantity and quality, aquaculture in the form of stocking and competition for freshwater', 'water abstraction and diversion, hydroelectric development, draining wetlands, and siltation and erosion from land use patterns'.[16] Moreover, resources under aquaculture are excludable, and therefore are private goods.

Since this research focuses on the governance of common-pool problems, we will exclude inland fisheries from our analysis and will focus on marine fishery, for which overexploitation is the major man-made (proximate) driver. We will examine how public and private regulation interact in establishing property rights and in overcoming the limitations of property rights in addressing overfishing at sea. The fact that there are many different drivers of depleting marine fish stocks obviously makes the choice of an adequate regulatory framework complex.

7 Together, they support livelihoods of about 10–12 percent of the world's population (employment and their dependants). Ibid., p. 10.
8 Ibid., p. 17.
9 Kolding, Bener & Bavinck 2014, p. 318.
10 Sharpe & Hendry 2009; FAO 2012a, p. 12; Delgado *et al.* 2003.
11 Around 3.23 million fishing vessels operated in marine waters in 2010. FAO 2012a, p. 10. The problem of overcapacity has long been highlighted. See Pauly *et al.* 2002, p. 692.
12 Cooper 2004–2005, pp. 5–6. The origins and growth of social and environmental certification programmes in the fisheries sector, p. 12.
13 Caddy & Seijo 2005.
14 Auld 2007, p. 12.
15 Andersen 1983; Gordon 1954; Hardin 1968.
16 Auld 2007, p. 13.

This chapter starts by giving a historical overview of the international laws, domestic laws, and private regulatory regimes that have been developed to address marine fishery overexploitation. It then goes on to describe how regulatory regimes addressing overfishing have evolved in four different countries (South Africa, Mexico, New Zealand, and Japan). After accounting for the choices of these case studies, each case study starts with a description of the most important problems concerning marine fishing. The focus is on the overexploitation of particular fish stocks, but related environmental problems such as loss of biodiversity and by-catch will also be addressed. Subsequently, the case studies describe how property rights have been established in a particular country and what additional public and private regulations have been implemented to prevent the depletion of fish stocks. The case studies show that each country has followed its own path in governing fish stocks and has its own set of stakeholders which is involved in the governance regime. The country-specific manner in which the governance regimes have evolved also determines how they play out in practice. After describing the evolution and functioning of the regulatory mixes, the case studies evaluate whether or not the preconditions for property rights' effectiveness are satisfied in the regulation of fish stocks – establishing property rights, enforcing, coordinating, collecting information, scaling governance – and whether or not public and private regulation overcome the limitations of property rights in countering overfishing. Each case study ends with a summary of the aspects in which the regulatory mix works well or badly. The chapter ends with a comparison of the four case studies in terms of their transaction costs in managing fish stocks.

4.2 Fishery governance: international law, domestic law and private regimes

This section provides a brief historical overview of the international law, domestic law and private regulatory regimes that have been developed to address marine fishery overexploitation.

4.2.1 International law

Regulating fisheries under international law has a long history, which dates back to the nineteenth century when a few landmark cases were adjudicated and conventions were adopted, such as the Pacific Fur Seal Arbitration, the Geneva Conventions on the Law of the Sea 1958, the Fisheries Jurisdiction Case and so on.[17] In the current legal regime, the UN Convention on the Law of the Sea 1982 (UNCLOS) plays a fundamental role in setting rights

17 Sands & Peel 2012, pp. 399–402.

and obligations of states in fishery conservation. The customary law developed in the 1970s is now consolidated under the UNCLOS, which extends the right of the coastal state to the Exclusive Economic Zone.[18] Under the UNCLOS, the coastal state has 'sovereign rights for the purpose of exploring and exploiting, conserving and managing the natural resources'.[19] The open access to high seas remains subject to limited conditions under the convention.[20] The control of fishing on high seas hence still relies largely on the flag state.[21] Member states are required to cooperate in establishing (sub)regional fishery organizations.[22] In 1995, the UN Agreement for the Implementation of the Provisions of the United Nations Convention on the Law of the Sea of 10 December 1982 relating to the Conservation and Management of Straddling Fish Stocks and Highly Migratory Fish Stocks (Fish Stocks Agreement) was adopted.[23] It codified the precautionary approach and ecosystem-based management to fisheries, and widened the inspection competency of non-flag states on the high seas.[24] Another recent shift is the adoption of the FAO Agreement on Port State Measures to Prevent, Deter and Eliminate Illegal, Unreported and Unregulated (IUU) Fishing in 2009. It tries to harmonize port state measures, enhance international cooperation and prevent IUU fishing.[25] In addition to these legally binding conventions, soft law also shapes international fishery governance, such as the 1995 FAO Code of Conduct for Responsible Fisheries and the 2005 FAO Guidelines for the Eco-labelling of Fish and Fishery Products from Marine Capture Fisheries. Many regional fisheries organizations have been established, which are usually mandated to collect information and to do research and to perform regulatory functions. The International Council for the Exploration of the Sea, the International Pacific Halibut Commission, and the Northwest Atlantic Fisheries Organization (NAFO) are examples of such regional fisheries organizations.[26]

4.2.2 Domestic law

4.2.2.1 Introduction

Fisheries have been a concern for domestic regulatory systems for a long time. In pre-industrial time, early fishing technologies had already started to put marine fisheries under stress. In some European countries, 'fishing

18 Stokke 2012.
19 Article 56, 1(a).
20 Article 116.
21 Stokke 2012, p. 105.
22 Article 118.
23 34 ILM 1542 (1995) in force 11 December 2001.
24 Stokke 2012, pp. 103–104.
25 www.fao.org/fishery/topic/166283/en.
26 Sands & Peel 2012, pp. 411–417; Warner, Gjerde & Freestone 2014.

activities have been regulated through professional organizations with conditional decision power or community-based groups'. Since the seventeenth century,[27] local use rights, regulations and religious taboos regulated fishing activities in certain areas.[28] With the rise of sovereign national states and the process of industrialization, such 'pre-capitalistic community-based management' of fishery rights has shifted towards a more 'capitalistic and market-driven centralized management' of those rights.[29] Early regulation focused on economic growth and the achievement of optimum yield. Since the 1980s, overfishing and overcapacity have become serious concerns and conservation issues started to play an important role in domestic regulation.[30]

At domestic level, four types of measures are often used to regulate fishing activities: technical measures (mesh sizes, gear requirements, seasonal closure of fishing ground, etc.), input controls (limited entry by permits, effort quotas), output controls (total allowable catch (TAC), catch quotas) and economic measures (tax exemption, subsidies).[31] Such measures, if classified according to the typologies of governance instruments in this research, can be command and control (legally binding), incentive-based (subsidies, determining fishing rights and sometimes making them transferable) or suasive instruments (governmental advice and recommendations).

Traditional regulation often relied on technical measures, input controls, sometimes combined with a TAC.[32] The traditional approach, however, often fails to regulate fishing activities effectively. Technological innovations have often rendered technical measures and input controls obsolete.[33] Input control, when combined with a TAC, creates 'incentives for fishers to race against each other for a share of the catch before the total harvest limit is reached'.[34] Therefore, an incentive-based approach has gained popularity. Various fishing rights, either of an individual or collective nature, have been established to restrict access to fisheries resources and to incentivize fishermen to better conserve fish resources.[35]

4.2.2.2 Fishing rights

As discussed earlier, four types of property rights regimes can be established for common-pool resources: open access, private property,

27 Garcia, Rice & Charles 2014a, p. 5.
28 Ibid.
29 Ibid.
30 Ibid., pp. 5–9.
31 Stokke 2012, p. 101; Garcia, Rice & Charles 2014b, p. 19.
32 Garcia, Rice & Charles 2014b, 19; Huppert 2005, p. 202; Grafton *et al.* 2006, p. 700.
33 Squires 1992.
34 Grafton *et al.* 2006, p. 700.
35 Huppert 2005, p. 202; Garcia, Rice & Charles 2014b, p. 19; Grafton *et al.* 2006, pp. 701–702.

communal property and state property.[36] For fishery resources, the high seas remain largely open access with limited conditions imposed by international law.[37] The development of customary law in the 1970s and the UNCLOS extended the sovereign right of the coastal state to the Exclusive Economic Zone (EEZ).[38] Therefore, the resources in the EEZ and territorial seas are nominally state property. The fish in the EEZ and territorial seas are actually harvested by different actors, including industrial fleets, small-scale fishing and traditional users. These users usually do not have ownership of the fish resources in the sea, but have different extents of access to and rights to harvest the fish resources. The scope and boundary of such rights may differ among jurisdictions and even fisheries. Various rights arrangements can make the nominal state property (fish in EEZ and territorial sea) de facto an open access, private regime or communal property.[39]

Debates exist in defining the access of actors actually engaged in fishing activities. Some view it as property rights (e.g. use rights rather than ownership), while some others regard it as 'privileges'.[40] This research will not discuss its legal nature, but will use the term 'fishing rights' to refer to the rights of fishing actors to access and harvest fish. Different types of fishing rights are used in practice, including limited entry permits, individual quotas, communal quotas or territorial use rights.[41] Under the limited entry system, the number of permits or fleets is limited, whether or not accompanied by a TAC. But there is no individual quota for permit holders. Therefore individuals compete with each other to harvest. With individual quotas, individuals have a share of a TAC to harvest and sometimes such quotas are transferable (individual transferable quotas, ITQs). The quotas can also be granted to a group instead of individuals, especially fishing communities or cooperatives. Under such collective quotas (communal property), communities decide how to manage fisheries and allocate rights to individual fishers. The individual and collective quotas define the fishing rights according to the quantity of fish they can harvest. Another approach is the territorial use right (TUR), which defines the rights based on fishing areas. Under such a scheme, right to access or harvest in a specific area is limited to members of a community, other groups or individuals according to custom or statutes. This chapter focuses on three commonly used fishing rights: limited entry, individual quotas and TUR.

36 See *supra* section 2.1.
37 E.g. Article 116, UN Law of Sea 1982 (UNCLOS).
38 Stokke 2012, pp. 99–101; Article 56, 1(a) UNCLOS.
39 Young 2001, p. 286.
40 E.g. US. See Huppert 2005, p. 205.
41 Huppert 2005, p. 207.

The limited entry system was first proposed to replace the open-access scenario[42] and has become widely used since the 1970s.[43] Under this system, the access to fisheries resources is restricted, usually via licence requirements. The licences can be transferable or non-transferable.[44] In other words, the fishing rights are limited via input: the number of fishermen, vessels, the type, size or engine power. This scenario is closest to the public property right. The licensed actors can access or harvest the resources, but the right is not exclusive and without quantity limits. The government decides how the fisheries are managed by public regulation, such as imposing a TAC, adding conditions to the licences regarding fishing efforts, technical issues and so on. The literature holds that the limited entry system has been effective to reduce overfishing under some particular restrictive conditions. One is that the entry itself should be restrictive (as far as the number of allowed fishers are concerned), but also that the type of fishery to be regulated would not be too complex.[45] In addition it is held that for a successful limited entry system it is necessary that the management of the system is based on a particular social and political attitude.[46] In the short run, entry limitation can reduce the entry rate and slow down the fleet/effort expansion. However, without clear individual rights, the fishermen still have incentives to compete with each other to harvest the resources until the total cap is reached (if there is a cap). This provides them no incentives to delay current catches and to choose environmentally friendly fishing methods.[47] Moreover, in the long run, the benefits of limited access can be compensated by the growth of licensed actors and the increased efficiency of fishing activities. Therefore the overcapitalization of the fishery sector and the exploitation of fishery resources cannot be solved.

Individual quotas have been introduced to end the 'race for fisheries' under the limited entry system. Compared to a limited entry system, clearer individual rights to fish resources have been established with more specific content. This is referred to in the literature as 'rights-based fisheries management'.[48] Under the individual quota systems, the fishermen hence no longer compete with each other in harvesting and have better incentives to manage the resources sustainably. However, such rights are

42 Sinclair 1961.

43 Townsend 1990, p. 359.

44 Ibid., p. 360.

45 The extent to which a limited entry system can protect the fishery resources is influenced by 'the specific technical and operational features of the fisheries'. 'For example, in a terminal salmon gillnet fishery, a limit on net length and vessel size may eliminate most, if not all, avenues for overcapitalizing the fishing operation'. See Huppert 2005, p. 206. If the benefits from restricting one technical aspect can easily be compensated by increasing efforts in the other unrestricted aspects, the limited entry system can hardly be effective.

46 Townsend 1990, pp. 371–372.

47 Ibid., p. 372.

48 Arnason 2012.

not exclusive either. Many fishermen work in the same area and hence cannot decide how the resources are managed individually. The government still plays an important role in deciding when, where and how to harvest and manage the resources and in enforcing the rights. Therefore, individual quotas have the combined characteristics of private and public property rights. One type of individual quota system, the individual transferable quota system (ITQ) has received intensive academic attention for its capacity to allow for flexibility and to improve efficiency.[49] The first ITQ fisheries came into being in the late 1970s. Until 2009, 249 species in eighteen countries are under ITQ management, covering 10 per cent of the total marine harvest.[50] The ITQ system is believed to be able to solve the overcapitalization of the fishery sector, rationalize the fleet structure, increase the economic gains of fishermen, reduce the administrative costs of the government and end the race for fish.[51] However, it has limitations as well. ITQ provides only property rights over specific species, but not over other species or the ecosystem in the same marine area. Therefore the ITQ holders have no incentives to internalize the remaining externalities (reducing by-catch and protecting the habitat).[52] Even for the targeted species, the ITQ system may lead to a high level of discarding low-quality catch and an inefficient harvest pattern when the stock is heterogeneous.[53] To overcome the inefficient allocation of fishing efforts over space, coordination between fishermen is crucial. Coordination can be conducted by self-governing institutions in case of communal property rights (such as the territorial user rights discussed below).[54] In addition, fishermen need to engage in costly searches 'when stock densities vary in an unknown way over space or time'.[55] Information sharing can help to reduce such search costs. Some self-governing institutions can also make pooling arrangements of their proceeds and coordinate market behaviour, hence reducing the competitive incentives among individual fishers and allowing for information sharing.[56] Under an individual quota system, however, the self-interest of fishermen may limit the incentives of information sharing and the optimal dispersion of fishing efforts.[57]

49 E.g. Grafton 1996; Squires *et al.* 1998; McCay *et al.*1995; Branch 2009.

50 Chu 2009, pp. 217–218.

51 Chu 2009; Griffith 2008; Symes & Crean 1995.

52 Branch 2009; Cancino, Uchida & Wilen 2007.

53 Under individual quotas, the volume of fish one fisherman can exploit is limited. In case of spatial heterogeneity, the fishermen tend to compete for the most profitable stocks and overexploit them. The less profitable stocks may be under-exploited. See Cancino, Uchida & Wilen 2007; Costello & Deacon 2007.

54 Cancino, Uchida & Wilen 2007, p. 399.

55 Costello & Deacon 2007.

56 Cancino, Uchida & Wilen 2007, pp. 401–402.

57 Cancino, Uchida & Wilen 2007; Costello & Deacon 2007; Arnason 2012.

Another approach is to vest fishing rights in communities rather than in individuals. This usually takes two forms: the collective quotas or TURs. The latter are especially interesting, since they are area-based but not quota-based. The TURs are usually established for multiple species, especially benthic, sedentary species in coastal areas, given their suitability for defining rights in a specific area. In some cases, the TURs can be established on all the species in a certain area.[58] Usually, a self-governing institution is established to decide the allocation of the resources to its members, the management and protection of resources and to coordinate the marketing strategy. Therefore, the TURs are communal property rights. Though the adoption of TURs is still limited compared to the limited entry and ITQs, an incomplete overview shows that there are at least 1,000 fisheries managed under the TURs in 41 countries. Given the more complete property rights they cover and the existence of coordination institutions, the TURs can overcome some deficiencies of the ITQs. They can partially internalize the externalities left by ITQs, such as allowing considerations of multiple species, accounting for species heterogeneity and creating more incentives to protect the habitat. The ability to achieve those, however, depends on the suitability of establishing TURs for the species and areas and the ability of self-governing institutions to set the rules and enforce them.

4.2.3 Private regimes

4.2.3.1 Introduction

In spite of the comprehensive governmental regulation at both international and domestic levels, many fishery-management challenges remain, such as the overcapacity of fishing fleets, large quantities of by-catch, degradation of habitat and IUU fishing.[59] In response to these failures, social movement[60] and consumer concerns triggered the development of private governance regimes, such as eco-labels, buyers' guides and certification schemes. Early attention was focused on marine mammals and sea turtles as by-catch of fishing activities. Some eco-labelling regimes were established for single species, such as the Dolphin Safe Tuna Certification established under the Agreement on the International Dolphin Conservation Program[61] and the Dolphin Safe label established by the Earth Island Institute.[62] Such schemes, however, focus on protecting animal rights and

58 See *infra* case study of Seri fishery in Mexico.
59 FAO 2012a; Gulbrandsen 2010, p. 115.
60 For example, the public concern over the substantial dolphin by-catch by tuna fisheries triggered the development of the dolphin-safe labelling scheme.
61 Baird & Quastel 2011.
62 Oosterveer 2010, p. 9.

endangered species, rather than preventing overfishing and fish stock depletion.[63] Another approach is to develop seafood buyer guides, providing information for consumers to choose fish products from sustainably managed fisheries. The Audubon Seafood wallet card, the Seafood Watch guides, and the Seafood Choice Alliance programme are examples.[64] Such programmes have enhanced consumer awareness, but do not involve standard setting and third-party inspections.

Based on the earlier efforts, some more comprehensive, sector-wide certification schemes started to develop. Inspired by the success of the FSC, the WWF formed a partnership with Unilever, the largest seafood buyer at that time, to develop a certification scheme for the fishery sector in 1996.[65] Based on studying the governance structure of the FSC, the Marine Stewardship Council (MSC) was established in 1997. Differently from under the FSC, the MSC does not have a membership-based governing body, but leaves the decision-making power to the Board of Trustees. It also adopts centralized standards for fishery management and the chain of custody, rather than having national affiliates to develop their own national standards.[66]

Another certification scheme with broad geographical application and substantial coverage is the Friend of the Sea certification scheme (FOS). The FOS was founded in 2006 and included standards for both capture fish and aquaculture. It has also a substantial coverage of the world capture fish (10 per cent).[67] Unlike the MSC which bases its assessment on field study, the FOS adopts a strategy to rely on existing published data.[68] Therefore, it has a much shorter assessment period and less costly assessment fees.[69] This allows a substantial coverage in small-scale fisheries in developing countries.[70]

In addition to these two transnational certification schemes, some countries also have their own national schemes, such as KRAV certification applying mostly in Sweden, Naturaland in Germany and the Marine Eco-Label in Japan.[71]

63 Allison 2001, p. 945.
64 Gulbrandsen 2010, pp. 115–116.
65 Gulbrandsen 2010, pp. 117–118.
66 Ibid., pp. 118–125.
67 Washington & Ababouch 2011.
68 Oosterveer 2010, p. 7.
69 The assessment under MSC takes around fourteen months (pre-assessment and full assessment together), while the assessment under FOS takes only one day to two weeks. See Karlsen, Hermansen & Dreyer 2012, p. 1124. The full assessment cost for MSC is estimated to be around US$10,000 for a small fishery and US$250,000 for a large one, while the FOS certification costs are only around US$2,000. Washington & Ababouch 2011, p. 39.
70 www.friendofthesea.org/fisheries.asp?ID=71; www.friendofthesea.org/about-us.asp?ID=10.
71 Oosterveer 2010, pp. 6–9; Thrane, Ziegler & Sonesson 2009.

4.2.3.2 The choice of private regulatory scheme: the MSC

Of the various certification schemes, only the MSC and FOS have global coverage.[72] FOS, although it expanded quickly since its establishment, has been criticized for a lack of professionalism[73] and relying on published data for assessment has also been said to reduce the added value of the certification scheme.[74] The MSC, though also criticized in several aspects, is generally accepted as the most legitimate wild-capture certification scheme. The longer history and public availability of certification reports allow for a more meaningful analysis. Hence the following research on private regimes focuses on the MSC.

The MSC holds two standards: the Fisheries Standard, and the Chain of Custody Standard. The first version of the Fisheries Standard (the MSC Principles and Criteria for Sustainable Fishing, hereinafter MSC P&C)[75] was developed on the basis of the 1995 FAO Code of Conduct and stakeholder consultation between 1996 and 1999.[76] In developing the standards, the MSC decided to restrict the scope to environmental issues and wild-capture fisheries. Unlike the FSC, social issues, like fishing rights, local communities and labour issues, are not included. The MSC P&C includes three principles, focusing on target fish stocks, environmental impact and the fishery-management system and a few more specific criteria.[77] The assessment and certification are conducted by accredited certification bodies according to MSC standards. The assessment procedure includes a mandatory, confidential pre-assessment which determines whether it can proceed further to a full certification assessment, for which the reports are published.[78] The MSC P&C only provides general requirements, and no specific standard for different nations/regions has been developed like the FSC did. Expert teams appointed by certification bodies used to develop 'an assessment tree' comprising detailed indicators and scoring guides for assessment in a specific fishery.[79] However, such an approach was criticized as allowing too much room in interpreting the principles and criteria and creating inconsistent assessment among similar fisheries, especially concerning principle 2.[80] To

72 Each covers around 10 per cent of world captured fish. www.msc.org/business-support/key-facts-about-msc; Washington & Ababouch 2011, p. 25.

73 Oosterveer 2010, p. 7.

74 Ibid.

75 www.msc.org/documents/scheme-documents/msc-standards/MSC_environmental_standard_for_sustainable_fishing.pdf.

76 Gulbrandsen 2010, p. 122.

77 www.msc.org/documents/scheme-documents/msc-standards/MSC_environmental_standard_for_sustainable_fishing.pdf.

78 Gulbrandsen 2010, p. 125.

79 Ibid.

80 Marine Stewardship Council and Governance Review Commission, 'Simpler, Faster and More Consistent: MSC Launches New Fishery Assessment Methodology', Press release; Marine Stewardship Council 2008; Ward 2008.

enhance consistency, the MSC adopted the Fishery Assessment Methodology in 2008, which introduced a 'default assessment tree' for all fisheries to be assessed to replace the individual assessment trees developed by certifiers for each fishery. Specific indicators and scoring guides are incorporated in the new methodology.[81] This consistent approach has been consolidated in the new Fisheries Standards that came into force in 2015.[82] The new Fisheries Standards include a few important changes which include standards on the protection of vulnerable marine ecosytems (VMES), which operationalize the rules under the UN General Assembly Resolution 61/105.[83] The effect of MSC certified fisheries on by-catch will be evaluated cumulatively.[84] A new assessment method, the Risk-Based Framework, has been introduced for those types of fisheries where adequate data inter alia on the fish stock were lacking. Those fisheries have difficulties in acquiring certification under the just-mentioned default assessment tree, precisely because of this lack of data.[85]

During assessment, all indicators for the three principles are assessed according to scoring guidelines. To obtain certification, a fishery must at least obtain an average of 80 for each principle and 60 for each indicator. For indicators scoring between 60 and 80, the certification body will issue conditions requiring increased scores within a determined time frame.[86] By October 2015, 275 fisheries had been certified according to MSC standards and a further 103 fisheries had been undergoing assessment. The certified fisheries produce a catch of around 9 million metric tons annually, comprising roughly 10 per cent of the world capture harvest.[87] Owing to the assessment procedure and high costs associated with it, most certified fisheries are located in developed countries.[88] Despite the efforts of the MSC to promote development in developing countries, only 8 per cent of the certified fisheries are located in such countries.[89] With large fish product retailers increasingly engaged with the MSC, and the limited number of fisheries certified, a few fishery improvement projects (FIPs) have been developed to help the fisheries improve towards pre-assessment or from

81 www.msc.org/documents/scheme-documents/msc-scheme-requirements/methodologies/Fisheries_Assessment_Methodology.pdf/view.

82 MSC Fisheries Standard and Guidance v. 2.0. available at: www.msc.org/documents/scheme-documents/fisheries-certification-scheme-documents/fisheries-standard-version-2.0.

83 www.fao.org/in-action/vulnerable-marine-ecosystems/en/.

84 www.msc.org/newsroom/news/marine-stewardship-council-publishes-updated-fisheries-standard.

85 www.msc.org/about-us/standards/fisheries-standard/msc-risk-based-framework.

86 Foley 2013, p. 296.

87 www.msc.org/business-support/key-facts-about-msc.

88 Pérez-Ramírez *et al.* 2012; Ponte 2012.

89 www.msc.org/about-us/credibility/working-with-developing-countries/fisheries-involved-in-the-developing-world-programme.

pre-assessment to full assessment.[90] The fishery improvement projects usually involve multiple stakeholders, such as fishermen, managers, suppliers, buyers, funders, and NGOs.[91] A typical FIP involves four steps: an MSC pre-assessment when a gap is identified between the current status and MSC requirements; the development of an action plan for improvement to close this gap; the implementation of actions and progress tracking and finally a full assessment of the fishery.[92]

4.3 The selection of countries

4.3.1 Characteristics of fishery governance

Before explaining the choice of countries for case studies, this section first discusses several characteristics of fishery governance, specifically how public regulation and fishery certification influence fishing rights and the interaction between public and private regulation.

The above analysis shows that fishing rights can be vested in individuals or communities. However, unlike forest rights, such rights are not ownership and are less complete. The rights holders can access and harvest specific fishery resources. Under the individual quotas or TURs, they can also execute part of the rights to manage resources. Compared to forest resources, however, governments have more direct control over the management of fishery resources. Under the individual quota systems, many individuals have non-exclusive rights over the same fishery. They cannot decide how to manage the fisheries on their own. Under the TURs, usually the self-governing institutions have more say in fisheries management. However, the TURs usually cover a limited type of species in small-scale fisheries. Decisions need to be made at a higher level regarding the management in broader areas. Therefore, sometimes it is governments rather than rights holders who make plans or establish standards for fishery management.

In contrast to forest certification, the MSC standards focus on environmental issues and do not have criteria specifically on fishing rights. Their influence on fishing rights is less direct: the MSC requires legal compliance, including the lawful rules on fishing rights.[93] Moreover, the MSC can inject 'a new form of exclusive rights or privileges into common-pool resource situations already complicated by access and property privileges',[94]

90 Bush *et al.* 2013, p. 289; www.msc.org/about-us/credibility/all-fisheries/tools-for-fisheries-improving-towards-msc-certification/tools-for-improving-fisheries-towards-msc.

91 www.sustainablefish.org/fisheries-improvement; www.msc.org/about-us/credibility/all-fisheries/tools-for-fisheries-improving-towards-msc-certification/fips.

92 www.msc.org/about-us/credibility/all-fisheries/tools-for-fisheries-improving-towards-msc-certification/fips.

93 Principle 3, Management System Criteria 4, in MSC P&C.

94 Foley & McCay 2014, p. 27.

strengthen the rights claims of the certified and weaken the others.[95] As for the small-scale fisheries (SSF) and community-based fisheries, some argue that the MSC weakens their property rights due to the long and costly certification procedure and standards which are more favourable to industrial fishing.[96] Others argue to the contrary, that the MSC has the potential to empower the SSF and communities, by strengthening their rights claims and enhancing their role in decision-making.[97] Which argument holds true needs to be revealed through empirical research.

Compared to forest governance, the interaction between domestic law and certification schemes for fishery governance is more obvious. States usually control and manage access to marine fisheries and sometimes they are the clients of certification.[98] Other clients of MSC certification can be a fishermen's association, an industry association of quota holders or an organization of processors. Even when private actors are clients, the government needs to have a certain level of control over the fishery.[99] This, to a large extent, depends on the design of fishing rights under domestic law. The Fisheries Standards require the maintenance of fish stocks and the ecosystem (Principles 1 and 2). To achieve these, data collection (e.g. stock assessment) and monitoring are necessary, which can only be done by the government.[100] Principle 3 of the MSC Fisheries Standards requires effective management. Many fisheries are actually directly managed by governmental agencies, which develop management plans. Therefore certification can to some extent be regarded as an endorsement of public regulation. However, since the MSC has no standards directly related to fishing rights, the discussion on the interaction between the MSC and public regulation focuses on the practical level but not on the theoretical (standards) level.

4.3.2 The choice of countries

This research examines the interaction between public regulation and private regimes and their potential to address overfishing, especially the interaction between fishing rights and the MSC. Four countries have been chosen for further examination: South Africa, Mexico, New Zealand and Japan. All four countries have substantial marine fish capture and are all ranked in the top 25 in the world.[101] Two are developed countries where the

95 Ponte 2008.

96 Pérez-Ramírez *et al.* 2012.

97 Lopuch 2008, pp. 317–320; Novy-Hildesley & Short 2003, pp. 167–170.

98 Gulbrandsen 2010, p. 125. E.g. Alaska Department of Fish and Game is the holder of the certificate for the Alaska salmon fisheries.

99 The state has 'ultimate legal responsibility over controlling and managing access to marine fisheries'. Foley 2013, p. 301.

100 Foley 2013.

101 FAO 2012b, available at: ftp://ftp.fao.org/fi/stat/Overviews/CaptureStatistics2012.pdf.

public fishery regulator has comparatively better implementation and enforcement capacity than the two developing countries. Different fishing rights are used across the countries. Individual quota and limited entry systems co-exist in South Africa. In Mexico, both limited entry and TUR systems apply. New Zealand presents a typical example of an ITQ system and Japan has the best established TUR networks. Certification plays a role in all these countries, but differs in coverage. More fisheries have been certified by the MSC or are under full assessment in New Zealand and Mexico. Only one fishery has been certified in South Africa, which provides an opportunity to examine how it performs in comparison with other fisheries in the country. Although in Japan, an important developed fishing country, only two fisheries have been certified by the MSC. One reason for this is that industry and government have responded strategically to the MSC and have established a competing domestic certification scheme.

4.4 Case study: South Africa

4.4.1 Fishery problems

In 2012, 0.7 million tons of fish were caught in South Africa, ranking second in Africa.[102] Around 50 per cent of the main fish stocks are considered to be unhealthy and around 22 per cent are depleted or heavily fished while 28 per cent are heavily depleted.[103] The concern is especially true for inshore resources, due to the accessibility of 'a wild range of marine user groups, including traditional linefishers, recreational fishers, and inshore trawl and longline fisheries'.[104] Abalone and linefishing are two examples that are subject to serious illegal fishing and collapse.[105] The status of offshore resources is better, with only 14.8 per cent of species overexploited and another 29.6 per cent with uncertain status. Hake and small pelagic fisheries are the two most important fisheries in South Africa. The deep-water hake is considered to be overexploited while the shallow-water hake as well as small pelagic fish are not.[106]

Commercial, recreational and small-scale sectors all operate in South Africa. For the commercial sector, industrialized fleets mainly work in

102 FAO 2012b.

103 South Africa Department of Agriculture, Forestry and Fisheries, Status of the South African Marine Fishery Resources 2014, p. 1, available at: www.nda.agric.za/doaDev/sideMenu/fisheries/03_areasofwork/Resources%20Research/STATUS%20OF%20THE%20SOUTH%20AFRICAN%20MARINE%20FISHERY%20RESOURCES%202014%20WEB.pdf.

104 WWF 2011, p. 9, available at: http://awsassets.wwf.org.za/downloads/wwf_a4_fish_facts_report_lr.pdf.

105 Ibid., pp. 11–12.

106 Ibid., pp. 14–16.

offshore fisheries while traditional and less capital intensive fishers work mainly in near-shore fisheries.[107]

4.4.2 Good governance

South Africa is an upper-middle income country with annual GDP at US$350.1 billion in 2014.[108] According to the World Bank WGI project, South Africa is rated 68.47 (out of 100) regarding importance and accountability, 43.20 regarding political stability, 65.38 regarding government effectiveness, 63.94 regarding regulatory quality, 63.94 regarding the rule of law and 54.33 regarding the control of corruption.[109] The just above average World Bank WGI indicator on control of corruption is also confirmed according to the extent of corruption control from Transparency International. According to Transparency International, South Africa was ranked 61 out of 167 countries.[110]

4.4.3 Property rights as well as public and private regulation

4.4.3.1 Public regulation

Fishing is mainly regulated under the Marine Living Resources Act of 1998 (MLRA), which aims to achieve three goals: ecological sustainability, economic viability and equity.[111] The fisheries in South Africa are mainly regulated under two instruments: the total allowable catch (TAC), the total allowable effort (TAE) or a combination of them.[112] The competent authority determines the TAC or TAE applying to a particular area, particular species or particular equipment, fishing methods or vessels.[113] It also allocates the proportion to subsistence, recreational and commercial sectors and furthermore to each rights holder.[114] To use a fishing vessel, a fishing vessel licence must be applied for[115] and a permit needs to be obtained to

107 Ibid., p. 9.
108 http://data.worldbank.org/country/south-africa.
109 http://info.worldbank.org/governance/wgi/index.aspx#reports.
110 www.transparency.org/cpi2015/#results-table.
111 Section 2, MRLA.
112 'Total allowable catch' means the maximum quantity of individual species of fish or groups of species made available annually, or during such other period of time as may be prescribed, for combined recreational, subsistence, commercial and foreign fishing; 'total applied effort' means the maximum number of fishing vessels, the type, size and engine power thereof or the fishing method applied thereby for which fishing vessel licences or permits to fish may be issued for individual species or groups of species, or the maximum number of persons on board a fishing vessel for which fishing licences or permits may be issued to fish individual species or groups of species; section 1, MRLA.
113 Section 14(1)(3).
114 Section 14(2).
115 Section 23.

execute the fishing right.[116] The competent authority also declares the fisheries' management area and approves the plans for 'the conservation, management and development of the fisheries'.[117] The Operational Management Procedure (the OMP) is one type of plan. The OMP covers management of resources in times of scientific uncertainty. The important fisheries in South Africa are under the OMP.[118] In addition, the MLRA and corresponding regulations also adopt some other technical conservation measures, such as gear restrictions, seasonal closures and so on.[119]

Since 1994, the fishery management was centralized and put under the Sea Fisheries Chief Directorate,[120] which was renamed as the Department of Marine and Coastal Management (MCM) in 1998.[121] Its authority was also extended from fishery research to broader fishing management issues such as fishing rights distribution, monitoring and so on.[122] Since 2010, the authority of MCM was transferred to the Department of Agriculture, Forestry and Fisheries (DAFF), which has different branches responsible for fisheries research, fishing rights allocation, marine resources management (e.g. determining TAC, establishing management procedures and plans[123]) and monitoring.[124] For each fishery, two groups are advising the DAFF on decision-making. The Scientific Working Group (SWG) provides scientific input regarding fishery management. The Resource Management Working Group (RMWG) advises the DAFF on quota and permit allocations.[125] These groups are consultative forums providing a bridge between the DAFF and the fishing industry.[126]

4.4.3.2 Property rights

According to section 18 of the MRLA, any person undertaking commercial or subsistence fishing needs to have fishing rights granted by the competent authority. As discussed earlier, the competent authority determines the TAC or TAE to the fisheries and further distributes them among the rights holders. Therefore two types of fishing rights exist in South Africa: the rights with individual quotas and the limited entry. Currently, most coastal

116 Section 13.
117 Section 15.
118 Plagányi *et al.* 2007, p. 626; www.olsps.com/fisheries/know29.html.
119 Witbooi 2006.
120 Hauck & Kroese 2006, p. 75.
121 Van Sittert *et al.* 2006, p. 103.
122 Hauck & Kroese 2006, p. 75; Van Sittert *et al.* 2006, p. 103.
123 Article 6(a), MLRA.
124 www.daff.gov.za/daffweb3/Branches/Fisheries-Management/Marine-Resource-Management.
125 Nielsen & Hara 2006, p. 47; www.daff.gov.za/daffweb3/Branches/Fisheries-Management/Fisheries-Research-and-Development.
126 MSC 2015, p. 96.

fisheries are managed under the TAE. The biggest fishery by volume, the one for small pelagics, is managed by individual quotas and the most important fishery by volume, the one for hake, is managed under both TAC and TAE.[127]

Fishing rights can be short term (up to three years) which can be extended once or long term (up to fifteen years), which will automatically terminate and revert back to the state for reallocation.[128] Both the commercial and subsistence fishing rights are transferable, but the transfer needs to be approved by the competent authority.[129]

In addition to the introduction of current fishing rights systems, the redistribution of fishing rights in the post-apartheid period is worth noting as well. The South African coast has been fished by traditional artisan fishermen for a long time. However, during the apartheid period, the government granted fishing rights exclusively to around 300 white-owned companies, with a few large ones dominating. Black and coloured people and traditional fishing communities were excluded. Owing to the high concentration of the fishery sector, 'the industrial fisheries developed as highly vertically integrated entities, combining harvesting, processing and marketing', which makes it difficult for new entrants to get established.[130] During the post-apartheid period, the government started to redistribute fishing rights to historically disadvantaged (HD) individuals and groups. The redistribution has been conducted both externally and internally. Under the first approach, HD individuals and groups have been granted access rights, bringing the number of rights holders to 5,837 in 2004.[131] Established industries, unwilling to give up their rights, tried to frustrate the transformation and to preserve the status quo by using litigation and creating an alliance with labour unions.[132] They claim that transformation has been implemented through internal changes, such as 'shared ownership, shareholding schemes, change in leadership/management structures, transfer of managerial and technical skills and joint ventures/partnership agreements'.[133] Such efforts are viewed by many as window dressing, and even strengthened the position of established companies in maintaining or increasing their quota.[134] The transformation process has also been criticized as the quotas allocated to new entrants are often too small to make them economically viable. This is especially true when the fishery industry in South Africa is vertically integrated. The new

127 Attwood & Farquhar 1999; Nielsen & Hara 2006; FAO, p. 9, available at: www.fao.org/fishery/facp/ZAF/en.

128 Section 18(6)(6A).

129 Article 19(2), 21, MLRA.

130 Nielsen & Hara 2006, p. 44.

131 Van Sittert *et al.* 2006, p. 102.

132 Isaacs, Hara & Raakjaer 2007, p. 304.

133 Ibid., p. 305.

134 Ibid.

entrants with small quotas cannot afford to establish their own processing facilities and can only resort to the existing ones with processing branches. Therefore, the small quota holders often sell their quotas to established companies, increasing concentration.[135] Besides, other scholars argue that the openness of the redistribution to all HD individuals and groups actually favoured newcomers who had good connections with politicians, while marginalizing the bona fide traditional fishers.[136]

In spite of the history of exclusion, it is estimated that there are still at least 30,000 fishers relying on fishing for their livelihood.[137] The reform has introduced the legal category of the subsistence fisher, which is defined narrowly as those fishing for personal consumption rather than for commerce.[138] This definition excludes fishers who occasionally sell some of their catch but mainly fish to satisfy subsistence needs.[139] This category of fishers can be admitted as 'a limited commercial sector', subject to similar regulation as the industrial sector. Only a small percentage of SSF fishers were granted rights for commercial fishing due to the complex application procedure and the small available quotas for the limited commercial sector.[140] Moreover, even when fishing rights are granted to SSF fishers, they are usually granted to individuals rather than to communities. This situation is contrary to existing customary systems that still operate in many coastal areas.[141] The Policy for the Small-Scale Fisheries Sector was published in 2012, trying to empower this marginalized sector. It is more community-oriented and tries to establish territorial user rights.[142] However, it should be taken into account that this is basically how the system of fishing rights of the SSF should function on paper. That obviously does not provide any specific insight concerning its effectiveness in practice. Although there is a limited access to the SSF fisheries, there is a large historical dependence on fishing and the local population often lacks an alternative livelihood. As a result, even when no formal rights are allocated, many fishers continue to catch fish even without formal rights, but simply in an informal sector.[143]

4.4.3.3 Private regulation

Only one fishery is certified through MSC certification: the South Africa hake trawl. It is one of the earliest fisheries certified in the developing world. As will be discussed later, the management of deep-sea hake fishing

135 Nielsen & Hara 2006.
136 Van Sittert *et al.* 2006, p. 105.
137 Sowman 2011.
138 Article 1, MLRA.
139 Young 2013, p. 291.
140 Sowman 2011.
141 Sowman *et al.* 2014, p. 35.
142 Department of Agriculture, Forestry and Fisheries Republic of South Africa 2012, pp. 27–29, 33–36.
143 Gezelius & Hauck 2011, pp. 456–458.

provides an example of adequate management in South Africa. However, it should be remembered that this only applies to deep-sea hake. It is therefore not representative of how well other fisheries are managed in South Africa. For many such fisheries, because of rampant illegal fishing and poor management, there is slim chance of getting certified. Nevertheless, comparing the certified deep-sea hake fishery with the non-certified inshore fishery helps to understand why fisheries' protection varies so much within South Africa. Therefore, general fishery management is introduced briefly before the analysis of hake fishery governance.

4.4.4 The protection of fishery resources in general

Illegal fishing is widespread in South Africa, which leads to an annual loss of US$815 million.[144] The no-take zones are controlled inadequately and resources, like lobster and abalone have suffered large-scale poaching.[145] Particularly inshore resources are generally poorly protected in South Africa and many species are depleted. This is especially true of abalone and linefish.

The commercial fishing of abalone started in 1949. At that moment, the fishery was an open-access system, with hardly any regulation except for caps on the catching of abalone. The abalone harvest rose to unsustainably high levels in the 1960s (a record high of 2800 tons in 1965), resulting in a shift to quota management in 1968.[146] The annual TAC was reduced in the following years to allow the restoration of the resources and stabilized at around 615 tons until the mid-1990s.[147] However, the situation has deteriorated dramatically since the mid-1990s. The TAC dropped to 75 tons in 2007, leading to a total closure of the fishery in 2008.[148]

There are many causes of the collapse of abalone. First, the price of abalone increased sharply in the 1990s due to the increased demands from East Asia. This led to widespread poaching and even highly organized illegal fishing networks. Besides, the ecological situation also changed, with predators settling into traditional abalone fisheries.[149]

Another important contributor is the beginning of fishing rights redistribution in the mid-1990s. As in many other South African fisheries, abalone fishing rights were historically concentrated in five white-owned companies with local fishermen as employees.[150] Fishing rights were redistributed to local fishermen in the form of subsistence rights and

144 Sjöstedt & Sundström 2015, p. 82.
145 Ibid.
146 Raemaekers *et al.* 2011, p. 434.
147 Ibid.
148 Ibid., p. 435.
149 Hauck & Kroese 2006, pp. 76–77.
150 Raemaekers *et al.* 2011, p. 436.

limited commercial rights.[151] This led to a sharp increase in the number of rights holders. The increase of rights holders and the decrease of the TAC led to many fewer quotas allocated to the existing commercial sector, which was subsequently incentivized to overfish to compensate for the loss of profit.[152] Moreover, the credibility of the redistribution process is questioned due to the fierce competition to obtain the limited resources. The process has been criticized as long, cumbersome and with community elites dominating.[153] Therefore, many of those excluded from formal rights still continued to fish, leading to an increasingly organized illegal fishery.[154]

Responding to the fast deterioration of abalone resources, the government introduced territorial user rights to the abalone fisheries in 2003.[155] Under the TURF system, the individual is granted rights to manage and harvest a certain area of the abalone fisheries. However, this policy has had limited success. First, the zones assigned to fishers are not always exclusive, leading to conflicts between those living in adjacent areas. Second, rights are allocated to individuals but not to collective organizations. The rights holders could not exclude the members of their communities who did not manage to obtain fishing rights. The lack of formal organization of abalone fishers also weakens their position.[156]

The MCM also made efforts to strengthen law enforcement, by cooperating with the police to fight against organized crime, delegating enforcement authority to the local municipality and establishing an environmental court to adjudicate illegal fishing cases. However, some of the efforts did not persist due to a lack of funding and shifting of political priorities. Corruption and lobbying from industrial fishing groups in policy-making and implementation are still pervasive.[157]

Commercial linefishery in the Cape started to develop in the early nineteenth century. At that moment, there were few regulatory restrictions, and boat-based linefishing became a thriving industry.[158] 'In spite of a long history, the first comprehensive management framework for the South African linefishery was only introduced in 1985'.[159] Since then, a permit system has been introduced and the legislation tried to cap the fishing effort at that level.[160] However, by then the fishing industry had already been overcapitalized and most of the damage to the fish population had

151 Ibid.
152 Ibid., p. 437.
153 Ibid.
154 Ibid., p. 438.
155 Hauck & Gallardo-Fernández 2013, p. 8.
156 Hauck 2009.
157 Raemaekers *et al.* 2011, pp. 439–441.
158 Griffiths 2000, p. 81.
159 Ibid.
160 Ibid., p. 104.

already been done. Moreover, during the permit granting procedure, almost all applicants were successful, maintaining the fleet size at an unsustainably high level of vessels (2000).[161] In addition, 'since 1985 many single boat licences have been split to enter two smaller vessels in place of one large one, which has had the effect of increasing effort'.[162] The technological advances also helped to increase the fishing capacity. Therefore a limited entry system failed to prevent the collapse of linefishing.

4.4.5 Interaction between public and private regulation

4.4.5.1 Case study

The deplorable state of the governance of inshore fishery is in clear contrast to the offshore hake fishery. Hake fishery is the most important fishery in South Africa in value, accounting for 45–50 per cent of the landing value.[163] It is based on two species: deep-sea hake and shallow-water hake. Europe is the main market for the hake exports, with Spain, Italy and Portugal dominating in the early 2000s while northern Europe has recently become more important as a market.[164]

Four sectors operate in hake fishery: deep-sea trawl, inshore trawl, longlining and handlining, with the deep-sea trawl dominating.[165] The deepwater hake is mainly caught by deep-sea trawlers and longliners and the shallow-water hake is mainly caught by inshore trawlers and the handline sector.[166] The deep-sea trawl is capital and labour intensive and is organized via the South Africa Deep Sea Trawling Industry Association (SADSTIA).[167] The inshore trawl industry also has a long history and is organized under the South East Coast Inshore Fishing Association (SECIFA). Longlining and handlining sectors started to develop much later, in the 1990s and early 2000s, on a much smaller scale and without adequate organization.[168] The deep trawl, inshore trawl and longlining sectors are mainly managed by the TAC while the handlining sector is managed by the TAE.[169] Since 2008, the two trawling sectors have been subject to TAE limits as well. Their fishing days are capped according to the TAC allocation and the vessel size and power.[170]

161 Ibid.
162 Attwood & Farquhar 1999, p. 430.
163 Nielsen & Hara 2006.
164 Ponte 2006, p. 25.
165 Ibid., p. 27.
166 Ibid.
167 Nielsen & Hara 2006, p. 45.
168 Ponte 2006, pp. 30–31.
169 Ibid., p. 28.
170 MSC 2015, p. 25.

As mentioned above, every fishery has an SWG and RMWG. The SWG is a scientific consultant forum. With the involvement of DAFF and SADSTIA scientists, it advises a TAC level based on the OMP to the DAFF. After the approval of the TAC, the RMWG advises DAFF on the allocation of rights. Then a DAFF fishery manager meets with the industry to prepare annual fishing plans and permit conditions.[171]

In the early 1990s, fewer than 20 licence holders operated in the hake fishery (deep trawl sector), with 80 per cent of TAC concentrated in the hands of two groups. The post-apartheid reform required redistribution of fishing quotas to HD individuals and groups. The redistribution process in the hake sector has been slow since it has been often challenged by incumbent companies in courts. Incumbents who lost their rights as a result of the redistribution filed court cases which obviously slowed down the process. After the medium-term allocation (quota allocations for the following four years) in 2001, only 25 per cent of TAC was redistributed to HD individuals and groups.[172] The government planned to allocate the long-term rights (fifteen years) in 2005/2006. Under this background, the SADSTIA applied for MSC certification in 2002 and got certified in 2004. Officially, the motivations for the SADSTIA to apply for MSC certification are to maintain competitiveness in the international markets, to promote market access, to match buyer demands and to achieve a price premium.[173] However, scholars have noted a few informal motivations, including entrenching interests of major trawl companies,[174] avoiding external transformation which further reallocates the quotas of established companies to new entrants and marginalizing the longliners who are the competitor for deepwater hake.[175]

4.4.5.2 The interaction

This section examines the role of government in certification by looking at the full process of certification. Using the first assessment in 2004 as an example, one of the expert team members is a former scientist for the regulatory agency: MCM.[176] During the assessment, several meetings were held where MCM staff have participated in discussing issues like stock assessment and fishery and environmental management.[177]

The first assessment required seven conditions to be met for the fishery to remain certified, including issues concerning by-catch, structure of

171 Ibid., p. 98.
172 Ponte 2008, p. 166.
173 Ibid.
174 The major companies have advanced processing chains to obtain MSC chain of custody certification, which is the precondition for the products to be sold as MSC certified.
175 Ponte 2008, pp. 166–167.
176 MSC 2004, pp. 32–33.
177 Ibid., p. 33.

stock, ecosystem relations, and effects of trawling on benthic habitats, external review, compliance monitoring and the impact of trawling on seabird populations.[178] Though SADISTIA and SECIFA members are actual rights holders, it is the DAFF and its working groups that decide how the fishery is managed. Therefore, compliance with the conditions that were imposed during the assessment in order to achieve certification required the joint efforts of the regulatory agency, MCM and industry. For example, the by-catch management plan initiated by MCM has to be in place one year after the certification, to include appropriate reference points and stock rebuilding measures;[179] the implementation of stock and ecosystem modelling and related research is also required, which is also under the authority of MCM.[180] As shown in the surveillance report from 2005 to 2008, both MCM and the certification applicant SADSTIA have made efforts to implement the actions required in the conditions that were imposed as a result of the first assessment.[181]

The South Africa Hake Fishery has been recertified twice: in 2010 and 2015. A few benefits have been noted. For example, its export market has been diversified to include the UK, Germany, Switzerland and Australia, which have a high demand for MSC products. The MSC certification is argued to have provided the South African industry with a competitive advantage, at the cost of traditional fishers, who have been hit hard by the economic crisis.[182] Price premiums have also been reported.[183] MSC certification has also been linked to a significant reduction of seabird deaths. However, many challenges still remained, some of which had already been indicated during the first and second assessment of the certification process. For example, although a rebuilding plan has been put in place since the second assessment, there are still concerns with respect to the sustainability of the stock in the deep sea. Other concerns include by-catches, uncertainty of impact on benthic habitat and so on.[184] It is worth noting that the hidden goals of the client in seeking certification have also been achieved. In the reallocation of long-term fishing rights in 2006, 'the proportion of TAC allocated to the longlining industry has not increased' and 'no new entrants were assigned quotas and some of the smaller existing quotas were not renewed'.[185] The argument of established companies

178 Ibid., pp. 40–42.

179 Ibid., condition 1.

180 Ibid., conditions 2 and 3.

181 www.msc.org/track-a-fishery/fisheries-in-the-program/certified/south-atlantic-indian-ocean/south-africa-hake-trawl-fishery/assessment-downloads.

182 Lallemand *et al.* 2014, pp. 6, 9, available at: http://ir.library.oregonstate.edu/xmlui/bitstream/handle/1957/55162/Purves%20392%20IIFET.pdf?sequence=1.

183 Ibid., p. 13.

184 www.msc.org/track-a-fishery/fisheries-in-the-program/certified/south-atlantic-indian-ocean/south-africa-hake-trawl-fishery/sustainability-notes.

185 Ponte 2008, p. 168.

that 'fewer players are better for conservation than many players' has been accepted by the MCM.[186] It is even argued that the hake fishing industry may become more concentrated in the medium term.[187]

4.4.6 How does the interacted system address overfishing?

The above analysis examines the fishery management in South Africa in general and the governance of the hake fishery in particular. It shows that illegal fishing is widespread in South Africa and many inshore resources are depleted. However, the deep-sea hake fishery is comparatively better managed with the interaction between public regulation and the MSC. This section tries to clarify why the deep-sea hake fishery is regulated more effectively than others.

4.4.6.1 Definition of property rights

As discussed earlier, two types of fishing rights exist in South Africa: individual quotas and limited entry. For the limited entry system, access may be limited in different forms, such as vessel number, size, power and fishing days. The establishment of such fishing rights is closely related to public regulation: the TAC/TAE and permits. To obtain fishing rights, one needs to get permits from the competent authority. Whether the fishing rights take the form of individual quotas or limited entry depends on whether the fisheries are managed under TAC or TAE. The competent authorities first decide the TAC and TAE and then divide them among the rightful applicants. Though there are two types of de jure fishing rights, the property rights regimes play out differently in practice.

Many inshore resources are used by a wide range of actors, including commercial, recreational and traditional fishers.[188] Not all of them have official fishing rights. Like abalone, many of such resources were historically used by SSFs. However, during the apartheid period, white-owned companies held most fishing rights, while traditional fishers were excluded. Though the post-apartheid period tried to redistribute fishing rights to HD individuals, progress has been limited due to the narrow definition of the subsistence sector, the small number awarded and the individual-based reallocation process.[189] Many local communities and fishermen, who have limited alternative opportunities for livelihood, continue to fish informally. Some lucrative resources like abalone even attracted outsiders or foreign poachers, gradually developing into organized crime. Such resources, though officially managed under individual quotas or a limited entry

186 Ibid.
187 Ibid.
188 WWF 2011, p. 9.
189 See *supra* section 4.3.2.

system, are de facto open-access regimes. Linefishing presents another story. The effort limitation started very late, more than one century after the establishment of the commercial fishery. At that moment, the fleet size was already too large and resources were already overexploited. The newly introduced limited entry system was generous in granting fishing permits. The large number of rights holders, without a cap on allowable catch, failed to stop the resources from collapsing.

Two species of hake are fished: the deepwater one (*M. Paradoxus*) and the shallow-water one (*M. Capensis*), with the former being more profitable. The deepwater hake is not easily accessible to traditional fishers and only started to be commercially explored by trawl sectors in the twentieth century. Since the 1990s, most foreign fishing vessels have no longer been allowed to fish off the coast of South Africa. Therefore, most deepwater hake is now fished by the trawl sector, which is permitted by the competent authority. Unlike the abalone fishery which is an open-access system, hake fishery is managed by formal sectors. The MSC certification does not include standards for fishing rights. However, it de facto helped the trawl sectors to consolidate their fishing rights, while keeping out new entrants and marginalizing the longline sector.

4.4.6.2 Enforcement

Property rights need to be enforceable to ensure the protection of fishery resources. The fishery management fishery-management authority was centralized under the MCM. However, at the same time, staff numbers have sharply decreased (reduced by more than two thirds in a decade) and budgets were tight, leading to most resources focusing on quota monitoring and harbour management and none on visible policing in the late 1990s.[190] Since 2000, the enforcement capacity has increased. However, ineffectiveness in legal enforcement and non-compliance is still widely acknowledged.[191] The lack of visible patrolling,[192] popular corruption,[193] together with mistrust of enforcement agencies and legislative authority due to historic injustices against traditional fishing communities all create weak deterrence.[194]

The situation, however, is quite different for the hake fishery. Being the most profitable fishery in South Africa, the hake sector is a priority in DAFF enforcement. All the catches are monitored at offloading points to ensure

190 Hauck & Kroese 2006, p. 75.

191 Gezelius & Hauck 2011, p. 458.

192 For example, it is estimated that a fisherman is only likely to have his licence checked in Cape Town once every 22 years. See Sjöstedt & Sundström 2015, p. 82.

193 Corruption is also believed to disrupt fishermen's trust in public regulations and hence creates incentives for non-compliance. See Sundström 2013.

194 Ibid., pp. 458–459.

compliance with quotas and by-catch. Vessels are tracked via a VMS system in real time. Furthermore, there are four mobile patrol vessels.[195] Compared with the generally low possibility of getting checked, eight deepwater trawl rights holders and sixteen inshore trawl rights holders were checked in 2014 by the public authority.[196] The MSC certification adds an additional layer of monitoring. The auditors do not only rely on existing documents from the government and their client, but also conduct field inspections. The first and second assessment also required conditions regarding monitoring. The first assessment required a review of compliance provided within thirteen months of certification, followed by ongoing annual monitoring.[197] The second assessment required the establishment of an external review programme for fishery management.[198]

The different level of enforcement makes the de facto property regime treat the hake fishery and many others in South Africa differently. The lack of enforcement makes many fisheries de facto open access, while the hake fishery is formally regulated under individual quotas.

4.4.6.3 Coordination

Coordination is crucial for fisheries management because multiple fishermen normally operate in the same areas. As discussed earlier, it is the competent authority (MCM/DAFF) that determines how fishing rights are allocated and how the resources are managed. However, two consultative working groups exist to involve the stakeholders in decision-making: SWGs and RMWGs. These two groups are crucial in advising the competent authority in deciding TAC and the fishing rights allocation.

The competent authority 'chairs, convenes and sets the agenda' for SWG meetings and the representatives of rights holders attend meetings as observers. However, it is the established industry which typically attends the group meetings. They have been influential in making recommendations. The fact that they have highly qualified scientists puts the established industry in a particularly strong position in the SWGs.[199] The new entrants are poorly represented in SWGs due to the concerns of being dominated by the established industry in the meetings and the lack of their own scientists to engage in the scientific debate.[200]

195 The MSC certification 2015, p. 99.
196 Ibid. There are around 30 inshore trawlers fishing for hake. www.sadstia.co.za/the-fishery/the-cape-hakes/the-hake-fisheries/inshore-trawl; in the deepwater trawl sector, rights holders are organized into 11 clusters. www.sadstia.co.za/about-sadstia/members.
197 MSC 2004, condition 6.
198 MSC, Reassessment Report for South Africa Trawl Fishery, 2007, condition 4.
199 Nielsen & Hara 2006, p. 47.
200 Ibid.

The membership of RMWGs is also an obstacle for new entrants or small-rights holders. Using the small pelagic fishery as an example, its RMWG is limited to the 'representatives of recognized stakeholder associations', excluding individual participants.[201] In a highly competitive and vertically integrated industry, the small-rights holders have difficulties in surviving independently, but have to 'enter into complex catching and processing agreements with vessel and factory owners belonging to the existing recognized stakeholder associations. For the small-rights holders, it does not make sense to join these associations or even to form their own if they cannot actively participate in the industry independently.'[202] Therefore, such small-rights holders have no channel for participation in the RMWGs.

It is already difficult for new entrants and small-rights holders to participate in the decision-making on fishery management, let alone for other traditional fishers and communities who do not even have formal fishing rights. Not being able to influence the agenda setting, such marginalized parties have no choice but to fish illegally, depleting the resources.

The hake fishery is also managed by the DAFF with the advice of SWG and RMWG. The lack of representation of small-rights holders is also present here. The trawl sectors are well organized via SADSTIA and SECIFA. These two associations are very active in the working groups, either as observers or members.[203] The other sectors, particularly the longline and handline sectors, are marginalized. The MSC provides another forum for multi-stakeholders to negotiate and shape the fishery governance agenda. The first assessment, however, has been criticized by NGOs as not providing a forum for them to exchange ideas and comments.[204] In the recent assessment, many stakeholders have been identified, including SADSTIA, SECIFA and a few environmental NGOs and research institutions. However, other rights holders are not included.[205] Therefore, there is still plenty of room to improve the coordination of the management of South African fisheries.

4.4.6.4 Information

The lack of information on the biological features of the fisheries and the influence of fishing activities is a problem facing many South African fisheries.[206] This negatively influenced the decision-making regarding fishery management. In response to scientific uncertainty, the OMPs were

201 Hara *et al.* 2014, p. 185.
202 Ibid.
203 MSC 2015, p. 96.
204 Ponte 2006, p. 39.
205 MSC 2015, pp. 96–97.
206 Attwood & Farquhar 1999.

developed, providing procedures to determine the TAC and other management measures during uncertainty and a model to adjust them to newly available information.

For the hake industry, stronger governmental monitoring makes more information available. For example, the VMS provides real-time information about the vessels. The industry is well organized via SADSTIA, which has developed a scientific observer programme since 2002 to collect information on fishing activities and the status of fish stocks.[207] The MSC standards have intensive information requirements. The assessment also promotes information gathering and transparency. A lot of conditions have been required in the three assessments of the MSC concerning information and research needs.[208]

4.4.6.5 Scale of governance

As far as forest governance is concerned, governments regulate at the national level while individual producers are regulated by forest certification operators. However, when it comes to fishery, although governments can issue some general regulation at the national level, both government and certification authorities target the fishery level. Deep-sea hake is the only certified fishery. The hake fishery is large, both in terms of area and economic value and hake is a mobile fish species. Though transboundary problems exist,[209] the comparatively large fishing area already covers a substantial part of the activities influencing this resource. The large scale of hake fishing and the existence of a strong industry render the certification costs affordable and the information requirements attainable. This has enabled the certification of the deep-sea hake fishery.

4.4.6.6 The role of public and private regulation in overcoming the limitations of property rights

With clearer established property rights, the hake fishery presents a successful example of countering fish stock depletion. However, non-targeted species and the health of the ecosystem are still externalities that are not internalized by the establishment of individual quotas for hake. The MSC has contributed to protection in these regards. Many conditions have been issued, leading to improved performance of the fishing industry.

207 www.sadstia.co.za/the-fishery/management/monitoring-control-surveillance.

208 For example, in the first assessment, conditions 2, 3, 4 and 7 identified information gaps on structure of stock, ecosystem relations, effects of trawling on benthic habitat and on seabird populations. They required information collection and research in these areas. Responding to insufficient information and scientific uncertainties, the second assessment required the adoption of precautionary management (condition 5).

209 Some Namibian fishers are using the same resources.

4.4.7 Conclusion

As we have seen, South Africa presents a case where the governance of the MSC certified deep-sea hake fishery seems to be more effective than the governance of inshore fishing.

Initially fishing has officially been managed under an individual quota property regime. The large number of traditional fishers, who were marginalized during apartheid, have not been able to benefit from the post-apartheid redistribution of permits. Hence, they have continued to fish illegally for the informal market. Since the levels of oversight and enforcement are low, fish has de facto remained an open-access resource. The later adoption of a limited entry system for linefishing has not been able to stop this because of the lenient granting of permits.

However, the situation is different for deep-sea hake fishing. This fishery has only recently become accessible to a highly selective number of domestic commercial trawlers. Small fishers lack the means to catch deep-sea hake, while foreign vessels are no longer allowed to catch it. These domestic commercial fishers are held to territorial quotas. They have the capacity to collect and provide the information that is needed for governmental oversight and to obtain an MSC certificate. This certification adds another layer of monitoring. This, however, does not mean such a system is perfect, especially not in terms of distribution and fairness. The established industry is in a strong position and has managed to control the redistribution of fishing rights. It has also used the MSC as a strategy to consolidate its rights, effectively preventing reallocation of fishing rights to HD individuals and the longline sector. As for enforcement, due to the importance of the hake sector, it receives stronger monitoring than other fisheries. The MSC also contributed to strengthening further enforcement for the hake fishery.

In sum, the gap between the de jure fishing rights and the de facto ones is apparently smaller when it comes to the offshore fishing of the MSC certified hake fishery than it is otherwise in fishery governance in South Africa. While marginalized and traditional fishers relentlessly continue to compete for scarce fish resources because monitoring and enforcement are largely ineffective, commercial trawlers fishing for hake not only have the capacity to comply with the rules but also have been able to use these rules in order to appropriate property rights and to safeguard their dominant position by excluding newcomers.

4.5 Case study: Mexico

4.5.1 Fishery problems

In 2012, 1.47 million tons of fish were captured in Mexico, ranking it sixteenth in the world and third in Latin America.[210] The main species harvested include minor pelagic species, shrimps, tuna, oysters, clams, shark and crab. The Gulf of California is the most productive fishing area in Mexico, providing 70 per cent of the harvest by value.[211] A major part of the fish resources have been exploited to their maximum capacity or over-exploited, leading to no new permits being issued for most fisheries.[212] Efforts have been made to reduce overcapacity in the fishing sector, resulting in a 13 per cent and 8 per cent reduction in the number of fishers and vessels respectively.[213] In Mexico, 41 per cent of its population lives in coastal areas and 350,000 fishers work in the small-scale fishery sector. Together they harvest 40 per cent of the national catch.[214]

4.5.2 Good governance

Mexico is an upper-middle income country with an annual GDP of US$1.295 trillion in 2014.[215] According to the World Bank WGI project, Mexico ranks 47.78 in prestige and accountability, 20.87 in political stability, 61.06 in government effectiveness, 66.83 in regulatory quality, 37.98 in the rule of law and 26.44 in the control of corruption.[216] The World Bank WGI indicator on control of corruption was also confirmed according to the level of corruption control by Transparency International. According to Transparency International, Mexico ranked 95 out of 167 countries.[217]

4.5.3 Property rights as well as public and private regulation

4.5.3.1 Property rights

Commercial fisheries started to develop in Mexico in the 1920s and 1930s. At that time, a major part of Mexican fisheries were exploited by foreign fleets.[218] Since indigenous people who used to exploit inshore fishery resources largely diminished during colonization, the relatively new in-shore fishing communities started to develop.[219] The early settled communities were either organized as cooperatives or fishing as individual

210 FAO 2012b. Note however, that the actual catch in Mexico could be substantially higher. It is estimated that the actual catch from 1950 to 2010 was twice the amount officially reported. See Cisneros-Montemayor *et al.* 2013.
211 Cudney-Bueno *et al.* 2009, p. 208.
212 Ibid; OECD 2013, pp. 336–337.
213 OECD 2013, p. 334.
214 Espinosa-Romero *et al.* 2014, p. 291.
215 http://data.worldbank.org/country/south-africa.
216 http://info.worldbank.org/governance/wgi/index.aspx#reports.
217 www.transparency.org/cpi2015/#results-table.
218 Espinoza-Tenorio *et al.* 2011, p. 344.
219 Young 2001, p. 291.

fishermen. Such communities usually have no formalized management rules for local fishing activities. The members are nonetheless bound by blood and kinship ties.[220] Since 1938, cooperatives of fishermen have been granted 'exclusive access to the most important fish stocks, such as shrimp, abalone and lobster'.[221] The Fishing Law of 1947 further stated that industrial fleets can obtain fishing rights via permits and that artisanal fishermen, organized as cooperatives, can procure fishing rights via concessions.[222]

Investment in fisheries was not a priority and did not grow substantially until 1970 when the government 'strengthened co-operative rights to inshore fisheries and encouraged the expansion of fishing fleets'.[223] Viewing the fishing sector as an alternative for securing food safety and as a solution for employment problems caused by the agriculture crisis as well as with the intention of facilitating the export of fishery products, Mexico started to promote fishery development through centralized government institutions.[224] It established a fishery development bank (BANPESCA) to provide loans to the fishery sector as well as a state company (PROPEMEX) to regulate domestic prices and to develop export markets, build port facilities and collect and analyse fishery statistics.[225]

With this governmental promotion, the fishery sector developed quickly in the 1970s and peaked in 1981.[226] In the following years, however, though fishermen and fleets continued to increase, the catches did not. The fishery sector became overcapitalized and many fish started to be overexploited.[227] In the 1980s, the fishery sectors were affected by the following events: a strong El Niño phenomenon, the tuna embargo and the peso crisis. By the late 1980s, many cooperatives went bankrupt; BANPESCA and PROPEMEX were deep in debt.[228] The government blamed the small-scale fishery sector for failures and tried to improve the efficiency and to solve the financial problems by promoting privatization. The cooperatives lost their exclusive access to valuable resources, which was largely replaced by permit systems.[229] The new period witnessed the encouragement of private investment, trade liberalization and the reduction of government involvement.[230] Both PROPEMEX and BANPESCA were shut

220 Ibid.
221 Espinoza-Tenorio *et al.* 2011, p. 344.
222 Ibid., p. 345.
223 OECD 2011a, pp. 46–47.
224 Hernandez & Kempton 2003, pp. 508–509.
225 Ibid., p. 509; OECD 2011a, pp. 47–48.
226 Hernández & Kempton 2003, p. 509.
227 Ibid., p. 508.
228 OECD 2011a, p. 48.
229 Ibid.
230 Young 2001, p. 288.

down.[231] Many cooperatives were dismantled and some valuable fisheries, such as shrimp were taken over by private investors.[232]

By now, 'the industrial sector targeted high economic-value species, such as shrimp, tuna, and sardines'; while artisanal fishermen targeted other 'species in ecologically complex systems of multi-species fisheries'.[233] The industrial fishers were organized under the National Fisheries and Aquaculture Industry Chamber (CANAIPESCA). 'Artisanal fishers are usually organized in fishing cooperatives, grouped under the National Confederation of Fishing Cooperatives (CNCP)'.[234] But not all cooperatives are affiliated to the CNCP and many free fisheries do not belong to any cooperative or organization.[235]

> The political organization and power of the commercial fishers tends to be significantly greater than that of the cooperatives, with more coordinated policy positions supported by access to scientific expertise. However, the cooperatives retain an element of regional influence, particularly in the poorer coastal states where artisanal fishing plays a stronger social role in the rural economy.[236]

Currently, all marine resources are federal property.[237] Actors can gain fishing rights mainly in two ways: via permits or concessions issued by the fishery agency CONAPESCA. The permits allow holders to fish in certain areas in registered vessels and are usually valid for two to five years.[238] Concessions are area-based: the concessioners are allowed to harvest specified species (usually one species or groups of species) exclusively in a certain area for five to twenty years. Only in very limited cases, such as the concessions for the indigenous group, Seri, do exclusive fishing rights cover all marine species.[239] Another way to obtain exclusive fishing rights is by being designated a '*predio*' by the environmental agency, SEMARNAT to regulate the resources listed as 'under special protection'.[240] Permits are more widely used while concessions are limited to a few marine species of high commercial value on the west coast of the Baja California peninsula.[241] TAC is not widely used in Mexico and the limiting of fishing efforts is usually conducted through reducing the number of

231 OECD 2011a, p. 48.
232 Ibid., p. 49.
233 Hernández & Kempton 2003, p. 511.
234 OECD 2011a, p. 52.
235 Ibid.
236 Ibid.
237 Mexico Constitutional Law, Article 27.
238 OECD 2013, p. 335; Bourillón-Moreno 2002.
239 Bourillón-Moreno 2002.
240 Rights-based management in LA fisheries, p. 32.
241 Bourillón-Moreno 2002; Cinti *et al.* 2010, p. 331.

permits and vessels.[242] In other words, two types of fishing rights exist in Mexico, the limited entry system under permits and territorial user rights under concessions or *predios.*

4.5.3.2 Public regulation

As fishery policy developed, the regulatory institutions also changed accordingly. In the early days, the Forestry, Hunting and Fishing Division of the Secretariat of Agriculture and Development was the competent authority in charge of fishery regulation. The National Institute for Fisheries Research (INP) was created in the early 1960s to promote scientific research regarding fishery issues.[243] In the 1970s and 1980s emphasis was placed on fishery development and specific fishery agencies were created and upgraded to the secretariat level, the SEPESCA.[244] In the 1990s, environmental problems attracted wide attention, leading to the creation of several environmental institutions, such as 'the Mexican biodiversity commission (CONABIO), the environmental enforcing agency (PROFEPA) and the Institute of Ecology (INE)'.[245] At this moment, a 'sustainability discourse' started to be introduced to fishery management. To achieve this, the fishery agency, SEPESCA, was downgraded to the level of Under-Secretariat, and merged into a new Secretariat of the Environment, Natural Resources and Fisheries (SEMARNAP).[246] During this period, PROFEPA enforced fishery regulation.

The current system, however, redirects the focus of fishery management back to production and export rather than to environmental protection. Therefore, SEMARNAP was divided, with environmental governance competence remaining in the Secretariat. It was renamed as SERMARNAT (Secretariat of the Environment). Fishery governance competence has been moved to the Secretariat of Agriculture, Livestock, Rural Development, Fisheries and Food (SAGARPA) and is executed by the National Commission for Aquaculture and Fisheries (CONAPESCA).[247] With the separation of SAFAPRA and SEMARNAT in 2001, the environmental enforcement agency PROFEPA was no longer responsible for fishery regulation enforcement. A gap in enforcement existed until 2004, when an enforcement department was created under CONAPESCA.[248] The existence of multiple agencies creates challenges for coordination.[249] For

242 OECD 2013, p. 336.
243 OECD 2011a, p. 45.
244 Ibid., pp. 46–47.
245 Ibid., p. 49.
246 Ibid.
247 Ibid., pp. 49–50.
248 Ibid., p. 50.
249 Ibid., pp. 57–58.

example, although CONAPESCA has an enforcement department, it only undertakes fisheries surveillance and has to refer legal actions to the National Prosecutor for further action. It hence prevents rapid and effective response to fisheries violations. PROFEPA may also be involved when the issues concerned involve 'tortoises, endemic species and endangered species'.[250] The SERMARNAT is responsible for designating and managing Protected Natural Areas (ANPs). Therefore, for fishing in marine ANPs, CONAPESCA needs to coordinate with SERMARNAT.[251]

Efforts have been made to decentralize the regulatory system, by relocating the office of CONAPESCA from Mexico City to a major fishing region, establishing state offices for SAGARPA and CONAPESCA, and by transferring some authority to state governments and local municipalities.[252] However, these efforts have achieved limited success. The relocation of CONAPESCA only changed the location of decision-making but the decision-making power is still centralized.[253] Besides, 'The state offices for CONAPESCA are small in number, are co-located in SAGARPA state offices, and have low staffing and limited responsibility'.[254] The government also tried to organize some joint meetings with scientists and fishers to help set fishery management standards. However, fishers are not actively involved in standard setting due to the domination of powerful fishing industry lobbies and the insufficient consideration of the recommendations of such meetings by the public regulators.[255]

Currently, the General Law on Sustainable Fisheries and Aquaculture of 2007 is the basic law regulating fisheries in Mexico. Two common instruments that used to regulate fisheries are the National Fishery Chart and the Official Mexican Standards (NOMs). The National Fishery Chart provides information for decision-making, including indicators on fisheries status, species availability, equipment and conservation issues. The official Mexican standards body establishes regulations for specific fisheries, including applicable species, fishing seasons, capture systems and technical requirements.[256] It is worth noting that the process to establish NOMs has been slow. By 2000, only fourteen fisheries had an NOM. The requirement for all stakeholders to be consulted before enacting an NOM was believed to cause the slow progress that has been made.[257] By 2012, there were 45 NOMs in Mexico.[258] Many fisheries are still developing a fishery

250 Ibid., p. 51.
251 Ibid., pp. 50–51.
252 OECD 2011a, pp. 51, 54–57.
253 Though the current location is closer to fishing communities. Ibid., p. 55.
254 Ibid.
255 Ibid.
256 OECD 2013, pp. 335–336.
257 Hernandez & Kempton 2003, p. 515.
258 OECD 2013, p. 336.

management plan, as required by the General Law on Sustainable Fisheries and Aquaculture of 2007.

4.5.3.3 Private regulation

Fisheries in Mexico have been actively seeking MSC certification. Three fisheries have been certified: the Mexico Baja California Red Rock Lobster Fishery in 2004, the Pacific Sardine Fishery in 2011 and the Sian Ka'an and Banco Chinchorro Biosphere Reserves Spiny Lobster Fishery in 2012. Three other fisheries are under full assessment[259] and several others have experienced pre-assessment but do not seek full assessment yet such as the crab fishery of the Seri and later the Gulf of California Swimming Crab.[260] The Pacific Sardine Fishery is fished by industrial fleets which are permit holders. This is the largest Mexican fishery in terms of volume.[261] The certification is similar to the South African hake certification, in that the Gulf of California sardine industry is attempting to gain a stronger position than other sardine fleets.[262] The two other certified fisheries are managed under exclusive fishing rights.

4.5.4 The protection of small-scale fishery resources

As discussed earlier, both industrial fleets and small-scale fishermen operate in Mexican waters. However, governmental policies have largely been geared towards industrial, large-scale fisheries.[263] Small-scale fisheries, however, play a crucial part in the Mexican fishery sector: the number of SSF fishermen far outweighs the number of industrial ones and the SSF provides a substantial part of total landing. With much weaker bargaining power, the property rights of small-scale fishermen tend to be more vulnerable than industrial ones. This section focuses on how small-scale fisheries are governed.

Fishermen can engage in the small-scale sector in three ways: as permit holders, organized as a cooperative or as free fishers.[264] Under the permit system, a permit has to be obtained from CONAPESCA. The permit holders can only use 'the fishing equipment (boat, motor, and fishing equipment) registered in their permits'.[265] 'Only permit holders can legally land the catch and report it at CONAPESCA' and sell their catch in

259 www.msc.org/about-us/credibility/working-with-developing-countries/fisheries-involved-in-the-developing-world-programme.

260 www.sustainablefish.org/fisheries-improvement/crabs/gulf-of-california-swimming-crab.

261 Pérez-Ramírez, Ponce-Díaz & Lluch-Cota 2012.

262 Ibid.

263 Bourillón-Moreno 2002, p. 80.

264 Ibid., p. 84.

265 Basurto *et al.* 2012.

the market.[266] Though the permit system aims at limiting entry to the fishery resources, the limitation is usually unenforced.[267] Moreover, the procedure of applying for fishing permits is complicated and costly, creating a barrier for many free fishers who actually conduct fishing activities but have few resources and little expertise. As discussed later, many cooperatives were dismantled in the privatization reform from the late 1980s, leaving free fishers no choice but to gain access to fishing rights via cooperatives. Without holding a fishing permit or being a member of a cooperative, the free fishers have to engage in an 'informal working relationship with permit holders', and actually sell the products to permit holders.[268] Such a situation creates little incentive for either the permit holders or the actual fishers to behave responsibly. To maximize profit, permit holders tend to buy fish from illegal sources and 'legitimize' it for sale. The currently marginalized fishers, who are excluded from the decision-making process and benefits of fishing, have no incentive to protect the resources either.[269]

The fishermen could also organize as cooperatives to gain fishing rights via permits or concessions. The cooperatives used to have preferential exclusive access to some valuable fishery resources. However, cooperatives have encountered difficulties since the 1980s. Even when cooperatives have exclusive access rights, free fishers can engage in subsistence use, leading to overlapping access rights. Moreover, free fishers can also work for permit holders to fish the resources which were not exclusively reserved for cooperatives. However, it is difficult to patrol the mobile resources and the government also has few resources to monitor and enforce the regulation, leaving many fisheries with de facto open access. In addition, the government regulation was more favourable to permit holders than to cooperatives, with the former being allowed to sell their catch on the free market, while the latter were mandated to sell products to a state-owned company at a fixed price. The prices, however, were much lower than the market price, depriving cooperatives of a large part of their profits.[270] These situations, together with the natural hazards and economic crises in the 1980s, hit the cooperatives badly. The government blamed the small-scale fishery sector for 'low productivity and unsustainable use of nature resources'. Therefore, it withdrew cooperatives' exclusive access to fishing. Many cooperatives were dismantled in the 1990s.[271]

However, there were not always sad stories. In the privatization reform, the government partially retreated from direct intervention, creating 'new

266 Ibid., p. 603.
267 Ibid., p. 606; Bourillón-Moreno 2002, p. 85.
268 Basurto *et al.* 2012, p. 604.
269 Ibid., pp. 604–605.
270 Young 2001, pp. 286–287.
271 Basurto *et al.* 2012, p. 86.

spaces in which marine dependent communities can be mobilized to contest policies that erode local resource access rights'.[272] On a few occasions, the communities managed to develop their management rules and gained some self-enforcement capacity. The active involvement of NGOs and the MSC also created new opportunities to develop a working fishery-management system. The following section examines two of such examples: the Mexico Baja California Red Rock Lobster Fishery and the Gulf of California Swimming Crab Fishery. The Lobster Fishery was the first fishery certified by the MSC in Mexico. Nine cooperatives under FEDECOOP have fishery concessions and had developed self-management capacity before they were certified. The crab fishery has undergone MSC pre-assessment and a Fishery Improvement Project has been adopted to promote its movement towards full assessment. This fishery is mainly harvested by small-scale fishers. Some of them are permit holders. An indigenous group, the Seri, also live in this area, holding territorial user rights over certain areas. How public and private regulation interact to govern these two fisheries is explored in the next section.

4.5.5 Interaction between public and private regulation

4.5.5.1 Case study 1: the Mexico Baja California Red Rock Lobster Fishery

4.5.5.1.1 CASE STUDY

The red rock lobster can be found on the west coast of central Baja California and is harvested by 26 cooperatives.[273] Nine of these cooperatives capture 80 per cent of the lobsters and are members of the Regional Federation of Cooperative Societies of Baja California (FEDECOOP).[274] These nine cooperatives are concessioners which were granted exclusive fishing rights for lobster, abalone and snails in adjacent areas to the coast.

The areas these cooperatives operate in are relatively isolated 'due to the scarcity of water and poor infrastructure.'[275] Fishermen began to settle in those areas in the early twentieth century to work for the Japanese fishing enterprises, which were prohibited during World War II. In the 1930s, the government started to grant generous concessions exclusively to the cooperatives of local fishermen. By the 1970s, the local members started to gain control over the fishing sector, by owning the processing plants and integrating the businesses vertically. The cooperatives became actively involved in fisheries management in the 1980s.[276] The fishery reform in the late

272 Young 2001.
273 Pérez-Ramírez, Ponce-Díaz & Lluch-Cota 2012, pp. 24–25.
274 Ibid.
275 McCay *et al.* 2014, p. 50.
276 Ibid., p. 51.

1980s and the early 1990s also influenced the cooperatives in these areas. The concessions 'became competitive, conditional and time-limited'.[277] When the cooperatives obtained new concessions in 1992, they were limited to twenty years. The exclusive user rights cover only limited species. Other fishers could still fish for other species in the concession areas. Moreover, the concession areas awarded became much smaller, but close to the fishing communities.[278]

In Mexico, the government reserves the authority to manage natural resources and rights holders, such as the cooperatives, have no formal autonomy in managing fisheries.[279] However, a co-management system has emerged in the study areas. The co-management system did not emerge naturally with the creation of cooperatives of concessions. A strong El Niño occurrence hit Mexican fisheries in 1982–1983, leading to a sharp decrease in fish production. The government threatened to close the fishery in study areas if the cooperatives did not adopt strict austerity measures. As a result, the cooperatives agreed to take more responsibility for sustainable management and continued fishing under tighter regulations.[280] The 1992 concessions also required cooperatives to invest in fishery stewardship, monitor resource levels and employ scientists.[281] Under the current system, public regulators operate via NOMs and National Fishery Charts.[282] Though the NOMs include a lot of technical standards, such as minimum mesh sizes, the cooperatives and FEDECOOP are allowed to and often do adopt more stringent standards. They also decide the allocation of rights among members and monitor resources and behaviour.[283]

With the co-management of public regulators and concessioners, the lobster fishery had remained sustainable even before getting certified.[284] Literature has summarized a few factors contributing to its success, including the features of the fisheries and concessions as well as the capacity of the cooperatives and FEDECOOP. For example, the relatively small character of the fisheries and their clearly defined boundaries allowed cooperatives to manage their resources and to exclude outsiders. In other words, transaction costs were relatively low. At the same time, the nine small concessions are adjacent to each other and organized under a federation which coordinates efforts and provides a link with the government and the market.[285] The geographical isolation makes the fishery resources less

277 Ibid.
278 Ibid.
279 Ibid., p. 52.
280 Ibid.
281 Ibid.
282 Pérez-Ramírez, Ponce-Díaz & Lluch-Cota 2012, p. 25.
283 McCay *et al.* 2014, p. 52.
284 Ibid., p. 56.
285 Ibid., pp. 52–54.

vulnerable to outside poaching.[286] The cooperatives also create organizational incentives by allowing broad participation, profit sharing among members, and rotating leadership. Furthermore, the cooperatives foster self-management capacities by developing knowledge and skills and they invest in social capital by building infrastructure.[287]

An NGO, Communidad y Diversidad (Community and Biodiversity) (COBI), promoted the MSC in Mexico in the early 2000s and viewed the well-managed red rock lobster fishery as an ideal candidate for MSC certification. The main market for the lobster is Asia. Consumers in Asia are not very environmentally sensitive. As a result the possibility of obtaining a price premium as a result of being certified is unlikely to drive fishers aiming mostly at the Asian market. The main driver for FEDECOOP to seek certification was to obtain international recognition and to strengthen its bargaining power for government support.[288] The fishery was certified in 2004 and recertified in 2011, making it the first community-managed fishery that was certified in developing countries.[289]

It is worth noting that though the cooperatives under FEDECOOP harvested several species, only the lobster fishery was certified under the MSC. The 20-year concession period was believed to be able to motivate good stewardship for fast growing species such as lobsters. However, this period might not be long enough for species with a long life cycle such as abalone.[290] FEDECOOP is also considering certifying its snail fishery if there are market opportunities.[291]

4.5.5.1.2 INTERACTION

The first assessment of the Red Rock Lobster Fishery cost around US$60,000. Half of the cost was funded by the WWF and the Mexican government.[292] The government also played a role in nominating experts for the evaluation team.[293] During the assessment, a few governmental officials were interviewed.[294]

Two conditions were imposed on the first assessment: the development of a strategy for ecosystem impacts and a research plan for harvest strategies.[295] To implement the conditions, FEDECOOP developed an action

286 Finkbeiner & Basurto 2015.
287 Pérez-Ramírez, Ponce-Díaz & Lluch-Cota 2012, p. 26; McCay *et al.* 2014, pp. 54–56.
288 Phillips, Bourillón & Ramade 2008; Pérez-Ramírez, Ponce-Díaz & Lluch-Cota 2012, p. 26.
289 Carlson & Palmer 2016, p. 131.
290 Pérez-Ramírez, Ponce-Díaz & Lluch-Cota 2012, p. 25.
291 Ibid., p. 28.
292 Ibid., p. 26.
293 SCS 2004, p. 16.
294 Ibid., pp. 18–19.
295 Ibid., pp. 62–64.

plan, which included seeking collaboration with IMP, which is responsible for fishery research.[296] The joint efforts of FEDECOOP and regulatory agencies were noted in the following surveillance report. More specifically, the certification also led to changes in regulatory standards. A few standards were added to the lobster NOM to reduce ecosystem risks. The General Law on Sustainable Fisheries and Aquaculture of 2007 also 'requires that these issues about the ecological risks in a fishery be taken up in the development of fishery-specific management plans' and requires the management plan to 'include any additional research needs that may be identified for supporting the harvest strategy in the fishery'.[297]

A similar pattern can be found in the second MSC assessment. Five conditions were required in the certification report, concerning stock assessment, by-catch (including bait species) and influence on ecosystem components.[298] To fulfil the conditions, FEDECOOP again sought collaboration from the regulatory agency, IMP to conduct stock assessment, and to monitor the by-catch (bait) species.[299] Responding to the conditions and development, the government updated the National Fishery Chart with two specific requirements regarding lobster fishery: banning the use of specific bait species and limiting the number of traps allowed per season.[300] At the same time, the surveillance team also made an effort to promote the publication of a fishery-management plan for the lobster fishery.[301]

The fishery was already well managed before MSC certification. The certification brought limited economic benefits such as broadening the export market.[302] It did however play an important role in empowering FEDECOOP and the communities engaged in fishery. FEDECOOP obtained governmental support for fishery research, communication with stakeholders and negotiating quotas. The government also provided social programmes and investment grants to communities to improve their infrastructure.[303] Moreover, after the original concession expired in 2012, FEDECOOP got it renewed.[304] MSC certification may have contributed to the renewal of the concession, since concessions need to satisfy some renewal conditions, such as technical and financial capacity, legal compliance, and environmental, social and economic impact and so on.[305]

296 Ibid., Annex 8.
297 Chaffee 2008, pp. 8–9.
298 SCS 2011, pp. 27–58.
299 Ibid., pp. 59–60.
300 SCS 2014, p. 14.
301 Ibid.
302 Pérez-Ramírez, Ponce-Díaz & Lluch-Cota 2012, p. 27.
303 Ibid., p. 27.
304 Cunningham 2013, p. 4.
305 Pérez-Ramírez, Ponce-Díaz & Lluch-Cota 2012, p. 27; Philips, Bourillón & Ramade 2008.

4.5.5.2 Case study 2: The Gulf of California Swimming Crab Fishery

The Gulf of California Swimming Crab Fishery is an important fishery in Mexico and is especially important for small-scale fishers.[306] The US is its main export market. Ninety per cent of the catch originates from three states: Sinaloa, Sonora and Baja California.[307] In the Gulf of California, most fishermen obtain fishing rights via permits. It is estimated that there are around 139 fishing permits, with 2,193 fishing boats and 4,386 fishers active in this area.[308] However, parts of the area are managed under territorial user rights. The Seri fishery is such an example. With its self-governing institution, the Seri managed to maintain its fishery at a sustainable level and applied for MSC pre-assessment in the early 2000s. Since the crab the Seri fish comprises only a part of the crab species in the nearby areas, later efforts have been made to certify the fishery of all crab species in the Gulf of California. This part hence examines the two pre-assessments respectively. While discussing fishery management by the Seri, a comparison is made between the Seri and a neighbouring community, the Kino, who manage their resources under a permit system.

4.5.5.2.1 TWO COMMUNITIES: SERI AND KINO

The Seri and Kino are two neighbouring communities located in Sonora. The Seri are indigenous people who have lived in the north-west corner of Sonora for centuries.[309] Historically, they assumed 'a nomadic lifestyle based … on hunting, collecting wild plant foods, fishing and shellfish gathering'.[310] As a small nomadic group, the Seri had 'no political organization other than a local and temporary war chief'.[311] Self-governing institutions only started to develop in the twentieth century, responding to the need for permanent representation to negotiate with governmental agencies and the private sector.[312] Kino fishers emigrated to Kino Bay in the 1920s. In the 1930s, both Seri and Kino established their cooperatives. From then till the early 1970s, both communities were managing fisheries under a permit system. They fished for many species including swimming crab, penshell and others. As discussed above, under this permit system, fishers usually are not permit holders and have to commercialize their products through the permit holders (buyers). This created incentives for overfishing and led to a de facto open-access regime.[313]

306 Espinosa-Romero *et al.* 2014, p. 292.
307 www.sustainablefish.org/fisheries-improvement/crabs/gulf-of-california-swimming-crab.
308 Ibid.
309 Bourillón-Moreno 2002, p. 17.
310 Ibid., p. 23.
311 Ibid., p. 63.
312 Ibid.
313 Basurto *et al.* 2012, p. 600.

The situation began to change in the 1970s, when the government granted formal rights to the Seri in parts of their historic territory, including communal land tenure in the mainland and permanent and exclusive withdrawal rights of marine resources in certain marine areas.[314] Such rights are territory based. Unlike the lobster concessions mentioned above, the rights of the Seri are permanent and cover all species in their territory. Governmental orders, however, are not very specific in determining the boundary of exclusive fishing zones for the Seri, leading to different interpretations and conflicts between the Seri and outsiders such as the Kino.[315] The Seri interpret the area extensively to include all areas they historically used.[316] However, they lack the capacity to patrol the whole area and can only control and patrol the activity within the Infiernillo Channel. The narrowness of the channel makes it easy for the Seri to control this area compared to the outside open waters.[317] The outsiders narrowly interpreted the exclusive Seri fishing zone as the waters within the Infiernillo Channel and continued to fish in other areas regarded by the Seri as their property.[318]

The creation of an exclusive fishing zone for the Seri does not mean the resources will be automatically managed effectively. In the first decade, the Seri did not make an effort to control their resources. Until the mid-1980s, the scarcity of fin fish increased the attractiveness of fishing benthic species in the Infiernillo Channel. In response to the increasing interests of outsiders in accessing Seri resources, the Seri started to develop measures to control access. They allowed other fishers to fish in their territory if the other fishers made payments to Seri government officials, hired a Seri as fishing crew and paid them the same amount as other crew, did not fish in culturally important areas and complied with the maximum allowable catch per day.[319] The Seri even established an informal armed guard, the Guardia Tradicional, to police their territory to prevent poaching and to ensure access fee collection.[320]

The open-access scenarios for Kino fisheries continued. This also led to much better protection of Seri fisheries than of the Kino's.[321] In addition to the wrong incentives for permit holders and free fishers, some other issues also contributed to the poor performance of Kino fisheries. First, the Kino fishing areas were inside a large open bay, which was less controllable than the Infiernillo Channel controlled by the Seri.[322] Second, there were very

314 Ibid., p. 601.
315 Bourillón-Moreno 2002, p. 101.
316 Ibid., pp. 102–103.
317 Ibid., p. 105.
318 Ibid., p. 89.
319 Basurto *et al.* 2012, p. 602; Basurto 2005.
320 Ibid., p. 601.
321 Ibid., pp. 602–603; Cinti *et al.* 2010.
322 Basurto *et al.* 2012, p. 602.

few regulations regarding the crab fishery besides issuing permits. Until early 2000, there has been no NOM over crab fishery in the Gulf of California. Only in Sonora was an administrative agreement reached between government, producers and processors and exporters, including some technical standards.[323] This agreement, however, is not enforced.[324]

4.5.5.2.2 MSC REASSESSMENT: FROM SERI CRAB FISHERY TO THE GULF OF CALIFORNIA SWIMMING CRAB FISHERY

Considering the efficient management of Seri fisheries, WWF and a local NGO, Communidad y Biodiversidad (CoBi) formed a partnership in 1999 to explore the possibility for the Seri crab fishery to move towards MSC certification. A few efforts have been made to acquaint the Seri with the certification process, share the biology and monitoring techniques as well as traditional knowledge.[325] The fishery started MSC pre-assessment in 2000. The certifier concluded that the fishery 'has strong potential to continue to the full assessment phase once issues requiring special attention are addressed'.[326] The identified shortcomings included: lack of information on the stock status; 'the unknown impact of ghost fishing on the marine environment and the jaiba (crab) population' as well as the lack of participation in management decision-making.[327]

To address the first issue, a study has been done to analyse whether the crab 'fished by the Seris is a sub-species of a broader meta-population' or is the same species as other crabs in the Gulf of California. In the former case, a local stock assessment would be sufficient. In the latter, a stock assessment for crab in the whole Gulf needed to be conducted.[328] Finding the crab fished by the Seris is not biologically different from the rest, a first stock assessment for crab was done in Sonora and Sinaloa in 2000. The MSC pre-assessment also triggered the first public management tool, the Crab NOM.[329] The NOM includes standards regarding minimum size requirements as well as restrictions on equipment type and quantity, egg-bearing females and bait type.[330]

Efforts have shifted to facilitate the certification of the whole crab fishery in the Gulf of California since 2008 when the demand for crab in the US increased. A fishery improvement project (FIP) was launched by an NGO Sustainable Fisheries Partnership and Ocean Technology Inc., the

323 Bourillón-Moreno 2002, pp. 177–178.
324 Ibid., p. 227.
325 Phillips, Ward & Chaffee 2008, p. 173.
326 Ibid., p. 174.
327 Ibid.
328 Ibid.
329 Lopuch 2008, p. 313.
330 www.sustainablefish.org/fisheries-improvement/crabs/gulf-of-california-swimming-crab.

main importer of crab products in the US.[331] Other stakeholders in the FIP include a domestic crab processor, INP, CoBi and some research institutes. Ocean Technology applied for MSC pre-assessment in 2009. According to the pre-assessment report, an FIP has been designed and implemented. A fishery and by-catch monitoring protocol was developed by INP, CoBi and other research institutes and has started to operate. Fishery stock assessment has been conducted in Sonora and Sinaloa. A fishery management plan for crab has also been concluded.[332] Improving management for the Gulf of California Swimming Crab Fishery is an ongoing process. This fishery has not been subjected to full MSC assessment yet. It is still unclear whether current efforts are sufficient to correct the overfishing incentives created by the permit system and to create sustainable fishery management.

4.5.6 How does the interacted system address overfishing?

The above section analyses how public and private regulation interacts in governing two fisheries: the Baja California Red Rock Lobster Fishery and the Gulf of California Swimming Crab Fishery. These two cases reflect two typical types of fishing rights in Mexico: the limited entry system created by permit in most areas of the crab fishery, represented by the Kino communities, as well as TURs represented by the Seri crab fishery and the lobster fishery. This section examines to what extent public and private regulation can reduce the transaction costs in establishing fishing rights.

4.5.6.1 Definition of property rights

In the lobster fishery, cooperatives under FEDECOOP hold concessions for specific benthic species in clearly identified areas. Unlike many other cooperatives which were dismantled in the late 1980s and the 1990s, these cooperatives survived the privatization reform and managed to organize under FEDECOOP. The Seris have exclusive fishing rights over all species in their territory, especially in the Infiernillo Channel.

Both fisheries have natural and economic characteristics making them suitable to be regulated under TURs. The species harvested by the cooperatives under FEDECOOP and by the Seri are benthic species, which are not highly migratory, making them suitable for smaller-scale management by the communities. Besides, both areas are relatively isolated, making them less vulnerable to outside poaching. The smallness of the concession areas and the closeness to the community make them easier to monitor.

331 Ibid.
332 Ibid.

Although MSC certification has no specific standards on property rights, it played an important role in the renewal of concessions for the lobster fishery. It also helped to empower the FEDECOOP, strengthening its self-management capacity.

The limited entry system presents a different story. Most fisheries in Mexico are regulated under a permit system, usually without a TAC. Under this system, fishing effort is limited via the number of vessels or equipment allowed. However, as shown in the Kino community, the de jure rights holders do not fish themselves. The actual fishers have no expertise and resources to apply for the permits and hence have to sell their catch to the permit holders. The detachment of de jure and de facto rights holders prevents them from fishing at a sustainable level. In Mexico, the government claims ownership of natural resources. Therefore government has the authority to manage the fisheries. However, for a long time, no public regulation of the crab fishery existed in addition to the permit issuing. The MSC pre-assessment of the Seri crab fishery triggered the publication of the first NOM for crab. The further pre-assessment of the Gulf of California Swimming Crab Fishery also contributed to the introduction of the fishery-management plan. A public management system began to be put in place. However, whether the efforts have been able to address the mismatch between permit holders and actual fishers under the permit system is unclear.

The importance of clear property rights has also been illustrated by the example of the Seris. The shift from the permit system to exclusive fishing rights aligned the de jure and de facto rights holders. The law does not specify the areas reserved for the Seris. However, as discussed above, the access rules developed by the Seris define relatively clear property rights. Both the Seri and their neighbours accept the waters in the Infiernillo Channel to be Seri territory. In this area, a common property regime gradually emerged leading to good management. However, other areas considered by Seri as their territory were contested by outsiders and thus have remained a de facto open-access regime. Seris are allowed to share the benefit by acting as crew members of external fishing teams. The access rule also limits property rights in the sense that forbidden areas have to be respected and maximum allowable catch has to be complied with.[333]

Interestingly, in addition to these formal access rules required by Seri governing institutions, certain informal access rules exist. For example, outsiders may access and harvest resources in the Seris' fishing grounds by hiring a Seri crew member (without paying the governing institutions' access fee), contracting with a Seri fishing patron, becoming part of the Seri kinship and so on. However, all these informal access forms have clear boundary rules defining how and to what extent the outsiders can use the

333 Basurto 2005, pp. 650–652.

Seris' fishery resources. Therefore, the informal property rights are more or less clearly established.[334]

The example of FEDECOOP and Seri management also shows that the legal creation of territorial user rights does not automatically lead to good management. In both areas, there were no customary rules deciding the allocation of property rights among community members and management standards. The law only granted fishing rights to the community as a whole. How to allocate the rights among its members and to conduct daily management was unclear. Only decades after the establishment of TURs, FEDECOOP and the Seri started to develop their own management system when facing external pressure. When allocating the fishery rights among members, both self-governing systems encouraged the participation of community members in decision-making and allowed benefit sharing. Such arrangements create incentives for effective management.

4.5.6.2 Enforcement

Currently the enforcement capacity of government agencies is limited. As discussed earlier, the frequent change of regulatory institutions led to a division between CONAPESCA, PROFEPA and the Navy. The lack of coordination even created a gap in enforcement for a few years. CONAPESCA is currently the main institution responsible for fishery issues. However, it has little power to enforce its regulations but has to refer to the national prosecutor.[335] The ability of the current judicial system to prosecute fishery offences adequately, however, is also questioned.[336] Regional Offices of CONAPESCA and SAGARPA have been established in the decentralization process. However, the regional offices are sparsely staffed and have limited responsibility to ensure enforcement. Moreover, the multi-level design adds to the complexity in coordination between agencies. The local offices of CONAPESCA have to report to the local SAGARPA, instead of to CONAPESCA, which 'compromises the lines of authority within SAGARPA and hampers effective monitoring and enforcement of fisheries regulations'.[337]

The government has made an effort to improve the enforcement, by for example, introducing vessel monitoring systems (VMS) and using observers in industrial fleets. However, with current levels of staff numbers and resources, CONAPESCA is unable to enforce regulation in many small-scale fisheries.[338] With weak enforcement, illegal fishing is widespread in Mexico, particularly with 'inshore fisheries (for example, shrimp) and

334 Ibid., pp. 652–655.
335 OECD 2011a, p. 60.
336 Ibid.
337 Ibid., p. 55.
338 Ibid., p. 60.

sedentary fisheries (such as abalone and queen conch)'.[339] It is estimated that the total catch from 1950 to 2010 in Mexico was nearly twice as high as the official statistics mentioned.[340]

The lack of public enforcement is especially problematic for fisheries under the permit system. In this limited entry system, many fisheries operate in the same area and their rights are not exclusive. Government therefore should play an important role in helping fishermen to enforce fishing rights and exclude outsiders. However, under the permit system in Mexico, the government has insufficient capacity for enforcement in SSFs. The permit holders and fishers, however, have no incentives to enforce their rights or to manage the fisheries sustainably either.

As far as concessions are concerned, exclusive fishing rights are granted to rights holders in specific areas. To what extent the resources can be protected depends on the self-enforcement capacity of the rights holders. Both FEDECOOP and the Seris have developed such capacities. The cooperatives under FEDECOOP collaborate with the regional branch of INP in joint monitoring.[341] 'The small size and well-defined boundaries of the concession areas' also add to the effectiveness of monitoring and enforcement.[342] Considering the large economic value of the lobster resources, the cooperatives made a considerable effort to patrol the fisheries, either by the members themselves or by hiring apprentices and employees. The annual investment in monitoring costs millions of dollars.[343] Without government support in patrolling their territory and to enforce their rights, the Seris even organized a team of guards to patrol their waters. Though the guard remains informal and is not recognized by the government, it has been very effective in the Infiernillo Channel.[344] The narrowness of this channel makes monitoring easy. Moreover, when the Seris grant access to outsiders, one condition is the hiring of a Seri crew member, which helps to monitor the behaviour of outsiders. Even if the observation of every boat in real time is difficult, the Seris can detect the exploitation pattern by monitoring the level of resources in prohibited sites in the channel. With local knowledge, the level of resources can be easily monitored through the 'subsistence/semi-commercial harvesting practices'.[345] The lack of enforcement in other areas claimed by the Seris, however, grants them open access.

Similar to the hake certification in South Africa, MSC certification contributes to enforcement by conducting a field study during the

339 Ibid.
340 Cisneros-Montemayor *et al.* 2013, p. 285.
341 McCay *et al.* 2014, p. 53.
342 Ibid., p. 55.
343 Ibid.
344 Bourillón-Moreno 2002, p. 105.
345 Basurto 2005, pp. 650, 655.

assessment and annual surveillance in Mexico. The two assessments for the lobster fishery required the establishment of a monitoring system regarding the influence of the fishery on the ecosystem[346] and by-catch.[347] In preparing the Seri crab fishery for MSC pre-assessment, the NGO, CoBi, provided courses on monitoring techniques for the Seris. A grant was also secured to hire ecologists for monitoring activities.[348] With the involvement of multiple stakeholders, the FIP for the Gulf of California Swimming Crab Fishery also contributed to the development and implementation of a fishery and by-catch monitoring protocol.[349]

4.5.6.3 Coordination

The management of fishery resources involves many stakeholders, different public regulatory agencies, various types of fishers and their organizations, research institutions, NGOs, certification schemes and so on. How the multiple stakeholders interact and coordinate has an important influence on fishery management.

To engage the multiple stakeholders in developing NOMs, since the 1990s, SEMARNAP, the fishery agency, started to 'organize joint meetings with fishers, enforcers from PROFEPA and scientists from IMP'.[350] However, such meetings are dominated by the fishing industry while small-scale fishers are not encouraged to participate.[351] Both local knowledge and scientific research are important in deciding the current status of fishery resources and how to protect them. However, because of bad communication, the research and advisory agency, IMP, does not cooperate well with fishers and academic institutions. This is because IMP is entirely funded by the federal government and is not allowed to supplement its income from non-governmental or private sources.[352]

In addition, the conflicts between multiple marine users are also well noted, such as those between small-scale and industrial fleets, between capture and aquaculture and between fishers and other marine users.[353]

Coordination is especially problematic in fisheries regulated under permit systems. Under such systems, only permit holders have formal status endorsed by the government. The actual fishers have no formal legal status and do not participate in the fishery management's decision-making.

346 SCS 2004, Condition for principle 3, indicator 1.2.
347 SCS 2011, Condition 2.2.3.
348 Phillips, Ward & Chaffee 2008, p. 173.
349 www.sustainablefish.org/fisheries-improvement/crabs/gulf-of-california-swimming-crab.
350 OECD 2011a, p. 55.
351 Ibid.
352 Ibid., p. 58.
353 Ibid., pp. 58–59.

In the concession system, territorial user rights are granted to the communities or other actors exclusively, thereby alleviating coordination problems. The lobster fishery in Baja California and the Seris' fishery are such examples. The small size of the fisheries and the existence of self-governing institutions make coordination easier.

As mentioned earlier, a co-management system was established in the lobster fishery. Although formally, it is the government which promulgates fisheries regulations, the cooperatives can shape the norms by presenting scientific evidence on fishing.[354] Strong organization also exists in the FEDECOOP and cooperatives, facilitating internal coordination. Unlike many other cooperatives which were dismantled during the privatization reform, the cooperatives being studied are well organized and vertically integrated. The cooperative officers consult with members in making decisions concerning fishery management. The cooperatives are affiliated with FEDECOOP, which provides 'marketing services, technical expertise for fisheries management', 'a venue for collective bargaining' and 'a key liaison with government'.[355] The MSC assessment process provides another platform for multiple stakeholders to engage in discussions. The certification also empowered FEDECOOP, and strengthened its capacity to negotiate with the government. It was also invited to the National Advisory Council of Fisheries and Aquaculture, as the only regional fishery organization.[356]

For the Gulf of California Crab Fishery, the MSC pre-assessment promoted the development of FMP. A participatory approach was adopted in developing the FMP. Though IMP led the process, it worked closely with the NGO CoBi and met various stakeholders. Visits were made to all 28 communities harvesting crab in Sonora and Sinaloa to engage local fishers in the process.[357]

4.5.6.4 Information

'Fisheries scientific information is multidisciplinary and broad in scope, and is characterized by a breadth of knowledge, historical depth, various scales, and a mix of sources, which in turn adds to the complexity of the information requirements for decision-making and policy.'[358] How to produce sufficient information for fishery management is therefore a challenge. Like in many other developing countries, government investment in fishery research is limited and fishery information is insufficiently generated in Mexico.[359] Current research and information focus on single,

354 McCay *et al.* 2014, p. 53.
355 Ibid., p. 50.
356 Pérez-Ramírez, Ponce-Díaz & Lluch-Cota 2012, p. 28.
357 Espinosa-Romero *et al.* 2014, p. 293.
358 Cano Chacón 2013, p. 16.
359 Espinoza-Tenorio, Espejel & Wolff 2011, pp. 732, 738.

commercial species and tend to ignore ecosystems and interactions with social and economic processes.[360]

Local knowledge of fishers is crucial to supplement the lack of information available to the government and the public. The Seris have fished crab for subsistence long before the fishery became commercialized and have accumulated a wealth of traditional knowledge. Such knowledge helped in choosing fishing methods and patterns,[361] as well as monitoring the status of fishery stocks.[362] FEDECOOP and its cooperatives also developed scientific monitoring programmes and hired biologists and engineers to strengthen information generation for fishery management.

MSC certification also plays an important role in information generation. During the assessment, a team gathers information from existing literature produced by fishery clients, scientists, government authorities and other stakeholders. It also enables international scientists to study specific fisheries and provides channels for stakeholder participation. The publication of final reports and of surveillance reports enhances transparency. Assessment reports also identify gaps in knowledge and issue conditions or recommendations accordingly, which provide drivers for information production.[363] The lobster fishery is such an example. In the two assessments, conditions have been issued requiring the provision of information concerning by-catch and ecosystem impact.[364] When preparing the certification of the catch of swimming crab, a stock assessment was also made and fishery/by-catch monitoring programmes were established.

4.5.6.5 Scale of governance

Mexican fisheries have actively engaged in MSC certification. Those fisheries that have been certified or are being assessed vary in size. The Pacific Sardine Fishery is a large fishery harvested by industrial fleets. The Gulf of California Swimming Crab Fishery covers broad areas harvested by SSFs. The Baja California Red Rock Lobster Fishery is the largest lobster fishery in Mexico, but it is still small in size, covering a few neighbouring concessions. These examples confirm the general phenomenon that most MSC certified fisheries are either small or large.[365] The large fishery has economy of scale and can afford the costs and expertise related to assessment. Small fisheries are easier to manage and maintain.

360 Ibid., p. 738.
361 Bourillón-Moreno 2002, pp. 150–214.
362 Basurto 2005, pp. 655–656.
363 Cano Chacón 2013.
364 See *supra* section 4.5.5.1.2.
365 Kalfagianni & Pattberg 2013, p. 130.

4.5.6.6 The role of public and private regulation in overcoming the limitations of property rights

The above analysis shows that in Mexico clearer property rights were established under the concession system than under the permit system, resulting in different extents of resource protection. Though the territorial user rights granted via concessions have managed to induce a positive performance in the studied examples, such rights have limitations as well. For example, the concessions granted exclusive rights to cooperatives in Baja California to harvest lobster, snails and abalone. The 20-year period incentivizes cooperative members to protect lobster stocks. However, they have no exclusive rights to other species. Therefore insufficient attention has been paid to the by-catch and bait species. Cooperatives also did not sufficiently consider the impact of the fishery on the ecosystem, which has an important non-market value. The MSC assessment process identified these gaps and issued several conditions regarding the by-catch/bait species and ecosystem protection. Such concerns were also incorporated into the public regulation via revising NOMs and the 2007 Fishery Law.

4.5.7 Conclusion

Two types of fishing rights exist in Mexico: limited entry (permit without TAC) and territorial user rights (concessions). The limited entry system encourages permit holders to compete with each other in harvesting. Moreover, permit holders often do not fish themselves, but buy fish from fishers who do not have a permit. This illegal market continues to exist because of weak public enforcement. As a result, a de facto open-access regime exists.

Concessions have the potential to create better incentives to safeguard fish stocks, but are limited to a few fisheries and rely on the self-enforcement capacity of the concessioners. A few good examples indeed exist, creating opportunities for MSC certification. The case of Red Rock Lobster Fishery and the Seris' fishery are such examples. However, the granting of exclusive fishing rights does not automatically lead to good performance. Self-governing capacity has only started to really develop in response to external pressures. The cooperatives under FEDECOOP and the Seris eventually managed to develop clear rules to define and to enforce their property rights. MSC certification played an important role in empowering the FEDECOOP and in consolidating the fishing rights. The pre-assessment and FIP are also crucial in triggering the establishment of the public regulatory system for the crab fishery.

4.6 Case study: New Zealand

4.6.1 Fishery problems

New Zealand has an exclusive economic zone (EEZ) of approximately 1.3 million square nautical miles, ranking fifth in the world.[366] Around 130 species are commercially fished in the EEZ, of which 66 per cent are caught in deepwater fisheries and the rest caught in inshore fisheries. Most commercially important species are deepwater species with the exception of rock lobster.[367] Ninety per cent of the profit in the fish industry comes from export, making it the fourth or fifth largest export earner in New Zealand.[368] In 2015, 377 significant stocks are managed under the quota management system, of which stock information was available for 183 stocks.[369] Of these stocks with available information, around 17.2 per cent were estimated to be overfished and 6 per cent were estimated to be depleted.[370]

Three sectors operate in New Zealand waters, the commercial fishing industry, recreational fishers and customary fishers. The commercial sector consists of the inshore and the deepwater industry. The deepwater sector started to develop in 1983 with the declaration of the EEZ by the New Zealand government. It is dominated by vertically integrated companies and targets important species like hoki, hake, orange roughy and squid.[371] The inshore sector has a longer history of more than 100 years of fishing and consists of both small-scale fishers and large companies. With the exception of the indigenous Maori, 'historically strong community-based inshore fishing regulation' is lacking.[372] The Maori have a long history of exploiting fishery resources, for subsistence/cultural use (customary use) and commercial use.[373] Recreational fishing is also an important sector in New Zealand, with 20 per cent of New Zealanders engaging in it.[374]

4.6.2 Good governance

New Zealand is a high-income country with an annual GDP of US$200 billion in 2014.[375] According to the World Bank WGI project, New Zealand ranks 97.5 in prestige and accountability, 99.0 in political stability, 98.6 in

366 OECD 2011b, p. 94.
367 OECD 2013, p. 352; Yandle & Dewees 2008, p. 917.
368 OECD 2013, p. 349.
369 http://fs.fish.govt.nz/Page.aspx?pk=16&tk=345.
370 Ministry for Primary Industries 2015, p. 5.
371 Yandle & Dewees 2008, p. 917.
372 Ibid.
373 Lock & Leslie 2007, pp. 26–27.
374 OECD 2013, p. 354.
375 http://data.worldbank.org/country/new-zealand.

government effectiveness, 99.0 in regulatory quality, 98.6 in the rule of law and 100.0 in the control of corruption.[376] The World Bank WGI indicator on control of corruption is also confirmed according to the rank of corruption control from Transparency International. According to Transparency International, New Zealand ranked fourth out of 167 countries.[377]

4.6.3 Property rights as well as public and private regulation

4.6.3.1 Property rights

The Maori started to exploit marine resources 800 years ago. They did not only fish inshore resources for food, but also traded the fish products with inland tribes and later with European settlers.[378] Maori communities have developed marine tenure and customary rules governing the utilization of fish resources. Under the customary Maori property rights system, 'fishing grounds and shellfish beds' are 'held as property in common by the tribe or sub-tribe controlling the adjacent land'.[379] The traditional property rights and controlling system were destroyed by the European settlers, who introduced the idea of mare liberum and open access in the late eighteenth century.[380]

In response to the overfishing of some oyster and mussel beds, the first fishery legislation, the Oyster Fisheries Act, was passed in 1866. The early policies oscillated between resource conservation and industry development.[381] The Fisheries Amendment Act of 1963 represented a turn towards promoting the fishing industry, by allowing entry by any New Zealander or by companies with more than 50 per cent domestic ownership.[382] This led to a regulated open-access system.[383] With governmental financial supporting policies, such as subsidies and loans, the fish industry expanded quickly.

In the early stages, offshore resources were utilized by foreign fleets. With the introduction of the EEZ, New Zealand became a 'marine superpower'. Since then, the domestic fleets have expanded, with the support of governmental policies to replace foreign fleets. However, by 1982, a government assessment showed that the profitability of the nascent industry was marginal. A reform to improve efficiency began, by introducing a transferable quota system. Under such a system, quotas were allocated for

376 http://info.worldbank.org/governance/wgi/index.aspx#reports.
377 www.transparency.org/cpi2015/#results-table.
378 Hersoug 2002, p. 15.
379 Ibid., p. 16.
380 Ibid.
381 Ibid., pp. 16–18; Gibbs 2008, p. 25.
382 Hersoug 2002, p. 19.
383 Bess 2005, p. 339.

a ten-year period and could be transferred to companies with more than 75 per cent domestic ownership. Quota owners needed to pay an annual resource fee and were held to an aggregated catch limit.[384] Already by 1986, this reform proved successful in restoring fish stocks.

In contrast with the rapid development of deepwater fisheries, many inshore resources experienced a dramatic decline in the 1970s. In response to the decline, the 1977 Fisheries Amendment Act granted more power to competent authorities to regulate the fisheries, by issuing licences and attaching controlled conditions to them. In 1982, a general moratorium on licences for inshore fisheries was announced. However, the restriction could not prevent the expansion of existing licence holders. Therefore, a new policy was issued in 1983 that allowed the removal of part-time fishers without compensation. This led to a loss of more than 2,000 licences and hit the Maori particularly hard.[385] How such policy triggered the reform of Maori fishing rights is explored further in the following sections.

The exclusion of part-time fishers did not reduce the fishing pressure, and overcapacity was still widespread. The command and control approach to solve this problem was found to be cumbersome. The pervasion of neoliberalism and the broad public administration reform in New Zealand at that moment favoured a more flexible, and market-based approach.[386] Together with the influence of successful experience in the offshore sector, a comprehensive quota management system (QMS) was established, which granted individual transferable quotas (ITQs) to rights holders. The QMS now applies to both offshore and inshore fisheries, covering most commercially significant species.

The following section will discuss the property rights of the commercial sector and the Maori respectively. Given the importance of the recreational sector (with more than 20 per cent engagement of New Zealand people), their property rights will also be examined.

4.6.3.2 Commercial fishing rights

The 1986 Fisheries Amendment Act brought in a comprehensive QMS, which covers not only the offshore species for which quotas have been allocated since 1983, but also 27 important inshore species.[387] Under the QMS system, fish species are divided into different stocks 'defined by Quota Management Areas (QMAs), each of which is managed independently to ensure sustainability of the stock'.[388] A total allowable catch (TAC) is

384 Hersoug 2002, pp. 22–23.
385 Ibid., pp. 23–25.
386 Ibid., pp. 141–142.
387 Hersoug 2002, p. 31; Bess 2005, pp. 340–341.
388 Lock & Leslie 2007, available at: http://papers.ssrn.com/sol3/papers.cfm?abstract_id=978115.

determined for each QMA by the government.[389] Initially, ITQs were allocated to qualified permit holders, according to their 'catch history during the best two out of three years: 1981/82, 1982/83 and 1983/84'.[390] In other words, the quotas were allocated in tons rather than on a percentage of total TAC. Therefore the aggregate catch might be lower or higher than the TAC. The government hence acted as a quota bank, to buy back quotas when the determined allocation was higher than the TAC and to sell additional quotas when the TAC was higher.[391] Such design helped to gain initial support from the industry. The ITQs could be sold, leased or sub-leased and the transfer needed to be registered with the ministry.[392] To avoid transfer leading to excessive aggregation of quotas, the law sets limits for the total share of TAC each quota holder can have for each species.[393]

The early removal of part-time fishers and the introduction of commercial species into the QMS was of concern to the Maori, who claimed that their customary rights were not properly protected and obtained a few injunctions to prevent further introduction of species to the QMS before the disputes were settled.[394] Moreover, the early trial of the QMS showed a few limitations. First, relying on catch history in allocating ITQs provided incentives for fishers of non-QMS species to increase their catch levels for a higher amount of allocation if that species were to be included in the QMS in the future. This led to a moratorium on new permits for the non-QMS species in 1992.[395] Second, the early years also witnessed an increase in by-catch and high grading, a classical problem facing multi-species fisheries managed under a quota system.[396] Last but not least, allocating quotas as a fixed tonnage rather than as a percentage of TAC placed the risk of decreasing stock abundance on the government.[397] The initial over-optimism about resource abundance has led to large payments from the government to reduce the quotas in the following years.[398]

In response to the limitations, the fisheries management has experienced constant refinement. Important changes include: first, customary and recreational needs are considered in determining the TAC. The TAC is determined according to the need to achieve maximum sustainable yield (MSY). Customary catch is given priority in dividing TAC among sectors.

389 At that time, the TAC only covered commercial catch. The catch from recreational and customary sectors is not reported and included in the determination of the TAC. See Lock & Leslie 2007, p. 9.
390 Bess 2005, pp. 340–341.
391 Hersoug 2002, p. 32.
392 Ibid., p. 33.
393 Ibid.
394 Lock & Leslie 2007, p. 16.
395 Ibid.
396 Hersoug 2002, pp. 35–36.
397 Lock & Leslie 2007, p. 17.
398 Ibid.

After the allowance is made for customary catch, the rest of the TAC is then allocated to the commercial sector (as total allowable commercial catch, TACC) and to the recreational sector.[399] Second, the quota allocations are no longer made in terms of a fixed tonnage, but as a share of the TACC. Therefore, the risk of resource abundance is diverted to the quota holders, who have better incentives to protect the resources. Third, the permanent ITQ is transferred to a specific amount of allowable catch in tonnage via annual catch entitlements (ACE). After the determination of TACC each year, the ACE is calculated based on the ITQ the rights holders have.[400] The ACE provides rights to harvest a specific amount of fish in a QMA in a specific year. Both ITQ and ACE are transferable. Fourth, if fishers do not have sufficient ACE to cover their catch, they need to pay deemed values, which are set at levels making overexploitation unprofitable.[401] Fifth, the criteria for allocating ITQs have also been changed. For species introduced into the system before 2009, 20 per cent of the quotas are allocated to the Maori and the rest are allocated based on catch history. For the species introduced into the QMS after 2009, quotas are allocated to the Crown (80 per cent) and to the Maori (20 per cent). The fishing industry can obtain fishing rights via a tendering process for the Crown-held quota.[402]

It is worth noting that, though the most commercially significant species have been managed under the QMS, there are also many species not managed under the QMS.

4.6.3.2.1 MAORI FISHING RIGHTS

The early European settlers and the Maori signed the Treaty of Waitangi in both the English and the Maori language in 1840. The treaty confirmed the property rights of the Maori over their lands and resources. However, significant differences exist between the English and Maori versions regarding whether the control of fisheries lies with the Crown or the Maori.[403] On the one hand, the treaty provides a basis for the Maori's claims to protect their fishing rights and on the other hand, the different expressions left the seeds for future disputes. Maori fishing rights have been acknowledged in the subsequent legislation, but with limited treatment to actually protect their rights.[404] The removal of part-time fishers and the introduction of the QMS raised the concerns of the Maori over their fishing rights in the 1980s. Many claims were made, blocking the introduction of further species into the QMS.

399 Ibid., pp. 8–9.
400 Ibid., p. 28.
401 Bess & Rallapudi 2007, p. 721.
402 Lock & Leslie 2007, pp. 19–21.
403 Ibid., pp. 26–27.
404 Ibid., p. 27.

Intensive negotiations have taken place between the Maori and the Crown, leading to an interim settlement, the Maori Fisheries Act in 1989. Under this Act, the Maori Fisheries Commission was created, to represent and administer Maori interests. The Commission was given US$10 million and was allocated 10 per cent of existing quota (or the equivalent dollar value for the missing quota).[405] The Commission is responsible for allocating the quota to Maori communities, the iwi. Before the transfer of quotas, the Commission leased out quotas on an annual basis.[406] Final settlement was reached in 1992, followed by the Treaty of Waitangi Settlement Act. At that moment, the largest seafood company in New Zealand, Sealord Products Ltd, was offered for sale. The government bought half of the shares of the company (and its quotas) and transferred it to the Maori. In addition, the Maori were guaranteed to receive 20 per cent of the quotas of species which were to be introduced into the QMS thereafter.[407] These two Acts have largely resolved the disputes between the Maori and the Crown regarding the Maori's commercial interests. Under the 1992 Act, the Maori Fisheries Commission was restructured and became the Treaty of Waitangi Fisheries Commission (TOKM). TOKM holds the settlement assets and manages them before they are allocated to iwi. It took twelve years for TOKM to work out an allocation plan acceptable to the Maori. The allocation plan was formalized in the 2004 Maori Fisheries Act. To receive the allocation, iwi needed to be organized as a 'mandated iwi organization' (MIO), which was set up as a company, trust or incorporated society.[408] As a result of the development, the Maori commercial fisheries were penetrated by capitalism. Iwi are organized in modern commercial forms to obtain quotas and to manage them. The centrally controlled assets, such as Sealord and Aotearoa Fisheries Ltd (AFL) play an important role in New Zealand fisheries. It is estimated that though the Maori only form 15 per cent of the total population in New Zealand, they controlled more than 30 per cent of the total commercial quotas.[409] Unlike many other countries where autochthonous communities have a different organization and culture from modern capitalist society, the Maori and commercial sectors are better integrated.[410]

In addition to the commercial claims, the Maori also exploit fish for subsistence and cultural use. These rights are called customary rights of the Maori and are also recognized under the Maori Fisheries Act 1989 and the Treaty of Waitangi Settlement Act 1992. The traditional self-governing

405 Given that the government was not able to buy sufficient quotas from the market to compensate the Maori, they provided equivalent monetary value for the missing quotas.

406 Ibid., pp. 29–30.

407 Ibid., p. 32.

408 Ibid., pp. 36–37.

409 OECD 2011b, p. 97.

410 De Alessi 2012.

institutions have been gradually dismantled since colonization. How to create 'a customary regime within the modern regime' became a challenge.[411] The Maori Fisheries Act 1989 tried to enhance customary Maori fishing rights by establishing taiapure reserves. Taiapure reserves can be established in waters that have been customarily used by Maori communities for food or cultural reasons.[412] Though Maori fishers have property rights in perpetuity, their rights are not exclusive. The commercial and recreational sectors can also harvest in the same areas. A local management committee composed of Maori community members can make management recommendations to the government, but the government retains the final say.[413] To fulfil the obligations under the Treaty of Waitangi Act 1992, the government published the Customary Fisheries Regulations in 1998 and 1999, providing the basis for recognizing customary rights. According to these regulations, Maori tribes need to appoint guardians to manage customary fisheries in their areas. The guardians decide who is allowed to fish for what, where, when and how.[414] They are accountable both to the tribes and the government, and need to supply them with information regarding customary fishing activities.[415] The guardians can also request the government to establish mataitai reserves within their areas. Unless reinstated according to the recommendation of the guardians, commercial fishing is excluded from the established mataitai reserves. Recreational fishing is still allowed in the reserves.[416] The guardians can issue by-laws in the reserves, which are binding for both customary fishers and recreational fishers.[417] Up to 2013, 29 mataitai reserves and eight taiapure reserves had been established.[418] However, given the fact that these reserves are not exclusively used by customary fishers, spatial conflicts still exist between the different actors using the same areas.[419] The guardians have limited information on the resource status and fishing activities. Therefore the local guardians often claim lack of control over their resources against outside harvesters.[420]

4.6.3.2.2 RECREATIONAL FISHING RIGHTS

Recreational fishing is an important sector in New Zealand. It involves 20 per cent of all New Zealanders and generates important economic

411 Hersoug 2002, p. 77.
412 Bess & Rallapudi 2007, p. 722.
413 Yandle 2007, p. 34.
414 Hersoug 2002, pp. 77–78.
415 Lock & Leslie 2007, p. 39.
416 Bess & Rallapudi 2007, pp. 722–723.
417 Hersoug 2002, p. 78.
418 OECD 2013, p. 354.
419 Bess & Rallapudi 2007, pp. 724–725.
420 Walshe 2010, pp. 10–11.

value.[421] The recreational fishers, however, are 'a rather diverse group, regarding background, participation, degree of investment and actual catch'.[422] They are therefore poorly organized. Some associations exist, representing interests in specific types of fishing activities though. Some of these associations are connected via an umbrella organization, the New Zealand Recreational Fishing Council (NZRFC), which lobbies for the interests of the recreational fishers. Recreational fishing is regarded as a birthright by many New Zealanders. Therefore, people have free access to the resources, without the need to obtain a permit or pay. Only limited technical control measures exist for the recreational sector.[423] This has raised concern among commercial and customary fishers who compete for the same resources with the recreational sector. Efforts have been made to reform regulation for the recreational sector by, for example, integrating the recreational sector into the commercial sector. However, the recreational sector has effectively lobbied against this. Other proposals, such as relying more on the self-management of the recreational sector, were not possible due to its weak organization.[424]

4.6.3.3 Public regulation

The public regulatory system has changed during the reforms in fishing rights in New Zealand. To cover the management costs for the commercial sector, New Zealand has shifted from a resources rental system to a costs-recovery system. At the same time, the government tried to devolve some management responsibility to the industry, such as the registry of quotas and trade, as well as research. Some self-management organizations in the commercial sector began to develop, contributing to a co-management system involving both the government and industry. These reforms concern mainly the commercial sector. However, as mentioned above, the fisheries sector consists of commercial, customary and recreational fishers, who often engage in spatial conflicts. To coordinate and engage the various actors, Fishing Planning is under development. Before examining this shift, the institutional change in the competent authorities is briefly sketched.

4.6.3.3.1 INSTITUTIONAL CHANGES

Fisheries reform started in the 1980s when New Zealand was experiencing 'a general overhaul of the whole public administration'.[425] The neoliberal

421 Hersoug 2002, p. 85. Another estimate shows that the percentage may even reach 30 per cent. See Bess & Rallapudi 2007, p. 721.
422 Ibid., p. 84.
423 Borch 2010, pp. 659–661.
424 Ibid.; Hersoug 2002, pp. 88–99.
425 Hersoug 2002, p. 141.

revolution led to a deregulated financial sector and labour market, the removal of most subsidies and tariffs and a major tax reform. The goal of the reform was to make the state 'meaner and leaner' and to adjust state management costs according to available resources.[426] The general background together with the introduction of the QMS called for reforms in the fishery management institutions.

Fisheries were the responsibility of the Marine Department from 1877 to 1972. The Marine Department was reorganized in 1972, with fishery competencies transferred to the Department of Agriculture, resulting in the establishment of the Ministry of Agriculture and Fisheries (MAF). The MAF experienced three internal reorganizations from 1972 until 1994. 1995 witnessed the setting up of the new regulatory institution, the Ministry of Fisheries (MoF).[427] Before the introduction of the QMS, the compliance unit focused on inspection and apprehension of offenders to ensure fishing activities to comply with technical requirements. The introduction of the QMS showed that misreporting is an important issue and hence shifted the emphasis to auditing, monitoring and information analysis.[428] Fishery research used to be conducted by scientists in the MAF, which the industry claimed was reflecting their own interests but which was not responsive to industry needs (the so-called provider capture). In 1995, the research function was transferred to an independent Crown research institution, the National Institute of Water and Atmospheric Research.[429] Efforts have also been made to make the research answerable and to devolve it to the industry. These issues will be discussed later. The MoF merged with the Ministry of Agriculture and Forestry and Food Safety Authority in 2012, forming the Ministry for Primary Industries (MPI).

4.6.3.3.2 FROM RESOURCES RENTAL TO COSTS RECOVERY

When the QMS was introduced in 1986, the quota holders needed to pay resources rental for their 'exclusive utilization of public property'.[430] The rate was initially set low with the state's intention to increase gradually to capture the full surplus of the fishery. This led to intense objections from the industry. Moreover, the idea of 'Crown ownership' was challenged by the Maori. However, a double standard scheme for the Maori and others was not feasible.[431] This led to a costs-recovery system in 1994. Under the cost-recovery system, the state can recoup the management costs for commercial fisheries. Debates took place between the government and

426 Ibid., pp. 141–142.
427 Ibid., p. 143.
428 Ibid., pp. 144–148.
429 Ibid., p. 149.
430 Ibid., p. 34.
431 Ibid., pp. 34–35.

industry regarding the extent of costs recovery. Finally the government adopted an approach that recovery of costs would be attributed to the industry itself, including services like management, enforcement, research and conservation.[432] Based on consultation with the industry, the government decided annually the scope of services to be provided and their price.[433] This approach has raised concerns of both environmental NGOs and the industry. The environmental NGOs feared that cost recovery might lead to the dominance of industry in fishery policy due to 'user pays, user says'. The industry, on the other hand is concerned with paying bills for services over which they have little influence.[434]

4.6.3.3.3 DEVOLUTION OF RESPONSIBILITY

With the goal of making itself 'meaner and leaner', the government tried to devolve some fishery-management responsibilities to the industry. Under the costs-recovery system, the industry was incentivized to take over more responsibility and items were removed from the list of recoverable costs.[435] The devolution of responsibility is backed by legislation, the 1999 Fisheries Act. According to the Act, the Governor-General can 'transfer any specified function, duty or power to an approved service delivery organization'.[436] After the devolution, the approved service delivery organization became responsible for the provision and quality of their services.[437]

Which services had to be devolved to the industry was heavily debated. When the new Fisheries Act was published in 1999, the scope of devolution was limited to quota registry.[438] Since then, the quotas, permit and trade registries and status reporting services have been transferred to a private company, Fish Serve.[439]

For the other services, the government is still responsible for their provision. However, instead of providing such services themselves, for the non-core services, the government can resort to direct purchase by the ministry or allow the industry to make purchases.[440] For example, the governmental purchases research via negotiable contracts: the National Institute of Water and Atmospheric Research, the industry and other research institutions are allowed to compete for research funding. However, the National Institute of Water and Atmospheric Research still

432 Lock & Leslie 2007, p. 51.
433 Hersoug 2002, p. 126.
434 Ibid., pp. 135–137.
435 Connor & Shallard 2010.
436 Section 296B(1), Fisheries Act 1999.
437 Section 296C, Fisheries Act 1999.
438 Ibid.
439 Hersoug 2002, pp. 159–160.
440 Connor & Shallard 2010, p. 352.

dominates the current research market.[441] In addition, some industrial organizations also conduct or purchase research services directly.[442] However, the government has adopted a more reluctant attitude in delegating more research responsibility to industry since it remains accountable for the services purchased by the industry.[443] There is also insufficient public trust in the services funded and purchased by commercial interests.[444]

In addition, 'the government of the day was not enthusiastic about outsourcing services involving a compliance role and decided that any direct purchase of services would need to be based on having an approved fisheries plan in place for the fishery in question'.[445]

4.6.3.3.4 CO-MANAGEMENT

The devolving of responsibilities meant that the industry started to gain more management power, gradually leading to a co-management system. Under the QMS, many quota holders operate in the same areas, making it difficult to exercise management. Those quota holders targeting specific species have therefore often organized themselves as quota owner associations (QOAs), taking over some management functions.[446] The Challenger Scallop Enhancement Company,[447] the former Orange Roughy Management Company,[448] the former Hoki Fishery Management Company (now part of the Deepwater Group)[449] and the Rock Lobster Industry Council[450] are such examples. These institutions are organized differently, and are engaged in fishery management to different extents. Their functions usually include: providing harvest or sale rules, monitoring fishery conditions and activities, undertaking enhancement and research activities, and sometimes only engaging in imposing penalties on rule breakers and settling disputes.[451] However, the establishment of a co-management system is still facing many challenges.[452] First, given the comparatively short history of commercial exploration especially in the deepwater sector, there is insufficient scientific knowledge about the species and ecology in the vast New Zealand EEZ to support the self-management

441 Lock & Leslie 2007, p. 52.
442 Harte 2008, pp. 330–331.
443 Ibid., p. 331.
444 Hersoug 2002, p. 150.
445 Connor & Shallard 2010, p. 352.
446 Hersoug 2002, p. 170.
447 Mincher 2008.
448 Clement, Wells & Gallagher 2008.
449 Aalders *et al.* 2008, pp. 151–153.
450 Yandle 2006.
451 Yandle 2008.
452 Townsend 2010; Memon & Kirk 2010.

institutions. Second, there is insufficient social capital for collaborative governance. To achieve a unified vision and trust among the different fishing sectors and environmental groups has been difficult, given the unsettled rights of the recreational sector, the insufficient capacity of customary institutions and the adversarial relationship between environmental NGOs and the commercial sector.[453] Besides, the institutional design of the QOAs is also criticized.[454]

4.6.3.3.5 FISHERIES PLANNING

Fisheries planning is an instrument used to reconcile the interests of different sectors. It was initially proposed in 1983, but was soon shelved due to the incompatibility of the centrally planned approach with the market-based approached represented by the introduction of the QMS.[455] Planning development was reintroduced into the Fisheries Act in 1999, which prescribed 'the Minister may from time to time approve, amend or revoke a fisheries plan'.[456] However, the law is not very specific about the content of the plan and the procedure to establish it. The government initially favoured an 'industry-led' approach, which let the stakeholders develop their own fishery-management plans. The ministry only acted as an enabler and auditor. However, in fact, only the eCommercial sector is equipped with resources to develop the plans, which raised concerns among other fishery sectors and the environmental sector. 'The lack of a cohesive strategy' caused a shift to a 'government-led approach' in developing fishery plans.[457] Currently, five national fishery plans have been or are being developed under the leadership of the MPI and the engagement of other sectors. These fishery plans are for inshore finfish fisheries, inshore shellfish fisheries, freshwater fisheries, deepwater and middle-depth fisheries as well as highly migratory species fisheries. Specific plans will be developed for important species[458] which set the objectives for fishery management for the following five years and require the engagement of different sectors.

4.6.3.4 Private regulation

The New Zealand Hoki fishery is the first fishery certified by the MSC in New Zealand. It is the largest fishery in New Zealand and one of the most valuable.[459] Obtaining a certificate in 2001 also made it the first certified

453 Hersoug 2002, p. 184; Memon & Kirk 2010, pp. 796–797; Aalders *et al.* 2008, pp. 151–153.
454 Memon & Kirk 2010, pp. 797–800; Townsend 2010.
455 Hersoug 2002, p. 185.
456 Section 11A(1), 1999 Fisheries Act.
457 Walshe 2010.
458 www.fish.govt.nz/en-nz/Fisheries+Planning/default.htm.
459 Aalders *et al.* 2008, p. 145.

large-scale fishery in the world.[460] In addition to the hoki fishery, four other fisheries have been certified in New Zealand: the albacore tuna troll, southern blue whiting, ling trawl and longline fishery as well as the hake fishery.[461] Two others are under full assessment: orange roughy and skipjack tuna.[462] They are all deepwater species and are fished by large-scale commercial enterprises. With the exception of two tuna fisheries, all fisheries are co-managed by the government and an industrial association: the Deepwater Group (DWG). This might not be surprising, since the deepwater species are more commercially valuable and are largely exported to environmentally sensitive markets such as Europe and North America. In addition, deepwater resources are dominated by commercial sectors. While inshore resources are fished by commercial, recreational and customary sectors, spatial conflicts need to be solved and multiple stakeholders need to be involved to ensure that the resources are properly managed and certified. Because the hoki fishery has a longer history of getting certified, this research uses it as an example to show how public and private regulation interact in New Zealand.

4.6.4 The protection of fishery resources in general

New Zealand has long been regarded as a forerunner and leader in fishery management. When the QMS was introduced in the 1980s, a large portion of inshore resources had been overfished or depleted and the offshore fishery had just started to develop. By then, New Zealand had just declared the EEZ and suddenly became a marine superpower. Information concerning species in the vast EEZ was lacking and stock assessment had just begun. Nearly 30 years after the introduction of the QMS, information is still only available for 183 stocks out of 377 significant stocks under the QMS. Of the stocks with known information, 17.2 per cent are still overfished.[463] A rebuilding strategy or TAC reductions are adopted in these overfished fisheries to allow for stock rebuilding. Owing to the lack of historical data, however, it is difficult to evaluate the trend of stock changes. Whether or not the health of fish stocks has been increasing or decreasing is debatable.[464] There were also concerns regarding dumping by-catch and high grading in the early years after the introduction of the QMS. This situation

460 www.msc.org/global-impacts/msc-annual-report/news/newsitem/new-zealand-hoki-fisheries-meet-international-best-practice-standard-for-sustainability.

461 www.msc.org/track-a-fishery/fisheries-in-the-program/certified/pacific/new_zealand_southern_blue_whiting_fishery/unit-of-certification.

462 www.msc.org/track-a-fishery/fisheries-in-the-program/in-assessment/pacific/tri-marine-western-and-central-pacific-skipjack-and-yellowfin-tuna/contacts.

463 http://fs.fish.govt.nz/Page.aspx?pk=16&tk=345.

464 Some research argues that ITQ has made a positive effect. See Hersoug 2002, pp. 55–56; Bess & Rallapudi 2007, p. 719; Annala 1996; Stewart & Callagher 2003. Some other literature shows, however, that improvement has yet to be achieved, see Gibbs 2008, p. 29.

has been addressed by the strengthened monitoring and gear selectivity.[465] Moreover, it is generally acknowledged that there are two important drivers for overfishing: overcapitalization and illegal, unreported and unregulated (IUU) fishing which have been largely controlled under the QMS system. Some concerns remained, for example, regarding the protection of some by-catch species, the impact on benthic habitats and ecosystems, as well as some overfished stocks.[466] The spatial conflicts especially in inshore fisheries between the commercial, recreational and customary sectors have also attracted wide attention. However, in comparison with the sustainability of fisheries in most other parts of the world, and in comparison with the situation before the implementation of the QMS in New Zealand, the present governance system in New Zealand is an example of 'best practice'.

The comparative success in New Zealand is closely related to the property rights regimes. As mentioned earlier, three sectors work in New Zealand waters. The property rights of the commercial sector have been clearly defined under the QMS system. Holding clearly demarcated ITQs in perpetuity provides incentives for rights holders to protect the health of fish stocks for future benefits. The difficulties in coordinating many different rights holders in the same fishery have been partially addressed by the development of self-management institutions (the QOAs). New Zealand is also a good example of protecting customary interests. Since the colonial period, the Maori have begun a 'transition from one set of social relations to another that included commoditization, industrialization and capitalism'.[467] This makes it easier for them to be integrated into the commercial sector and to adapt to the market-based QMA. As far as Maori commercial interests are concerned, clearly defined ITQs have been granted to them, to incentivize responsible behaviour. The traditional interests of Maori are also recognized in law. However, compared to the commercial part, they are less clear and secure.[468] The recreational sector has successfully lobbied against strict regulation. On the other side of the coin, however, is the poorly defined property right. The recreational fishery is basically an open-access regime, with limited technical regulation. The commercial, customary and recreational sectors can operate in the same fisheries, especially in inshore fisheries, leading to spatial conflicts and negatively influencing biological protection.

The above analysis shows that problems remain in defining the property rights of the recreational and customary sectors and coordination among sectors. But the rights-based approach in the commercial sector has largely been a success in terms of stock management. However, just like in many

465 Gibbs 2008, p. 24.
466 See *supra* discussion of hoki fishery.
467 De Alessi 2012, p. 408.
468 Bess & Rallapudi 2007.

other countries, the individual quotas system has led to quota concentration in the hands of a few big companies and the departure of many small-scale fishers. In many developing countries, this has aggravated 'the race for fish' sold on the informal market by small-scale fishermen who cannot obtain fishing rights and have few alternative employment opportunities. New Zealand, however, presents a different story. When the part-time fishers were removed from inshore fisheries and the QMS started to operate in the 1980s, New Zealand was experiencing a rapid development of offshore fisheries,[469] which provided alternatives for the fishers leaving the inshore sectors. Research shows that many leavers are in the middle of their careers and could find alternative employment.[470] Moreover, in the earlier days, the QMS was designed in such a way that the government bought back extra quotas. Such a design, together with the active trading market for ITQs, provides the opportunity for small-scale fishers to leave with economic compensation. Therefore the quota concentration did not cause social turmoil and devastating effects on the environment.

The above analysis shows the importance of property rights to the success of fishery management in New Zealand. However, property rights do not function alone. The remaining sections discuss how public and private regulation influence the functioning of property rights in New Zealand. As discussed earlier, the certified fisheries are all offshore fisheries. Therefore, the discussion focuses on offshore fisheries, especially on the hoki fishery.

4.6.5 Interaction between public and private regulation

4.6.5.1 Case study: New Zealand's hoki fishery

The hoki fishery is the largest and one of the most valuable fisheries in New Zealand. Hoki is mainly fished off the west coast of the South Island from late June to late August and in the Cook Strait from late June to mid-September when they aggregate to spawn.[471] The stocks spawning in different places are usually assessed and treated as two stocks, the western stock and the eastern stock.[472] This fishery was developed by foreign fleets in the 1970s. Foreign fleets were gradually phased out and replaced by domestic fleets after the declaration of the EEZ.[473]

The hoki fishery is mainly exploited by the commercial sector and the influence of recreational and Maori customary access is negligible.[474] The

469 Annala 1996, p. 60.
470 Stewart, Walshe & Moodie 2006, p. 333.
471 SCS 2001a, p. 4; Intertek Moody Marine 2012, pp. 13–14.
472 Intertek Moody Marine 2012, p. 15.
473 Aalders *et al.* 2003, p. 145.
474 Intertek Moody Marine 2012, p. 46.

hoki was introduced into the QMS in 1986. There are two quota management areas (QMAs) for hoki: HOK 1 and HOK 2. The boundaries of QMAs do not correspond with those of biological stocks. HOK 1 is a large area, including both western and eastern stocks. HOK 2 is only nominal.[475] A single TACC is set for HOK 1 by the government, without distinction between eastern and western stocks.[476]

In response to the shift towards co-management in the late 1990s, hoki quota holders began to organize as the Hoki Fishery Management Company (HFMC) in 1997/98, representing most quota holders.[477] The initial organization of HFMC was weak, due to the diversity of its members regarding catching and processing. However, some common ground exists: the pressure from environmental groups incentivized the introduction of an environmental code of practice. Having the US and the EU as important export markets, the HFMC also actively sought MSC certification in the early 2000s.[478] Some large hoki quota owners also possess quotas of two other important deepwater species: orange roughy and squid. Therefore, three QOAs merged in 2005 into one organization: the Deepwater Group (DWG), aiming to 'provide a collective voice for deepwater quota owners'.[479]

Over the years, a co-management system has been developed. The government still plays a crucial role in fishery management. It led the development of national fisheries plans, including a specific plan for hoki. Such a plan set five-year objectives to be met by hoki fishery management. The government also decides the TAC, purchases research services, conducts monitoring and issues penalties for non-compliance.[480] The DWG conducts self-management, such as dividing the general quotas between western and eastern stocks;[481] engaging in compliance by contracting out the observer programme;[482] and developing operational procedures to reduce mortality of juvenile hoki and by-catch.[483] The industry and the government also signed a formal partnership (MOU) to promote cooperation between them.[484]

The HFMC applied for MSC certification in 2000 and became certified in 2001. Shortly after the certification, an environmental NGO, the Royal Forest and Bird Protection Society of New Zealand (RFBPS) filed an

475 SCS 2001a, pp. 10–11.
476 Intertek Moody Marine 2012, p. 46.
477 Hersoug 2002, p. 176.
478 Ibid., p. 177.
479 http://deepwatergroup.org/about/who-we-are/.
480 Intertek Moody Marine 2012, pp. 30–31; 46–47.
481 Ibid., p. 46.
482 SCS 2001a, p. 59.
483 Deepwater Group Limited 2011, p. 8, available at: www.cport.net/assets/uploads/files/Sustainable%20Management%20of%20New%20Zealand%20Hoki%20120711.pdf.
484 Ibid., p. 5.

official objection, regarding especially the influence of fishing on sea birds and fur seals. This is the first formal objection received by the MSC. In response to this, the MSC established a formal dispute resolution process.[485] A final decision was made at the end of December 2001. By then, the dispute resolution institution decided that the Corrective Actions conducted by the HFMC were sufficient for it to remain certified and therefore rejected the objection.[486] Due to dissatisfaction with MSC certification, the RFBPS launched its own Best Fish Guide programme to rank the sustainability of fish resources in New Zealand.[487] The hoki fishery was reassessed and recertified in 2007. It faced objections from the RFBPS and the WWF again but managed to keep its certificate. In 2012, the hoki fishery passed its second reassessment.

In the early 2000s, the hoki stocks were dramatically reduced and the western stock was regarded as overfished for a few years.[488] The decrease was attributed to poor environmental conditions. It was also believed that stocks had been overestimated and the reduction of the TACC began too late.[489] A rebuilding strategy has been adopted to promote the recovery of the stock. Since the late 2000s, stocks have increased again. The current stock level is again above the management target.[490] There are debates regarding its influence on seabird and fur seals as well as on the environmental impact. The governmental and DWG information shows that these aspects are improving.[491] However, environmental groups generally believe that certification failed to improve performance.[492]

4.6.5.2 The interaction

The close interaction between the state and certification is also noted in the case of hoki certification. Using the first assessment as an example, one of the evaluators was the former director of fisheries policy and research within the MoF.[493] Governmental information is also an important source of information for the assessment. For example, the information regarding catches is based on the vessels' report, quota, management report and licensed fish receiver reports, all mandated by law.[494]

485 Aalders *et al.* 2003, pp. 152–153.
486 www.msc.org/track-a-fishery/fisheries-in-the-program/certified/pacific/new-zealand-hoki/assessment-downloads-1/NZ_Hoki_Panel_Decision.pdf.
487 Weeber & Wallace 2008.
488 Deepwater Group Limited 2011, p. 2.
489 Rees 2005, p. 138.
490 Deepwater Group Limited 2011, p. 2.
491 Ibid.
492 E.g. the hoki fishery is criticized by RBFS and Greenpeace: www.greenpeace.org/usa/oceans/sustainable-seafood/red-list-fish/; www.forestandbird.org.nz/what-we-do/publications/the-best-fish-guide-/fish-species-/hoki.
493 SCS 2001a, p. 35.
494 Ibid., p. 8.

The first assessment required ten conditions to be fulfilled, regarding target stocks, by-catches and ecosystem influence as well as regarding the management system. The implementation needs not only the efforts of the client, the HFMC, but also the active engagement of the government, particularly the MoF and the Department of Conservation (DoC). For example, the assessment report shows that the spatial structure of stocks is not well addressed under the single TACC for hoki (Condition 3). In response, the MoF proposed to split the catch between eastern and western catches. Although the HFMC worked out a division plan and implemented it, the MoF will intervene if the HFMC fails in the enforcement.[495] To address the concerns on ecosystem influence, the MoF and DoC also carried out relevant research (Condition 4) and engaged in the development of ecological risk assessment (Condition 5).[496] The assessment also mentioned the lack of comprehensive planning as a deficiency of the management system. In response, a hoki management plan started to be developed. The earlier attempt was led by industry. However, such an approach was not welcomed by the public and was later replaced by the efforts led by the MoF.[497]

Similar interactions can also be found in the reassessment. The effort of the MoF in developing the fisheries plan continued after the reassessment in 2007.[498] Responding to the conditions about compliance and enforcement, the MoF and DWG concluded an MOU. The compliance with voluntary operational procedures developed by the DWG is audited by the government. The failure to comply with these non-regulatory measures may also lead to regulatory controls and penalties.[499]

4.6.6 How does the interacted system address overfishing?

4.6.6.1 The definition of property rights

As mentioned earlier, commercial fishing rights have been well defined in New Zealand. ITQs are defined as a specific share of the TACC in clearly defined areas (QMAs) and are awarded to specific rights holders in perpetuity. The rights holders can therefore have clear rights to access and harvest the fishery resources. The transferability further allows the flexibility of rights holders to adjust their capacity more efficiently. Though individual rights holders cannot exclude each other, they have the de facto rights to exclude non-rights holders. Moreover, the shift from resources rental to costs recovery and devolution of some management responsibility

495 SCS 2001b, p. 12.
496 Ibid., pp. 13–16.
497 SCS 2007.
498 SCS 2001b, Condition 1.
499 Ibid., pp. 57–58.

from the government to the industry incentivizes the co-management model between the government and the industry. The established QOAs sometimes can play a role in further refining the definition and allocation of property rights and determining how the rights should be executed (by determining the daily management rules). Therefore, the commercial rights holders have high-quality property rights. They have the rights to access, harvest, and exclude and share the rights to manage with the government.

After long negotiations between the Maori and the government, a specific portion of TACC has been awarded to the Maori, creating a clear definition of their commercial rights. However, the customary rights of the Maori are not so clearly defined. Sometimes, they have to share access with recreational and commercial fishers. The recreational fisheries are essentially open access, without clearly defined rights. Therefore the imprecise definition of customary and recreational rights may lead to conflict between fishers of different sectors.

The New Zealand hoki fishery presents an example of how public and private regulation, together with self-management institutions, influence the property rights of rights holders. The state decides the QMAs and the TACC, and allocates the ITQs to individuals. However, as mentioned earlier, different biological stocks exist in the same QMA, which requires separate stock assessment and management. The first MSC assessment pointed out the insufficient consideration of spatial structure of the fishery and called for spatially explicit management. In response to this condition, the government and self-management institution HFMC worked together to find a solution. The HFMC decided to separate the ITQs between two stocks and implement their decision. If the HFMC failed to enforce this rule, the government would intervene. Therefore, the involvement of certification and the self-management institutions contribute to the more subtle definition of the property rights.

Compared to some inshore fisheries, the hoki fishery only involves commercial interests, including the commercial interests of the Maori. Unlike in many developing countries, where the indigenous people require a separate, more fine-tuned system, the Maori community has adopted the capitalist market system. Therefore they can easily accommodate the market-based QMAs.

4.6.6.2 Enforcement

Since the mid-1980s, New Zealand has moved from an input control system to an output control system, the QMS. In comparison to the former system, the new system is more demanding when it comes to reporting catch, the control of dumping and high grading, that is, selectively harvesting fish in order to bring only the best quality fish ashore, discarding and wasting other captured low-quality fish. Accordingly, the public regulatory system has been reorganized and monitoring and enforcement receive more emphasis. The quota holders are required to report their level of quota holding; vessel holders need to report fish caught and landed; and fish receivers need to be licensed and to give detailed information about their purchase.[500] Compulsory catch and land reporting allows the possibility of cross-checking information from different sources, contributing to the alleviation of IUU fishing in New Zealand.[501] The government adopts a sliding scale system of enforcement starting with education and warnings for low-risk offenders. The government adopts infringement actions for medium-risk (opportunistic) offenders and invokes criminal procedures for high-risk parties.[502] In this way, the government tries to rely on voluntary compliance by regulation therefore creating a sufficient deterrent. For example, fishing in excess of the ACE was not regarded as a criminal offence but the parties have to obtain sufficient quotas by trading or to pay deemed values (which are set sufficiently high).[503] Failing to fulfil such an obligation is not an offence per se, but may lead to the revocation of fishing permits, making any future catch illegal. On the one hand, it was assumed such a design reduces the incentive to dump when fishers exceed their quotas. On the other hand, it would still deter overfishing.[504] The adoption of an extensive vessel monitoring system (VMS), which relies on satellite tracking technology, also provides real-time information on the fishing vessels and contributes to solving the dumping problem.[505] Moreover, compared to many other countries, the regulatory authority, the MoF/MPI, has broad authority in enforcement, such as random entry, search and questioning. Furthermore, they cooperate with other public agencies in enforcement, contributing to the comparatively good enforcement in New Zealand.

In addition to relying on public enforcement, the possibility of transferring responsibility to the industry is considered as well. Compared to the transfer of registry and research services, the government was more cautious in transferring compliance functions in the beginning. But over the years, many self-management institutions started to take over some compliance functions.[506]

Enforcement in the customary and recreational sectors proved more difficult. As mentioned earlier, the customary management institutions

500 Hersoug 2002, p. 145.
501 Gibbs 2008, p. 29.
502 OECD 2013, p. 355.
503 Intertek Moody Marine 2012, p. 31.
504 Hersoug 2002, p. 146.
505 Ibid., p. 145.
506 Research shows that the role of self-management institutions in monitoring and issuing penalties has been increasing. See Yandle 2008, p. 136. The management of the Orange Roughy Management Company and Challenger Scallop Enhancement Company are such examples. See Clement, Wells & Gallagher 2008; Mincher 2008.

often lack control over their resources because of outside harvesters.[507] The recreational sector is poorly institutionalized. Without a permit and reporting system, the recreational catch is monitored or estimated through surveys.[508] General awareness and compliance with regulation are low.[509]

The hoki fishery traditionally relied on governmental control. The self-management institution, HFMC, was established in 1997/98. It issued an environmental code of practice and developed an observer programme and contracted the service to external companies.[510] In the beginning, its self-enforcement capacity was weak: there was no civil contract between the members to ensure compliance with their own rules. The potential sanctions were also limited to shaming and the threat of governmental closure of the fishery.[511] These insufficiencies were identified in the MSC assessment process and responding conditions were required. For example, the first assessment showed that the links between HFMC and its members were weak and there were insufficient enforcement procedures.[512] In the following years, the HFMC merged with other organizations and became the DWG. The members of the DWG concluded a civil contract to enforce their own management rules.[513] In response to the reassessment of 2007, the government and the industry concluded an MOU, including specific arrangements for improving compliance. The non-compliance with self-management rules may lead to governmental intervention and penalties.[514]

4.6.6.3 Coordination

Under the QMS system, quotas are allocated to many individuals who operate in the same areas and target the same species. The management of these resources needs coordination between many rights holders. Self-management institutions started to develop in the 1990s. These institutions share some management responsibility from the government, such as deciding the daily management rules, conducting/contracting for research services, engaging in monitoring and so on. Being closer to the fisheries and daily fishery activities, the self-management institutions have the potential to adapt the management to local ecologies, such as dividing the quotas between different stocks/areas and adopting voluntary closures of vulnerable areas.[515]

507 Walshe 2010, pp. 10–11.
508 Hersoug 2002, p. 87.
509 Ibid.
510 SCS 2001a, pp. 11, 21.
511 Rees 2005, p. 219.
512 SCS 2001a, Corrective Action Requirements, pp. 9–10.
513 SCS 2007, pp. 13–14.
514 Moody Marine 2008, pp. 57–58.
515 As shown in the example of hoki management.

However, the above analysis also shows that the development of a co-management regime was not easy due to the conflicting interests with other fishing sectors and environmental groups. Other fishing sectors may fish for the same resources in the same areas but have less clearly defined fishing rights than the commercial sector. These fishers cannot engage very positively in the policy-decision process. The relationship between the fishing industry and environmental groups is 'deeply mistrustful and adversarial'.[516] The environmental groups fear that the fishing industry is gaining too much power in determining fishing policy and object to devolving too much management power to the industry. To reconcile the interests of different groups, New Zealand resorts to a 'government-led' approach in determining the Fisheries Plans. Given the comparatively new history of such plans, it is difficult to evaluate their effect.

Certification provides another opportunity to engage different stakeholders. However, as discussed earlier, the certified fisheries are all offshore fisheries where customary and recreational sectors have little presence. The hoki fishery certification process did see the active engagement of environmental groups. The WWF is one of the initiators of the MSC, and its New Zealand branch plays an active role in promoting the spread of certification. The first and second hoki assessments have been confronted with objections from environmental groups, specifically RFBPS and WWF New Zealand. The NGOs have special concerns over the influence of fishing on by-catch (sea birds and fur seals) and the ecosystem. To reduce the risks mentioned, the industry has adopted many measures, such as developing and implementing the Ecological Risk Assessment Strategy, adopting operational procedures for seabirds and fur seals and establishing voluntary closed areas for bottom trawls.[517] In spite of the self-reported improvement in these areas, some environmental groups still argue that the industry performed poorly.[518] To evaluate which arguments are true falls out of the scope of this research and the capacity of the authors. However, the engagement of environmental groups at least raised concerns over the environmental impact of the fishing activities and led to the adoption of many protective actions, which might otherwise not exist.

4.6.6.4 Information

Information about the fished resources and fishing activities is crucial to decide the right management rules and to enforce such rules. However, New Zealand has huge EEZ areas with a short history of the fishing industry, especially in deepwater areas. Stock assessments only started to develop

516 Short 2008, p. 155.
517 See *supra* section 4.5.
518 See *supra* section 4.5.1.

in the 1980s and there is still a lack of information for many stocks.[519] Efforts have been made to evaluate more stocks and further research projects. The adoption of VMS and compulsory reporting requirements also help to make information regarding fishing activities available. The development of a co-management regime helps to incorporate local knowledge in fishery management. However, some literature shows that local information is still not sufficiently considered, especially regarding customary fisheries.[520] The dominance of quota owners in the self-management institutions is also criticized, since the actual fishers who possess local knowledge are marginalized in the decision-making.[521]

In the hoki fishery, the interaction between the government, industry and the certifiers has had a positive effect on the collection of information. The MSC assessment has identified many informational gaps, especially regarding the ecosystem,[522] the environmental impact of fishing activities,[523] and the risks to seabirds and seals.[524] In response, the industry has adopted measures, such as conducting ecological risk assessment, and seabird/sea snail action plans.[525] The recent MSC assessment has shown that such gaps have been largely filled.

4.6.6.5 The scale of governance

Under the public regulatory system, fishery plans have been or are being established for specific species at the national level. Such plans set the goals for fishery management and try to reconcile the interests between different sectors. Under the QMS system, species are managed at the QMA level, which can either be large[526] or small.[527] Many quota holders are organized under the self-management institutions. These institutions usually cover specific species at the national level, but some are established at regional level as well.[528] With the engagement of people conducting actual fishing, these self-management institutions can better accommodate considerations at a local level. Currently, only offshore fisheries are certified in New Zealand, and all of them are large fisheries. Their large scale of operation makes certification costs affordable and the information requirements achievable.

519 Memon & Kirk 2011, pp. 1003–1004.
520 Ibid., pp. 1004–1005.
521 Ibid.
522 SCS 2001a, pp. 4–5.
523 Ibid., CAR 6.
524 Ibid., CAR 7.
525 See *supra* section 4.5.2.
526 Such as the fishery.
527 Such as the lobster fishery, see Yandle 2006.
528 Such as the regional management institutions for lobster, see ibid.

4.6.6.6 The role of public and private regulation in overcoming the limitations of property rights

The above analysis shows how public and private regulation influence the transaction costs in establishing property rights. Clear commercial fishing rights (ITQs) have been established in New Zealand, contributing to comparatively good management. However, literature has also identified many limitations of ITQs. The holders of ITQs gained property rights over targeted species. They therefore have the incentive to maintain the health of targeted species for future interests (upstream externalities). However, the by-catch, habitat and ecosystem are still external to the ITQ holders. Hence these downstream externalities cannot be internalized under the ITQ system. In this situation, public and private regulation are crucial. National fishery plans include objectives in these regards. The Fisheries Act only imposes relevant obligations. Marine reserves have been established, covering 32 per cent of New Zealand's EEZ.[529] As mentioned above, the hoki fishery certification also includes many conditions regarding catch and ecosystem, leading to the adoption of many protection measures.

4.6.7 Conclusion

Compared to many other countries, New Zealand presents one of the 'best practices' in fishery management. The commercial rights have been well defined under a QMS system. The shift towards a cost-recovery system, enabling the state to recoup management costs for commercial fisheries, and the devolvement of management responsibilities, have promoted the establishment of self-management institutions. The co-management of quota owner associations and government, together with the influence of certification, have contributed to a subtle definition of property rights, enhancing enforcement, promoting coordination between various rights holders, and taking advantage of both scientific and local knowledge. Certain residual externalities are also addressed via public regulation and certification. However, the property rights of the customary and recreational sectors are not well defined, leading to spatial conflicts between customary, commercial, and recreational fishers from time to time.

4.7 Case study: Japan

4.7.1 Fishery problems

Japan has a vast Exclusive Economic Zone, the sixth largest in the world.[530] It is a large fishery country in terms of production, consumption and

529 OECD 2013, p. 357.
530 Makino 2011.

import. In 2012, Japan produced 3.6 million tons of marine capture fish, making it the sixth largest producing country in the world.[531] Japan is also a top fish-consuming country, with an annual per capita consumption of 56.9 kg, which far higher than other developed countries.[532] Its total consumption ranked the greatest in value and second in volume in 2007.[533]

Japanese fisheries include five sectors: coastal fisheries, offshore fisheries, distant-water (high sea) fisheries, marine aquaculture and inland aquaculture.[534] Since 1989, the fishery production in Japan has decreased.[535] Distant-water fishing used to be the most important sector in the 1960s and 1970s. However, the declaration and establishment of EEZs worldwide since the late 1970s led to the drastic decline of the distant-water fishing industry in Japan. Offshore fisheries started to develop since then and production peaked in 1984. However, the natural fluctuations in biomass of an important offshore species, the sardine, led to a severe decrease in the early 1990s. Compared to these two sectors, coastal fisheries and aquaculture production have remained relatively stable.[536]

With strong consumption demand and decreasing production, Japan relies on importing fish. The self-sufficiency rate has declined from 113 per cent in 1964 to around 60 per cent in the 2000s.[537] Therefore, to maintain a certain level of self-sufficiency, the sufficiency rate becomes an important target of fishery policy.[538] Fish export is significantly smaller than import and 90 per cent of Japanese production is oriented to domestic consumption.[539]

After a long, intensive harvest, the fishery stocks in Japan have been under stress. According to the abundance of resources, Japan classifies its fish stocks into three categories: high, middle and low. Out of 84 stocks evaluated in 2012, 41.7 per cent were classified as low level, 40.5 per cent as medium level and 17.9 per cent as high level.[540] Generally speaking, the status of coastal fish stocks is better than others.[541] This is already an improvement compared to 1995, when around 60 per cent of species were evaluated as being at a low level.[542] However, considering the large portion of stocks remaining at low level, substantial efforts still need to be made to protect resources. Many factors are believed to cause the decline in resources, such

531 FAO 2012b, p. 2.
532 Popescu & Ogushi 2013, p. 18.
533 Makino 2011, p. 48.
534 Ibid., p. 44.
535 OECD 2013, p. 317.
536 Ibid., p. 319; Makino 2011, pp. 44–45.
537 Makino 2011, p. 48.
538 Current target is set at 65 per cent in 2017. See ibid., p. 48.
539 OECD 2013, p. 322.
540 Popescu & Ogushi 2013, p. 33.
541 Matsuda *et al.* 2010, p. 899; Sarker *et al.* 2015, p. 33.
542 Ibid.

as changes in the marine environment and man-made factors, such as coastal land reclaimation causing loss of habitat and overfishing.[543]

The number of fishermen has been decreasing for decades and around 177,870 fishers were still active in 2011.[544] The aging of fishery labour also became an important problem, with 35.3 per cent of male fishers older than 65.[545] The licensed vessels also experienced a 30 per cent decrease from 1980 until 2010.[546] Most of the vessels were less than 5 gross tons or non-powered. In other words, 'most of Japanese fishers are small-scale operators in coastal areas'.[547] The remaining Japanese fishers face unfavourable business conditions. The fishery expenses, especially fuel costs, are high. The average income of fishermen, however, is much lower than the national average of all employees.[548]

4.7.2 Good governance

Japan is a high-income country with an annual GDP at US$4.601 trillion in 2014.[549] According to the World Bank WGI project, Japan ranks 79.3 in prestige and accountability, 84.5 in political stability, 97.1 in government effectiveness, 84.1 in regulatory quality, 89.4 in the rule of law and 93.3 in control of corruption.[550] The World Bank WGI indicator on control of corruption is also confirmed and Japan is ranked 18 out of 167 countries according to Transparency International.[551]

4.7.3 Property rights and public and private regulation

4.7.3.1 Property rights

Japanese coastal areas and EEZs are highly productive and people started to explore marine resources thousands of years ago. The first fishery regulation was compiled in the eighth-century Taiho Code. According to the Taiho Code, coastal areas were open to all for common use. Since then, local users started to manage marine areas themselves. In the Edo era (1603–1868), different management systems were established for coastal and offshore areas. The coastal areas were partitioned and allocated by feudal lords to local communities. These local communities developed their own autonomous management bodies and established rules for the

543 Makino 2011, p. 46.
544 OECD 2013, p. 317.
545 Ibid.
546 Ibid.
547 Makino 2011, p. 46.
548 Ibid., p. 47.
549 http://data.worldbank.org/country/japan.
550 http://info.worldbank.org/governance/wgi/index.aspx#reports.
551 www.transparency.org/cpi2015/#results-table.

exploitation of the areas. The autonomous bodies laid the basis for subsequent fishery societies (FSs) and current fisheries cooperative associations (FCAs). In the later Edo era, the increase in population and technological progress led to the monopolization of coastal fisheries by a few wealthy fishermen. On the other hand, offshore fisheries were freely accessible. Later on, large-scale fishery operators also began to establish guilds and rules for government. These regimes were protected by feudal lords.[552]

These feudal regimes were abolished in the Meiji period, when Japan began to modernize. The sea was nationalized and a centralized licence system was introduced. During this period, many individuals who had not fished before were able to obtain licences easily. Fishing efforts increased quickly, leading to overexploitation. After this institutional failure, Japan returned to the customary arrangement relying on local management.[553] In 1901, the first modern fishery law, the Meiji Fisheries Law, was enacted. It introduced two types of management systems for fisheries. In the coastal areas, four types of fishery rights were granted to the autonomous FSs or individuals. Offshore areas were managed under licences.[554]

After WWII, the General Headquarters of the Supreme Commander for the Allied Powers (GHQ) requested the reform of fishery institutions. In response, a new Fishery Law was enacted in 1949. During the reform period, a debate took place between capitalism (based on personal rights) and socialism (the existing approach based on fishermen's associations). It took more than two years for the GHQ to accept the importance of fishermen's associations and the new law was finally adopted, based on the existing regimes.[555]

Under the current Fishery Law, fishery rights are granted for coastal fisheries and fishery licences are granted for offshore and distant-water fisheries as well as other fisheries.[556] Prefecture governors have the power to grant fishery rights to local fishery-management bodies, FCAs or individuals, including common fishery rights, large-scale set-net fishery rights and demarcated (aquaculture) rights.[557] Under common fishery rights, the rights to harvest resources in a specific area are awarded exclusively to the FCAs. 'The FCAs are mandated to establish their own autonomous regulations for sustainable use by their members'.[558] These rights usually apply to sedentary resources, or to modestly migratory fish.[559] For fisheries using set-nets over 27m in depth, large-scale set-net fishery rights are granted. These types of rights can be granted to more FCAs and individuals, but the

552 Makino 2011, pp. 22–24.
553 Ibid., pp. 24–25.
554 Ibid., pp. 25–26.
555 Ibid., pp. 26–28.
556 Ibid., p. 29.
557 Ibid.
558 Ibid.
559 Yamamoto 1995, p. 27.

FCAs are prioritized. Demarcated fishery rights are also area-based fishery rights, but apply to aquaculture.[560] The common fishery rights are valid for ten years and the other two are valid for five years. Fishery rights are regarded as real rights but cannot be sold or leased.[561] These fishery rights systems are area-based management systems.[562] In the fisheries where rights are awarded to FCAs, the FCAs decide the management rules for specific areas. Therefore, these fisheries are managed under communal property rights.

In addition to fishery rights, FCAs and individuals can also engage in fishing activities by obtaining fishery licences. The competent minister in the Ministry of Agriculture, Forestry and Fisheries (MAFF) can issue licences to industrialized large-scale fisheries, which usually operate in offshore or distant waters. The prefecture governors issue licences to medium-scale fisheries, operating in coastal or offshore areas.[563] The validity period of fishing licences is usually five years with a possibility of renewal.[564] Unlike fishery rights, many licensees can access and harvest fishery resources in the same area. Fishing efforts are controlled by limiting the number of fishing licences, vessel size and equipment or closing of seasons and areas.[565] Therefore, the licence-controlled systems are limited entry regimes.

In Japan, fishermen play a crucial role in managing their own fishery resources. Local fishermen usually are organized as FCAs. The FCAs usually manage multiple resources in specific jurisdictions. Fishing households and small companies can be members of FCAs, which play a variety of roles, such as joint purchases of inputs, the administration of the market, the provision of financial support and the management of resources.[566] Typically, FCAs are associated with coastal fisheries, where fishing rights are awarded to FCAs. In these cases, FCAs have the authority to manage all fishery resources within their jurisdiction.[567] To manage multiple resources in specific areas, FCA members sometimes divide into smaller groups and form fishery-management organizations (FMOs), which typically manage single species within the larger jurisdictions under FCAs.[568]

In addition, many licensees of coastal or offshore fishery resources are also members of FCAs/FMOs.[569] For example, the minister licensed

560 Makino 2011, p. 30.
561 Yamamoto 1995, p. 27.
562 Ibid.
563 Makino 2011, p. 30.
564 Yamamoto 1995, p. 27.
565 Ibid.
566 Uchida & Makino 2008, p. 223.
567 Ibid.
568 Cancino, Uchida & Wilen 2007, p. 393.
569 Uchida & Makino 2008, p. 224.

fishermen in each major sector to establish their own FMOs, such as the All Japan Purse Seine Fisheries Association and the National Federation of Medium Trawlers.[570] The FCAs/FMOs play different roles in these cases, such as coordinating fishing efforts, pooling revenues,[571] and taking autonomous protective measures, such as no-take zones.[572] They sometimes cover most operators in specific areas, and determine who, when and how to access the resources.[573] In these cases, though the fishery is de jure managed under a limited entry regime, a de facto communal property right system exists.

The FMOs are autonomous organizations, consisting of groups of 'fishers who share the same fishing ground and/or operate in the same fishery and are collectively engaged in resource and/or harvest management according to mutually agreed rules'.[574] The FCAs are legal persons under the law, but FMOs usually have no formal legal status. FMOs and FCAs are closely inter-related. Most FMOs are operated directly by an FCA, its subgroups or its affiliate organization. Sometimes, when the resources migrate between the boundaries of several FCAs, fishers from multiple FCAs jointly manage the FMOs and resources.[575] The FMOs can take multiple types of management measures, such as resources management (e.g. stock assessment, TAC establishment), fishing ground management (e.g. habitat protection, enhancement, usage rule) and fishing effort control (e.g. fishing season, fishing methods).

Although FCAs play an important role in fishery management in Japan, they also face management challenges. Many FCAs were established during the 1950s, on a wave of democratization of fishing villages. At that moment the requirements for establishing FCAs were poorly defined, leading to the

570 Tanoue 2015, p. 14.

571 Such as managing the Sakuraebi fishery. See Uchida & Baba 2008, pp. 175–189.

572 Such as snow crab management, see Makino 2011, pp. 83–88.

573 E.g. in the Walleye Pollack fishery, the Hiyanma Walleye Pollack Long Line Association and its affiliated FCAs imposed many self-regulations such as deciding fishing seasons, no-fishing area and gear restrictions. It also divides its members into different groups and subteams. It conducts a rotation system for its members to fish in different areas, at groups, subteams and individual levels. See Uchida & Watanobe 2008, pp. 168–170. In the Sakuraebi system, the self-management institution determines on a daily basis in the fishing season '(a) whether or not to fish that day and if fishing (b) the time of departure, (c) vessels' locations, (d) total harvest, (e) landing volume for each port and (f), other operational items as necessary'. See Uchida & Baba 2008, p. 181. In the Sandfish fisheries, the government decides the TAC and allocates it to offshore and coastal sectors. Within each type, the government further allocates a certain share to the prefectural level (held by the prefectural level fishermen's organizations). No individual quotas are officially established. This organization, SRMC, however, further divides the shares among affiliated FCAs. And some FCAs opt for an individual quotas system, changing the types of property rights working in practice. See Suenaga 2008, pp. 195–196.

574 Uchida & Makino 2008, p. 224.

575 Ibid.

creation of many FCAs with weak economic foundations.[576] In the 1990s, 'the declining birth rate, aging population and financial difficulties' put pressure on merging municipalities and FCAs.[577] Since then, more than 1,100 local FCAs have merged into 976 in 2013.[578] The consolidation of FCAs has brought some short-term economic benefits and limited the need for the government to underwrite debts. However, the merger also led to an increased distance between the places of decision-making, service provision and local fishermen. In other words, the merger may have taken 'resource use decision-making and harvesting rights away from local people with whom responsibility over the resources resides'.[579] Responding to the financial insecurities of the fishery sector, proposals have been made to replace the community-based management system with an individual property rights system such as the ITQ systems in New Zealand and Iceland.[580] Such proposals have led to strong opposition from fishery communities[581] as well as from researchers.[582] So far, the IQ has been applied only in three fisheries in Japan.[583] In addition, fishery communities also face pressure from industrial development of coastal areas, such as nuclear power plants and the conflicts between commercial and recreational fishers.[584] As in New Zealand, the recreational sector is also important in Japan. However, no official property rights have been established for recreational fishery and the sector is not subject to a licence system. There is also no exclusive authority in managing this sector where little monitoring and weak sanctions exist.[585] The recreational sector is also hardly represented in fishery-management policy-making, adding to the difficulties in coordinating the activities of recreational and commercial fishers.[586]

4.7.3.2 Public regulation

The MAFF is the competent authority for fishery management in Japan. One of its affiliated agencies, the Fisheries Agency, is 'directly in charge of planning and implementing Japan's fisheries management'.[587] The Fisheries Agency consists of four departments (the fishery policy planning

576 Hokimoto 2009, p. 54.
577 Delaney 2015, p. 266.
578 Ibid., pp. 269–270.
579 Ibid., p. 276.
580 Tanoue 2015, p. 3.
581 Makino 2011, p. 151.
582 Tanoue 2015, p. 3.
583 Ibid.
584 Murota 2011.
585 Makino 2011, p. 38.
586 Ibid.
587 Popescu & Ogushi 2013, p. 21.

department, the resource management department, the resource enhancement promotion department and the fisheries infrastructure department) and six local offices (Fisheries Coordination Offices).[588] In addition, many administrative tasks are also delegated to prefectural governments, such as issuing fishery rights and licences and promoting local fisheries cooperation.[589] Many fishing entities can operate in the same marine area. To coordinate these activities, specific organizations have been established at various levels. At the prefectural level, Area Fisheries Coordination Committees (AFCCs) have been established, which advise prefectural governors on issuing fishery rights and licences.[590] Some highly migratory species require management at cross-prefectural level. Therefore three WAFCCs were established to coordinate resource use and management in the Pacific, Japan Sea and Western Kyusyu areas. At the national level, a Fishery Policy Council has been established as an advisory body to the Fisheries Agency.[591]

To provide rapid technical solutions to pressing fishery problems, the government also runs its own fishery research. At the national level is the Fisheries Research Agency (FRA). At the local level, prefectures also have their own fisheries research stations to conduct research adapted to local conditions. To direct the research, the Agriculture, Forestry and Fisheries Research Council was established under MAFF, which drafts research plans.[592]

The basic fishery law in Japan includes the Fishery Law of 1949 (FL), the Fisheries Cooperative Association Law of 1948 (FCAL), and the Act on Preservation and Control of Living Marine Resources of 1996 (APCL). The Fisheries Law prescribes the institutional structure and the rules concerning fishery rights and licences. The FCAL concerns the organization of FCAs. The APCL provides the basis as to how marine resources should be protected, including measures such as resources protection plans to be developed by the government, the total allowable catch (TAC), the total allowable effort (TAE) and so on.

According to the APCL, the MAFF 'shall establish a basic plan on the preservation and control of living marine resources (hereinafter referred to as "the Basic Plan") for preserving and controlling living marine resources in the EEZ, etc'.[593] The Basic Plan includes issues such as basic policy concerning the preservation and control of living marine resources, matters concerning the trends of each resource, the determination of the

588 Makino 2011, p. 10.
589 Ibid.
590 Ibid., p. 31.
591 Ibid., pp. 31–32; Popescu & Ogushi 2013, p. 22.
592 Makino 2011, pp. 16–17.
593 Article 3(1).

TAC and its allocation between fisheries managed by the minister and the prefecture, TAE and so on.[594] Each prefecture is also required to establish its own prefectural plans, which ought to include issues such as preservation and control of living marine resources in their jurisdictions and the management of TAC for fisheries under their jurisdiction.[595]

So far, the total allowable effort (TAE) system has been imposed on nine species, under which the fishing capacity is capped through limiting the number of fishing days or vessels accessing certain waters.[596] The total allowable catch (TAC) system was introduced in 1997. 'TACs have been set for seven commercially important species that account for around 35 per cent of domestic landings.'[597] These species are mainly managed under licence systems rather than via fishery rights systems.[598] The MAFF decides the TACs on the advice of the Fishery Policy Council. Then the TACs are allocated to the fishery sectors managed either by the min-ster or the prefectural governors.

Many fish stocks are still at a low level in Japan. To restore these resources, the 'Resource Recovery Plan' (RRP) was introduced in 2001. The RRPs are co-developed by the Fishery Policy Council, governmental officials, researchers and fishermen.[599] Various types of measures can be used to implement the RRPs, such as TAE/TAC, resource enhancement, preservation and rehabilitation of the environment.[600]

4.7.3.3 Private regulation

Japan is an important fishery country in terms of production, consumption and imports. The MSC opened an office in Japan in 2006. However, the development of the MSC in Japan is still slow. Only two fisheries have been certified by the MSC: the Kyoto Danish Seine Fishery Federation Flathead Flounder (certified in 2008) and the Japanese Scallop Hanging and Seabed Enhanced Fisheries (certified in 2013). One fishery lost its certification after the validity period expired (Tosakatsuo Suisan Pole and Line Skipjack

594 Article 3(2).
595 Article 4.
596 Popescu & Ogushi 2013, p. 25.
597 Ibid.
598 The criteria to select species to manage under TAC include: economic importance of the species (top 30); species considered overexploited and potential to be exploited by foreign vessels in Japan's EEZ. See ibid., p. 25. Fisheries managed under fishery rights are mainly located in the territorial seas and exploited exclusively by rights holders, usually FCAs. The coastal resources are usually also better protected and smaller in scale. This means that mainly resources which fall under the TAC system could be licence-managed resources.
599 Makino 2011, p. 35.
600 OECD 2013, p. 318; Popescu & Ogushi 2013, p. 25.

Tuna Fishery)[601] and another is under assessment (Japanese Pole and Line Skipjack and Albacore Tuna).[602] The expected price premium has not materialized in the certified fisheries.[603]

Many factors contribute to the slow development of an MSC in Japan. First, the MSC was initiated by an international environmental NGO, the WWF, and the transnational consumer goods company, Unilever. The Japanese fishing industry and government, however, are somewhat hostile towards ENGOs due to whaling issues.[604] The political decisions are mainly made by 'the iron triangle': politicians, bureaucracy and industry. ENGOs, such as the WWF, have little influence in policy making. The government, although it acknowledges the MSC in its documents, does not provide direct support or funding for certification. Compared to European countries, mistrust of NGOs is high in Japan.[605] Second, the seafood distribution system in Japan is less suitable for certification which requires traceability. Most producers are small-scale fishers who land 'small amounts of a wide variety of seafood'. The products are processed by small processing companies and retailers, adding to the difficulties in tracking the supply chain.[606] Third, there is no strong market demand for certified fish products. As mentioned earlier, most Japanese fish products are consumed domestically. However, the consumer awareness of certification schemes and their willingness to pay a price premium is still low.[607] Fourth, fishery cooperative associations play an important role in managing marine fisheries. The costs of MSC certification are, however, an impediment for many cooperatives. One of the main benefits of certification, being that there would be an increase in market demand, may be less appealing to the cooperatives because of their limited capacity to increase production.[608]

In spite of the current slow progress, studies indicate that there is potential for the MSC to gain support in Japan. The engagement between the WWF and fishery product retailers is increasing. Aeon and some other major supermarkets are increasingly committed to sustainable products.[609] Consumer awareness is also increasing. Research shows that if reliable information concerning the current fish resources and the content of certification are provided, many consumers tend to prefer certified food.[610]

601 www.msc.org/track-a-fishery/fisheries-in-the-program/exiting-the-program/withdrawn/tosakatsuo_suisan_pole_and_line_skipjack_tuna/assessment-downloads.
602 www.msc.org/track-a-fishery/fisheries-in-the-program/in-assessment/pacific/japanese-pole-and-line-skipjack-and-albacore-tuna.
603 Wakamatsu *et al.* 2010; Gondor & Morimoto 2011, p. 220.
604 Izawa & Makino 2005, p. 154.
605 Gondor & Morimoto 2011, pp. 224–225.
606 Izawa & Makino 2005, p. 154.
607 Gondor & Morimoto 2011, p. 221.
608 Ibid.
609 Ibid., p. 224.
610 Wakamatsu *et al.* 2010.

Scholars also indicated that, for international ENGOs, to establish connections with domestic NGOs and consumer advocacy groups, may be a good beginning in promoting the spread of certification.

Shortly after the MSC started to operate in Japan, a domestic certification system, the Marine Eco-Label Japan programme (MEL) was established in 2007. The MEL claims to be an affordable system, 'most suitable to the Japanese fisheries'.[611] It aims to promote 'its recognition in the Japanese market and gain wider acceptance' as well as promoting export.[612] Both government and industry play an important role in the establishment and operation of MEL. It 'was established out of the work of a preparation committee that included government officials. Its structure consists of a Council, an Audit Committee and a Board'.[613] Both the representatives from the competent authority, the Fisheries Agencies, and an administrative agency, the Fisheries Research Agency, have posts in these governing bodies.[614] The Council accredits certification bodies, registers industry organizations, and decides certification standards and procedures. The Audit Committee 'supervises the requirements of the certification bodies and the administration of the system'. The Board 'oversees basic administrative matters and serves as a general advisory board'.[615] A wide range of stakeholders plays a role in these governing institutions, including government, industry, 'community, academic, labour and women's groups'.[616] However, ENGOs have no position there. The Secretariat, which runs the daily administration of the scheme, is located in the national fishermen's organization, the Japan Fisheries Association.[617] Therefore, the MEL has 'a hybrid structure that requires that its activities meet the interests of both states and firms involved in the sector'.[618]

The MEL started to develop when many territorial certification schemes arose in response to the spread of MSCs, such as the systems in Iceland, the US and Canada.[619] Several factors are believed to motivate the development of these territorial certification schemes. First, the development of domestic regimes is a strategic response to the MSC. They try to reduce the high certification costs and to prevent a monopoly of the MSC in the market. Second, the sustainable seafood movement requires sufficient alternatives to provide sustainable products. Third, as a major international certification scheme, MSC threatens traditional state control. Therefore,

611 http://melj.jp/eng/index.cfm.
612 Ibid.
613 Foley & Havice 2016, p. 30.
614 Ibid.
615 Moye 2010, p. 551.
616 Ibid.; Foley & Havice 2016, p. 29.
617 http://melj.jp/eng/index.cfm.
618 Foley & Havice 2016, p. 30.
619 Ibid.

the state and industry try to regain authority and market access via establishing domestic alternatives.[620]

The MEL system develops faster than the MSC system. Twenty-three fisheries have been certified.[621] However, as an industry and government-led scheme, it has aroused a lot of criticism. NGOs like WWF claim that the MEL standards are not as stringent as the MSC's.[622] The close connection with industry is, by some scholars, regarded as a threat.[623] Though the system claims that it uses third-party certification bodies, these bodies are accredited by the MEL itself. Since 2008, the Japan Fisheries Resource Conservation Association has been the only certification body.[624] Moreover, unlike the MSC's, much less information is available concerning MEL's and few studies have explored the effects of MEL certification.

Certification is not very active in Japan. The effects of MSC and MEL certification have hardly been studied. This example focuses only on how public regulation influences the transaction costs in establishing fishing rights in Japan.

4.7.4 How does the interacted system address overfishing?

4.7.4.1 The definition of property rights

As mentioned earlier, two types of property rights systems exist in Japan: TURF in coastal areas and a limited entry system in offshore and far sea areas. In Japanese fisheries law, the term 'fishery right' is narrowly defined to refer to property rights granted to FCAs or individuals in coastal areas. The fishery sector operating in the offshore and far sea area is regulated under the licensing system. The rights of licence holders are not defined legally as 'fishery rights'. However, licence holders can access and harvest resources according to standards specified under the licence. The fishing effort is controlled by limiting the number of licences. Therefore, as far as offshore fishing and fishing in the high seas is concerned, a limited entry system is applied.

A co-management system has been promoted by the Fisheries Agency as a key approach to manage both coastal and offshore fisheries in Japan since the 1980s.[625] Both government and fishermen's associations, the FCAs play an important role in fishery management. The property rights in coastal areas are area-based rights. Exclusive fishery rights are granted to

620 Foley & Havice 2016, p. 28.
621 http://melj.jp/fishery/index.cfm.
622 Gondor & Morimoto 2011, p. 224.
623 Ibid.; Moye 2010.
624 http://melj.jp/fishery/index.cfm;www.conxemar.com/ponencias2/IForum/36-Masashi%20Nishimura.pdf.
625 Suenaga 2008.

FCAs or individuals in a specified area. FCAs are prioritized or exclusively awarded such rights. Therefore *clear geographical boundaries* exist for these types of property rights. Sedentary species account for most resources awarded under fishery rights systems. There are relatively few FCAs and FMOs that effectively manage fisheries. A relatively small number of migratory species are equally covered under the fishery rights systems. As a result, in that particular instance, the boundaries of the management units do not necessarily correspond to the ecological boundaries. This obviously creates challenges for management.

The offshore areas are managed under a licence system, where multiple vessel owners hold licences to harvest resources in the same area. The vessel owners are usually members of FCAs. Sometimes the local FCAs control most licences in a certain area and decide the harvest and management rules. Such self-management rules can be far reaching in that they determine who can access resources or what the individual quotas are. In these cases, though the fisheries are de jure regulated under limited entry systems, a de facto TURF system exists. However, if the FCAs only partially cover the vessels operating in a specific area, or if the FCAs do not specify clear rules regarding fishery management and equally lack strong monitoring capacity, the property rights for those particular fisheries will be less clear.

In both coastal and offshore areas, *the users of resources* are also well defined. In the coastal areas where property rights are vested in FCAs, the user group of the resources is limited via the membership of the FCA. To avoid arbitrary refusal of a new member, the Fisheries Cooperative Association Law prescribes, 'no eligible applicant for membership shall be refused by any Association without due cause'.[626] The eligibility conditions include: residence within the area of the Association and engagement in fishing activities for more than the minimum number of days determined by the Association (ranging from 30 to 90 days).[627] To be able to fulfil the minimum number of days condition, new membership seekers need to be accepted by the FCAs or hired by current fishermen. Therefore, the FCAs can effectively control the user groups of the coastal resources. As for the offshore resources, the user groups are limited through restricting the number of licence holders.

Though the validity period of the *fishery rights in coastal fisheries* is limited, they are treated as real rights. Therefore the rights holders, usually FCAs, have broad rights/authority to decide how to harvest and manage their resources. As discussed earlier, the FCAs have their roots in the feudal era. As exclusive rights holders, they have strong incentives to protect their own resources. The FCA members in coastal areas are often those who have

626 Article 25.
627 Article 18.

fished in their local areas for generations. Therefore taking care of resources is regarded as their own responsibility, which 'is connected to the household and related ancestor worship'.[628] Members of FCAs are often organized as FMOs to manage specific species. The FMOs can use different measures to manage the resources and many of these measures are more stringent than public regulation.

For resources managed *under licensing systems,* the TAC is an important instrument used by the government. The TAC defines the amount of fish that can be harvested by certain user groups for specific resources. In other words, it defines *the boundary of rights in terms of volume.* The TAC is determined by the minister based on allowable biological catch (ABC). The FRA and prefectural research stations conduct stock assessments every year and estimate the ABC.[629] The Fisheries Agency then decides the TAC based on ABC and potential social impact on the fishery industry.[630] Unlike in New Zealand, the government does not allocate the TAC to individuals via IQs. Instead, it differentiates the parts allocated to fisheries managed by the ministry and by prefectures. For ministry-managed fisheries, the quotas are further allocated to different sectors. To implement the TACs' obligations, the fishermen's organizations in these sectors usually establish a committee for TAC enforcement. The committee decides how to allocate the TAC among areas and seasons.[631] Other quotas are allocated to prefectures. The prefectures establish management plans to ensure the total catch falls within the quotas allocated. To implement the prefectural plans, some FCAs adopt autonomous management initiatives, including measures such as 'individual quotas, seasonal closures, area closures, size limits and restoration of fishing ground'.[632]

The TAC system, however, has many problems. One often noted criticism is that the TAC is often set higher than the ABC determined by the scientists.[633] These high TACs cannot prevent the overfishing of resources. This may be related to strong market demand for fishery products and high dependence on import. Therefore the government sets a self-sufficient rate as a policy goal and has no incentives to impose stringent restrictions. Moreover, the TAC is often set for specific species, without differentiating between stocks. More tailored TACs can better reflect the biological limits.[634] In addition, because no individual quotas are established to prevent the total catch from exceeding the TAC, the control of the fishery depends on the autonomous measures that have been adopted

628 Delaney 2015, pp. 273–274.
629 Makino 2011, p. 33.
630 Popescu & Ogushi 2013, p. 25.
631 Ibid., pp. 33–34.
632 Tanoue 2015, p. 17.
633 Popescu & Ogushi 2013; Matsuda *et al.* 2010, p. 902.
634 Tanoue 2015, p. 96.

by the sectoral FMOs (for ministry-licensed fisheries) and on local FCAs/FMOs (for prefectural-licensed fisheries). Sometimes, the quotas are allocated to big groups with heterogeneous interests, complicating the coordination and management of quotas. Some FMOs are also criticized for lack of transparency of their management practices.[635] On the other hand, in the successfully managed cases, the FMOs are often rooted in the same geographic regions, share common interests and have a relatively small number of members. In these cases, better autonomous management can be achieved.[636]

4.7.4.2 Enforcement

The co-management system also contributes to the enforcement of property rights. Under the public regulatory system, the Fisheries Agency and coastguard cooperate in monitoring fishing activities. 'The Fisheries Agency has seven branches around the country with around 39 patrol boats and several aircrafts for deployment.'[637] They have been effectively detecting both Japanese and foreign vessels involved in illegal fishing. In addition, since 2012, Japanese vessels operating in offshore and distant waters are obliged to use vessel monitoring systems (VMS), which provide a real-time record of fishing activities.[638]

In addition, many FCAs and FMOs are also engaged in monitoring activities.[639] With their own interests at stake, the FCA/FMO members have incentives to monitor each other. Research also shows that many FCAs have developed their own sanctions and conflict resolution mechanisms.[640]

4.7.4.3 Coordination

Coordination is important for the management of fishery resources, not only among fishermen but also between fishermen, regulators, scientists and the public. As mentioned earlier, FCAs play an important role in managing coastal resources, where members negotiate and decide how their fishery rights are executed in practice. FCAs therefore act as a coordination institution. The Board members and chairmen hold leadership positions and consult regularly with general members in formulating operational rules.[641] Moreover, some FMOs also have pooling arrangements to share economic benefits among all members. Such arrangements replace

635 Ibid., pp. 97–98.
636 Ibid., p. 98.
637 Popescu & Ogushi 2013, p. 27.
638 Ibid.
639 Uchida & Makino 2008, p. 225; Sarker *et al.* 2015, p. 35.
640 Sarker *et al.* 2015, p. 35.
641 Ibid., p. 34.

'individualistic competition with behaviour more similar to a cooperation'.[642] These institutional arrangements hence diminish the incentives to race for fish and help to control coordination costs.

Coordination can be more difficult for licensed fisheries. In these cases, property rights are awarded to individual vessel owners rather than to FCAs. In most cases vessels are members of the same local FCAs, which have the capacity to control their members. However, in many cases, not all vessels operating in the same area belong to the same FCAs, leading to difficulties in coordination.

In addition, other stakeholders may also determine the management of fisheries or have a stake in the management. For example, the government also regulates fisheries and many coordination committees have been established at different levels. The determination of fishery rights and licences at the level of the prefecture is a case in point which shows how government and local fishermen coordinate their activities in deciding who is allowed to fish for certain species and how. As mentioned earlier, prefectures are the competent authorities to issue fishery rights and licences. Those fishery rights and licences are issued according to a prefectural coastal fisheries management plan (CFMP).[643] To develop the CFMP, first, the FCAs draft their own management plans and submit them to the prefectural government. The prefectural government synthesizes these FCAs, collects related material and forwards them to the coordination institution, the AFCC. The AFCC is constituted of fishermen and experts nominated by the government. It prepares the draft CFMP and forwards it to the government. The government then holds a public hearing before finalizing and announcing the plan.[644] Therefore, though the granting of fishery rights and licences are governmental decisions, it is the outcome of coordination between the government and self-management institutions. It is worth noting that there are some species that migrate beyond prefectural boundaries. WAFCCs (Wide-Area Fisheries Coordinating Committees) have been established to coordinate behaviour at higher levels. However, 'they are rarely used because of the lack of supporting scientific information'.[645]

The above analysis shows that institutions have been designed to promote coordination between commercial fishermen and between fishermen and the government. The involvement of other stakeholders, such as the recreational sector, environmental NGOs and the general public, however, is more limited.[646]

642 Uchida 2010, p. 3.
643 Yamamoto 1995, p. 27.
644 Ibid., p. 31.
645 Uchida & Makino 2008, p. 226.
646 Ibid.; Makino 2011, p. 38; Murota 2011; Wilson *et al.* 2006.

4.7.4.4 Information

One advantage of co-management systems is their capacity to bridge knowledge accumulated at different scales. Local fishermen are better informed about local issues, such as the status of resources, local livelihood demands and behaviour, 'whereas the state has a regional and national vantage point and a repertoire of tools and techniques not normally available to local institutions'.[647] To build a robust bridge between those complementary types of knowledge is beneficial to resources management. In Japan, case studies have shown some beneficial bridges.

As discussed earlier, research institutions have been established at both national (FRA) and prefectural levels (prefectural fisheries research stations) responding to the pressing issues facing fisheries. These institutions consult in broad issues related to fishery management. For example, the FRA is divided into three sections: the marine resource section conducts stock assessment and is involved in defining TACs; the marine environment section conducts research concerning environmental conditions, marine geology and fish species ecology; the aquaculture section concentrates on the development of aquaculture.[648] Moreover, to convey the knowledge acquired through research to local fishers, 'extension Officers are stationed at main fisheries sites in each prefecture'. The prefectural government also supports some local fishermen (called 'fishery mentors') to act as bridges between research institutions and local fishermen.[649]

At local level, many FCAs and FMOs are also engaged in knowledge accumulation. For example, around half of the FMOs were engaged in stock assessment in 2003.[650] Some case studies have shown that local fishers and researchers collaborate in managing local fisheries. For example, when managing snow crab in the Kyoto prefecture, the local FMO took many autonomous measures such as establishing voluntary protected areas. Deciding the areas to promote resource recovery and ecosystem protection is a demanding task. The decision on the specific areas and protective activities is therefore based on the advice provided by the local research institutions.[651]

4.7.4.5 The scale of governance

The fishery rights system mainly applies to sedentary species and modestly mobile species. Governance at a smaller scale such as FCAs/FMOs can

647 Berkes 2009, p. 1694.
648 Popescu & Ogushi 2013, p. 54.
649 Makino 2011, pp. 17–18.
650 Uchida & Makino 2008, p. 225.
651 Uchida & Makino 2008, p. 217.

address them sufficiently. However, for more mobile species, usually those regulated under licence systems, it is more problematic to rely on small-scale management systems. As mentioned earlier, sometimes FMOs co-manage resources across municipalities. However, few of them operate across prefectures.[652] The WAFCC, which is designed to address highly mobile species, has limited application in practice.

The above analysis shows that different types of fisheries perform differently regarding the various issues of transaction costs. For the definition of property rights, comparative geographical and user groups' boundaries have been established for both coastal and offshore fisheries. The fishery rights of FCAs in coastal fisheries are treated as real rights and FCAs/FMOs have broad authority to decide how the rights are implemented. For offshore fisheries, the government has more direct control. The TAC defines the volume of fish products the licence holders can harvest. However, the government often sets the TAC above the scientifically determined limits. The fishermen's organizations for offshore sectors are also not always well established. Differences also exist regarding coordination costs and scale issues. The coastal fisheries are more suitable for small-scale FCAs/FMOs' management and also help to keep the coordination costs low. The mismatch between these self-management organizations and the biological boundaries of offshore species create challenges in management and coordination. The co-management between fishermen and government in both types of fisheries has contributed to monitoring and to the collection of information. To evaluate the total transaction costs, however, is a challenge. A (n anecdotal) case study of fishery management in the Kanagawa prefecture shows that the transaction costs, including both the costs encountered by the prefecture and the FCAs in these areas, are around 27 per cent of the total production, which is regarded as efficient.[653] Generally speaking, however, the coastal fisheries perform better than offshore fisheries. This is in accordance with the status of the fishery species: the coastal species' populations are generally healthier than others.

4.7.4.6 The role of public regulation in overcoming the limitations of property rights

In Japanese fisheries, especially in the coastal areas, a TUR system has been established. The literature argues that a TUR system can internalize some externalities which cannot be solved by an individual quotas system. It is believed that spatial heterogeneity, multi-species interactions and ecosystem protection can be better addressed under TUR systems.[654]

652 Makino 2011, p. 187.
653 Makino & Matsuda 2005, p. 441.
654 Cancino, Uchida & Wilen 2007, p. 403.

Under the ITQ systems, fishers tend to concentrate their harvests in the hot spots or where resources are more valuable. In cases of spatial heterogeneity, the distribution of fishing efforts will not be optimal according to biological conditions. Under the TURF systems in Japan, the FMOs can coordinate the activities of their members and mitigate the inefficient allocation of effort, by, for example, rotating access or pooling revenues.[655]

ITQs are awarded for specific species and are hence criticized as not being able to tackle multi-species issues. In Japanese coastal areas, FCAs are often granted fishery rights for multiple species and hence have better capacity to address the species interaction issues. However, in practice, fishermen manage their resources via subdividing FCAs to FMOs, which are usually single-species oriented. Only in a few cases is the inter-species relationship considered sufficiently by FMOs.[656] The single-species oriented management is also insufficient in addressing ecosystem issues. In these cases, governmental intervention is crucial.

Currently fishery regulation, especially fisheries law, was developed after WWII to improve productivity and meet domestic food shortage. Therefore it is geared towards targeted species. Since 1999, ecosystem conservation has started to be taken seriously, with the issuing of the Fisheries Basic Policy Plan and the Basic Act on Fisheries Policy.[657] In addition, the protection of the marine ecosystem needs the involvement of multiple parties. However, the official coordination institutions involve only commercial fishermen. Other parties are not involved in policy-making.[658]

4.7.5 Conclusion

Japan is a major fishery country, in terms of production, consumption and import. It has implemented a co-management system. Both government and fishermen's associations, FCAs, play a central role in fishery management. The country has the best established network of TUR fisheries: its coastal areas are managed by FMOs and fishery rights are acknowledged by the state. The offshore areas are regulated under a limited entry system, though sometimes, de facto TUR fisheries exist. This co-management system is characterized by significant formal and informal enforcement, coordination between commercial fishers and government, the combination of local, regional and national information, and the effective management of sedentary fisheries. However, due to high demand and dependence on fish imports, the TAC is often said to be larger than the allowable biological catch; the management of mobile species poses problems, and the involvement in fishery management of the recreational

655 Ibid., p. 399.
656 Ibid., p. 400–401.
657 Makino 2011, p. 105.
658 Makino 2010, p. 293.

sector and environmental NGOs is limited. Owing to the distrust of NGOs, the peculiar fishery distribution system, the FCA system and the lack of consumer awareness, the MSC is not very active in Japan. A domestic, industrial and government-led certification scheme, MEL, has been established as a strategic response to the MSC. However, little information is available concerning the influence of MEL on fishery management in Japan.

4.8 Comparison and conclusion

Fishery resources are a typical type of common-pool resources. The exclusion of people from exploiting them is costly and the resources have ecological limits. Without appropriate property rights to limit access, resources users may tend to overexploit fishery resources. This can lead to 'the tragedy of the commons'. Open access to the high seas remains under the current international regulatory system. Coastal areas are subject to the sovereign rights of coastal countries. In these areas, different types of property rights have been established to limit access and to promote sustainable management.

Three types of commonly used property rights have been discussed in this chapter: limited entry, individual quotas and territorial user rights. Under the limited entry system (the total allowable effort system in South Africa, the permit system in Mexican and Japanese offshore areas), access to resources is limited, but there are no specific limits to volume that each rights holder can harvest. Competition for fish still exists between multiple rights holders. In this situation, government retains the most power in regulating the fisheries. Individual quotas (the TAC system in South Africa and in New Zealand) are often promoted as a 'rights-based approach' to replace limited entry. In this scenario, specific quotas are allocated to individual rights holders, diminishing the incentives to race for fish and promoting the incentives to protect the resources. Rather than via volume of catch, another way to define fishing rights is based on sea areas. The right to explore (some) species in specific areas can be vested in communities and individuals. These TUR fisheries (concessions in Mexico, coastal areas in Japan) have also been used in many countries. In most cases, TURs are usually granted to communities, who decide how to allocate them to their members and to manage the resources.

Those property rights, however, do not operate alone. Both public and private regulation influence their functioning to a large extent. As far as public regulation is concerned, the regulatory institutions and specific regulatory instruments, such as TAC, TAE, licences, funding mechanisms and the devolvement of responsibilities all have the potential to influence property rights. The discussion of private regulation focuses on the MSC, a broadly used and acknowledged global certification scheme. In Japan, industry and government have responded strategically to the spread of

MSC and have established a domestic certification scheme, MEL. Though neither of these certification schemes includes standards directly concerning fishery rights, the case studies show that fishery certification has the potential to help refine the contents of fishery rights and to consolidate property rights of certificate holders.

This chapter shows that public and private regulation do not operate separately but are closely linked. Compared to forest certification, the role of government in fishery certification is more obvious. States usually 'have ultimate legal responsibility over controlling and managing access to marine fisheries', and they therefore are sometimes the clients of certification. Even when they are not the clients, government can also provide financial and technical assistance for the assessment.[659] Sometimes, the experts on evaluation teams selected by MSC certification bodies are qualified scientists or experts for government agencies. During the assessment, governmental data and interviews with government officials provide an important source of information. After the assessment, governments may 'also be called upon by certification applicants to facilitate and implement management changes necessary to maintain fishery certification standards'.[660]

The public and private interaction can influence the necessary preconditions for property rights' proper functioning and the overcoming of the externalities not addressed by the property rights.

Table 4.1 provides a comparison of the satisfaction of the necessary preconditions of property rights' effectiveness among the discussed countries.

4.8.1 The definition of property rights

The four countries studied present different types of property rights. In South Africa, fisheries are regulated under the TAC or TAE systems. Under the TAC systems, quotas are further divided among individuals (IQ system). Under the TAE system, the total fishing efforts are capped. In other words, the number of permits issued to vessels is limited, suggesting a limited entry scheme. Two types of property rights also co-exist in Mexico: limited entry and TURs. Most fisheries in Mexico are regulated under a permit system. In these fisheries, the number of permits is limited and the TAC is not often used. During privatization, many cooperatives were dismantled. A few surviving cooperatives and indigenous communities obtained concessions to harvest resources within specific areas. The Baja California Red Rock Lobster Fishery and Seri fisheries are examples of these TUR fisheries. A comprehensive individual transferable quotas system has been established in New Zealand. In Japan, coastal areas are

659 Foley 2013, p. 301.
660 Ibid.

Table 4.1 National comparison of the satisfaction of the necessary preconditions for property rights' effectiveness in countering overfishing and overcoming their limitations

		Definition of property rights	*Enforcement*	*Coordination*	*Information*	*Scale*	*Externalities*
South Africa	General	De jure limited entry or individual quotas; but many inshore fisheries are de facto open access	Lack of staff, budget and visible patrolling; corruption and mistrust of enforcement agencies	SWGs and RMWGs are two important channels to engage stakeholders. However, they are dominated by a large existing fishing industry, small-rights holders and non-rights holders are excluded	Lack of information	Managed at fishery level	
	Hake fishery (certified)	Individual quotas system; fewer stakeholders involved, MSCs consolidate the rights of certified parties	Public enforcement is prioritized in hake fishery; real-time track via VMS; certification adds an additional layer of monitoring	Similar situation applies to hake fisheries; MSCs provide another forum for stakeholder engagement, but small-rights holders also have a weak position	More information available with stronger public monitoring; industry has also developed its own observer programme. MSCs contribute to information availability	Public regulation and certification operate at the same level; hake fishery is large, making certification easier	Insufficiency regarding ecosystem protection; positive role of certification

Table 4.1 continued

		Definition of property rights	*Enforcement*	*Coordination*	*Information*	*Scale*	*Externalities*
Mexico (focus on SSF)	Limited entry (more broadly used, represented by the Kino community)	Under permit systems, the rights holders hire fishers; lack of public regulation to decide how fishing rights should be executed. MSC incentivized the creation of public regulatory systems	Lack of capacity; neither permit holders nor the fishers they hire have incentives to comply	Lack of coordination among multiple and multi-level regulatory institutions; coordination institutions dominated by fishing industry and SSF not motivated	Public investment limited		
	TURs (represented by lobster fishery and Seri fisheries)	Clear exclusive fishing rights; suitability of resources and users for TURs; MSC consolidates the rights of certified parties	Depends on self-enforcement capacity; FEDECOOP and Seri have successfully developed such capacity; MSC certification further strengthened it	Coordination is easier since exclusive user rights are established; self-governing institution with strong coordination capacity; MSC assessment also provides a coordination platform	Local knowledge valuable. MSC promotes information generation	Certified fisheries are either large or small	TURs are not sufficient in by-catch ecosystem protection; certification contributes to it

Table 4.1 continued

	Definition of property rights	*Enforcement*	*Coordination*	*Information*	*Scale*	*Externalities*
New Zealand (ITQ)	Clear property rights established for the commercial sector; commercial interests of the Maori integrated into commercial sector; customary and recreational rights are not so clearly established Hoki: only concerns commercial interests; MSC promotes refinement of property rights	QMS called for strong reporting system; the government adopts a sliding scale system of enforcement, VMS; broad authority of competent authority in enforcement; self-management institutions take over some monitoring functions; enforcement in customary and recreational sector is more difficult Hoki: MSC promotes the strengthening of enforcement capacity of self-management institutions	Self-management organizations share some responsibility with government; fisheries plan to coordinate the interests from different sectors; engagement of ENGOs in the certification process	Vast EEZ and short history of stock assessment; government efforts to generate more information; certification promotes information generation	Public regulation: national or QMA level; self-management – for specific species; all certified fisheries are large in scale	ITQs have limited capacity in dealing with ecosystem issues; positive role of certification

Table 4.1 continued

	Definition of property rights	*Enforcement*	*Coordination*	*Information*	*Scale*	*Externalities*
Japan (TURs)	Coastal areas: TURs; clear geographical boundaries, user groups and contents Offshore (licence system): limited entry; TAC defines the boundary in terms of volume. But TAC is often set too high	Public regulation: cooperation between agencies; real-time monitoring Self-enforcement of FCAs and FMOs	FCAs and FMOs coordinating activities of fishermen Public coordination institutions bring government and fishermen together Involvement of recreational, environmental sectors is limited	Co-management system brings governmental and local information together	Management system more suitable for coastal areas with small-scale and sedentary species; more challenging for mobile resources	Better incorporating spatial heterogeneity issues but not ecosystem protection

managed under TURs and offshore areas are subject to limited entry systems.

The limited entry system is implemented in three of the four countries examined: South Africa, Mexico and Japan. In South Africa, historically, fishing rights were concentrated in the hands of a few white-owned companies. Many small-scale fishers had no fishing rights. With little alternative employment, they continued illegal fishing. The reform failed to lead to a fair and sufficient redistribution of fishing rights. Illegal fishing is still rampant in many South African fisheries. The de jure limited entry fisheries are de facto open access. The permit regime in Mexico separates the permit holders from the fishers. Most fishers lack the expertise and the resources to apply for a permit. Permit holders tend to maximize production by buying fish from illegal fishers. This structure does not incentivize anybody to protect resources. In Japan actors obtain rights to harvest offshore resources through licences. The licence holders are often self-organized through FCAs and FMOs. Sometimes, these FCAs and FMOs cover most vessels working in a specific fishery and have strong capacity to coordinate the activities of their members. In these cases, de facto TUR fisheries exist, with clearly defined geographical boundaries, user groups and user rules. The TAC has been established for economically significant species, defining the boundaries of rights in terms of volume. However, due to the high fishing pressure in Japan, the government tends to define the TAC above the allowable biological catch.

South Africa and New Zealand provide two examples of an individual quotas system. In South Africa, the transferability of quotas is limited. As is the case with the limited entry fisheries, the loss of opportunities for small-scale fishers results in illegal fishing on a large scale, and a de facto open access system. This is especially true in coastal areas where small-scale fishermen traditionally harvest fish. It is less problematic in offshore fisheries, such as hake fishery, where few SSF interests are present. New Zealand presents a different story. The introduction of a QMS system has also led to some SSF exiting the fishery industry. However, the availability of an alternative livelihood, the grandfathering approach to allocating fishing rights and the possibility to buy back quotas from the SSF who can no longer afford to operate in the competitive market have helped them to leave the industry, without engaging in illegal fishing. The Maori people have adopted commercial ways of fishing on which the market-based QMS is based. Moreover, a co-management system exists in New Zealand, where fishing-sector organizations (QOAs) share management power with the government in deciding how fishing rights are executed.

Concessions in Mexico and coastal areas in Japan are managed under TURs. The self-management system in Mexican fisheries started to develop after the government recognized the rights of communities/cooperatives and the rights holders faced pressure from outside poachers or resource depletion. However, many cooperatives have been dismantled during the implementation of privatization in the 1990s. Therefore, few TUR fisheries remain in Mexico. In Japan, the TUR fisheries are established based on hundreds of years of development. In coastal areas, FCAs/FMOs have been well established. In both countries, TURFs cover mostly benthic coastal resources, which are most suitable for local management. The management units are usually small. They can, however be nested in an overarching structure. For example, individual cooperatives are organized under FEDECOOP in Mexico to coordinate lobster management. In Japan, bigger FCAs are sometimes divided into different FMOs, managing various resources.

The certification scheme does not define property rights itself. Nonetheless, the hoki fishery in New Zealand shows that it can promote the refinement of property rights by further dividing quotas between eastern and western stocks. The hake fishery in South Africa and the lobster fishery in Mexico also show that certification may help certified parties to consolidate their fishing rights.[661] However, the role of certification cannot be exaggerated. After all, property rights are a domestic law issue. The law decides who, when and how to assess and harvest fishery resources. Usually, only large-scale fisheries have the capacity to be assessed and certified. The more problematic cases, where property rights are insecure, have little chance of becoming certified. Therefore, certification contributes little to the satisfaction of property rights effectiveness.

4.8.2 Enforcement

The case studies show the importance of enforcement. In South Africa and Mexico, public regulatory agencies have limited staff, budget and capacity to monitor fishing activities and to ensure the enforcement of property rights. Therefore, some de jure limited entry and individual quotas systems are de facto open access, resulting in rampant illegal fishing and overexploitation of resources. However, in these two countries there are also better examples, such as the hake fishery in South Africa and the lobster and Seri fisheries in Mexico. The government prioritizes monitoring the hake fishery, the most valuable fishery in South Africa. The rights holders have also stronger self-enforcement capacity, ensuring the protection of their own resources.

The good governance level provides an indicator for the enforcement capacity of regulatory agencies. New Zealand and Japan rank high on the Governance Indicators Index, with high scores for government effective-

661 This, however, does not mean that the consolidation is not always socially desirable. For example, the Hake certification in South Africa is used as an instrument to marginalize the longline sector. Pure equity issues are not examined in this research. Here we focus only on the role of certification in contributing to establish a clearer and stable property rights regime.

ness and low scores for corruption. Mexico and South Africa rank lower on this index. In addition, the authority of competent agencies and the coordination between multiple agencies also importantly influence the effectiveness of enforcement. For example, the MoF/MPI in New Zealand has broad authority in enforcement, such as random entry, search and questioning. It also cooperates with other public agencies. Close cooperation is also found between the Fisheries Agency and the coastguard in Japan. These arrangements enable the competent authority to detect illegal fishing and respond quickly. On the contrary, in Mexico, the main competent authority, CONAPESCA has little power to enforce its regulation, but has to refer to the navy and prosecutor for compulsory measures. The frequent adjustment of regulatory institutions and the lack of coordination have left many regulatory gaps.

In addition to the enforcement of public regulatory agencies, self-enforcement can also play an important role. Under limited entry and individual quota systems, fishing rights are not exclusive. Monitoring and excluding others by individual rights holders are difficult. In some cases, the rights holders are self-organized, adding their enforcement capacity. For example, the trawl sector in the South African hake fishery established its own organization. QOAs were also developed in New Zealand to share some management responsibility with the government. Under TURs, property rights are often granted directly to communities. The self-management institutions play an important role in enforcing their own rights.

The existence of MSC provides an additional layer of monitoring. To assess the fisheries, the evaluation teams need to conduct field studies. They also identify the weaknesses in monitoring and issue corresponding conditions. The pre-assessment of the Seri crab fishery and the Gulf of California swimming crab fishery in Mexico shows that the involvement of MSC can lead to improvements even before the full assessment starts.

4.8.3 Coordination

Many actors have interests at stake in marine areas and resources, including the commercial fishery industry, indigenous people and other small-scale fishers, the recreational fishery sector, actors conducting coastal construction, environmental NGOs and so on. The coordination of these interests is a challenge in many countries.

A self-governing institution is often established in TUR fisheries. These institutions have their own procedure for involving their members in decision-making. Fishermen's organizations can also serve this function, as in the case of the hake fishery in South Africa and in the co-management system in New Zealand. These institutions, however, typically involve only formal rights owners.

Sometimes, the government also establishes coordinating institutions to engage stakeholders in fishery management and policy-making. For example, in South Africa, the government retains most authority in fishery management. In specific fisheries, two consultative working groups are established: SWG and RMWG, which involve stakeholders in scientific and management-related decision-making. However, only the big commercial fishing rights holders are involved. Entrants are poorly represented. Coordination institutions also exist in Japan, such as the AFCC and the WAFCC. Similarly, only commercial fishermen are involved, and other parties, such as the recreational sector, environmental NGOs and the general public are excluded.

In many countries, the same regulation applies to both the industrial fishery sector and to small-scale fishermen or indigenous communities. For example, in South Africa, commercial utilization by small-scale fishers is regarded as the 'limited commercial sector', which is subject to similar requirements as applied in the industrial sector. Such regulatory measures are often onerous for the SSFs, hence few of them obtain formal status in law, let alone being involved in policy-making. It also proves challenging to provide sufficient consideration to indigenous interests in policy-making, given their ways of organizing and culture differ from highly commercialized fishers which most regulation targets. New Zealand is an exception. Commercialism has long penetrated the Maori community, which can be well accommodated in the QMS. They have bargained intensively for their rights for both commercial interests and customary interests vis-à-vis the government and have successfully obtained one of the most favourable agreements for indigenous people in the world.[662] Certification provides an additional platform for coordination.

The recreational fishery sector is also important in some countries such as New Zealand and Japan. However, unlike the commercial sector, it is poorly organized and has a weak presence in policy-making. In both countries, no specific fishing rights have been granted to them. Regarding recreational fishing as a birthright, they successfully lobby against any access regulation. However, their weak organization makes both internal coordination (among fishermen) and external coordination (between the recreational sector and other stakeholders) difficult.

The interaction between ENGOs and industry/government also has an important impact on how fisheries are managed. For example, the support of NGOs in Mexico has helped fisheries to obtain certification or to prepare for it. In Japan, the relationship between ENGO and industry/government is more adversarial due to intense conflicts about whaling. The ENGO are poorly represented in policy-making and the development of MSCs has been slow. In response, the industry established its own domestic certification regime, MEL. An adversarial relationship between ENGOs and industry also exists in New Zealand. The hoki fishery was one of the

662 With 15 per cent of the population controlling 30 per cent of the commercial fishery interests in New Zealand.

earliest certified fisheries in the world. But it has encountered consistent objections and criticism from ENGOs.

4.8.4 Information

The availability of information regarding the status of resources, fishing activities and their interaction is crucial for fishery management. Local users, government institutions and certification assessment can all act as information generators and dispersers. A co-management system has the potential of bridging different information sources, as shown in the examples of lobster management in Mexico, the development of QOAs in New Zealand and coastal management in Japan.

The government enjoys a large economy of scale in collecting and processing information. However, obtaining information about vast marine areas remains a challenge, especially when the capacity of regulatory agencies is weak, such as in South Africa and Mexico. The declaration of the EEZ made New Zealand a marine superpower overnight. To obtain information in the vast EEZ is challenging and the introduction of the QMS also requires effective detection of misreporting. Thus high reporting requirements are imposed on quota holders, vessel holders and fish receivers. Cross-checking ensures the accuracy of information. Some new technologies, such as VMS, are also widely used in Japan, New Zealand and some important fisheries in South Africa. The availability of real-time information helps to strengthen enforcement.

The self-management institutions, especially those under TUR systems, play an important role in generating local information. Relying on local resource users, cheap and tailored information can be collected and processed quickly. In New Zealand, this research function is not devolved by the government. Besides, businesses often purchase research services themselves, tailored to their own needs.

MSC standards are demanding in terms of generating information. For a fishery to obtain certification, it needs to provide sufficient information regarding the stock status, ecosystem and management system. The case studies show that assessment often leads to conditions regarding more information generation. In this regard, the application of certification helps information creation and publication.

4.8.5 Scale

Fishery management can take place on different scales. Public regulatory measures can be directed at the national scale (such as the fisheries plan in New Zealand) or at fishery scale (such as the management of fisheries in South Africa). The case studies show that certification is more broadly applied to either big or small-scale fisheries. The big fishery has economy of scale, which makes the improvement of performance and certification costs more affordable, such as the hake fishery in South Africa and the hoki fishery in New Zealand. Small scale makes fisheries easier to manage and to monitor the sustainability standards of certification, such as the lobster fishery in Mexico. For the different types of fishing rights, limited entry and individual quota systems can be applied to either big or small fisheries. TURs are usually adopted for smaller-scale fisheries, such as in Mexico and Japan. The smaller management units can be nested in broader institutions and hence promote coordination at a larger scale. A challenge with fishery management is that the management scale, usually administrative units or self-management units, sometimes does not coincide with the biological boundaries of the species, especially for the highly migratory ones.

4.8.6 Externalities

An important goal of property rights in addressing common-pool resources problems is to internalize externalities and to limit the incentives of free-riding. Three types of property rights were discussed in this chapter. Via limiting fishing input, limited entry systems have been used to replace open-access scenarios. Literature shows that in limited situations, such as restrictive regulation, a simple fishery and a favourable social and political environment, the limited entry system can be successful. However, it cannot reduce the incentives for fishermen to race for fish. With specific quotas allocated to individuals, the IQ system can prevent the incentives to race for fish. However, in order to safeguard resource heterogeneity, fishermen need to coordinate their activities, which is not always easily achievable. Moreover, no property rights have been established on by-catch, habitats and ecosystems. Therefore, these issues are not considered sufficiently by rights holders. Under the TURs, property rights are often awarded to communities, which can coordinate behaviour among their members and account for resource heterogeneity. They are sometimes awarded rights over several species (e.g. coastal fishery rights in Japan) or even all the species in a given area (e.g. the Seri fishery in Mexico). In this case, communities can manage entire ecosystems rather than single species. Sometimes, habitat enhancement measures are adopted. However, generally speaking, non-target species and ecosystems are still external for them. Given the limitation of property rights in addressing remaining externalities, public and private regulation intervenes. One of the MSC three principles is ecosystem protection. The case studies show that the conditions required by MSC often concern ecosystem protection or by-catch reduction.

5 Comparative analysis

The above chapters discussed the governance systems for two types of common-pool resources: forests and fishery stocks. Nine case studies have been conducted which showed different degrees of fulfilling the five preconditions for the proper functioning of property rights. This chapter provides a comparison between the governance systems for forestry and fishery in the different countries studied; it will do so by following the analytical framework introduced in Chapter 2. This chapter is structured as follows: first, a comparative summary of the case studies is provided (5.1), followed by the comparison between them according to the framework presented in Chapter 2. Section 5.2 compares the characteristics of the two types of resources and the accordingly adopted property rights. This is followed by a comparison of public regulation and private regulation used to manage forest and fishery resources (5.3). Section 5.4 investigates the similarities and differences in the interaction between public and private regulation. The five preconditions for the proper functioning of property rights are reviewed in section 5.5. Section 5.6 concludes.

5.1 Summary of the case studies

This part provides a short summary of the nine case studies and then conducts the analysis. Recall that five case studies (Indonesia, Bolivia, United States and Canada, and Sweden) focused on forestry, whereas four others (South Africa, Mexico, New Zealand, and Japan) focused on fisheries. The summaries of the case studies provide a general appraisal of the system of public and private regulation in relation to the preconditions determining the proper functioning of the property rights in their protection of the common-pool resources, more particularly forestry and fish stocks. The idea in these summaries is to provide a general appraisal of the strengths and weaknesses of the regulatory system in each particular country based on the earlier analysis.

5.1.1 Indonesia

Traditionally forests in Indonesia were managed by local communities according to customary law. However, during the colonization period, the colonial government controlled the forests in Java and left forests on outer islands subject to customary law. This approach was accepted by the government after independence. During the Suharto period, the forest management authority was concentrated in the hands of the central government, which claimed control over forests in the whole country. But it could only exert direct control over forests in Java and outsourced forests in other areas to big timber companies with good political connections. The rights of smaller loggers and customary communities were not recognized.

Only after the Reformasi period did rights of the marginalized actors start to become recognized to a limited extent. Another important process which was initiated during the Reformasi period concerns decentralization. The government tried to transfer decision-making power to local governments, closer to where forestry activities take place. However, the self-interest of different levels of government and corruption led to a tug of war between local and central authorities. Departmentalism also caused conflicts between governmental agencies responsible for different sectors related to forestry, agriculture and environmental protection. These conflicts between different levels and sectors of government led to contradictory spatial planning and overlapping or conflicting permits. Therefore the decentralization process became a source of contested property rights which hampered effective forestry management.

International pressure since the 1990s and democratization after the ending of the Suharto government incentivized the active development of forest certification schemes in Indonesia. Several schemes have started to operate since the 2000s, of which the FSC and LEI are the best known. However, so far the coverage of forest certification in Indonesia is still limited, especially for the community-managed forests.

In summary, three layers of forest governance rules exist in Indonesia: customary law (adat), national law and private certification schemes. The interaction between the different levels of the systems has not resulted in the establishment of clear forest tenure and the prevention of deforestation. There are multiple reasons for this.

First, inconsistencies exist in defining property rights over forest in the multi-layered system. Traditionally, most forests were managed by local communities under the adat system. The communal rights, however, have been replaced by state ownership in modern Indonesia. Forest certification schemes require secure land tenure and respect for customary rights. However, the recognition and protection of customary rights under national law have been tortuous. Customary rights were excluded during the New Order regime and got limited acknowledgement in the BFL 1999.

The independence of customary forests from state forests has only been acknowledged recently. Moreover, even state forests have continued to be plagued by unclear boundaries, due to a lack of coordination between different levels of government and departments. Private property rights over forests have also continued to face uncertainties due to large numbers of unregistered forests. Certification schemes, which demand clear forest tenure as a criterion for certification, have hardly been a driver to address the problem of the incoherent definitions in law in Indonesia.

Second, the enforcement of property rights also presents challenges. Traditional self-governing institutions have been replaced by public regulation. However, public regulation has not been able to fill the gap left by the dismantled adat systems because of the limited capacity of governmental authorities and corruption. The costly and slow judicial system has also prevented effective private enforcement. Certification schemes have provided an additional layer of oversight, but have limited coverage and have their own limitations as well.

Third, traditional communities used to coordinate their activities via self-governing institutions. Excluding customary communities and promoting commercial interests in the forests traditionally controlled by communities have led to many conflicts between different actors, thereby increasing coordination costs.

In the current governance system, deforestation in Indonesia continues on a large scale.

5.1.2 Bolivia

The first Forestry Law of Bolivia in 1974 declared state ownership of all forests and required the users of forests to obtain permits from the state. Permits were granted only to registered enterprises and local and indigenous communities were excluded. A land and forest law reform was enacted in the mid-1990s due to pressure from indigenous movements, a deteriorating economy, privatization and growing environmental concerns. Under the new land and forest law, the state still owns all forests in Bolivia, but the land on which forests grow could be owned by the state, private parties and indigenous groups (via TCOs). Three types of forest tenure were established on public land: long-term contracting, concessions and ASLs. Reforms that have been implemented since 2006 have further strengthened the rights of indigenous people and have put a halt to issuing new concessions to timber companies.

Similar to Indonesia, a decentralization process and departmentalism have been implemented in Bolivia. As a result, land tenure and forests have been regulated by different agencies authorized by different laws. For example, TCOs and ASLs are granted by different agencies. Out of fear that claims for TCOs endanger the availability of municipal forests reserves, municipalities sometimes oppose the granting of TCOs, further complicating the conflicting land regularization process. The forest law also includes performance standards for both industrial and community-managed forests. Compliance with such standards requires a scientific approach which big companies conducting large-scale forest management have been able to apply, but which has proven challenging for local communities due to organizational, cultural, technical, and funding constraints.

The discourse that has triggered the forest law reform has also influenced the development of forest certification (CFV) in the late 1990s. This explains the similarities between the 1999 Forestry Law and the CFV standard. Only regarding issues like labour rights, community organization and consensus building, the CFV has more stringent standards. Highly compatible standards and governmental support made the certification scheme develop quickly earlier on, reaching a coverage of 2.21 million hectares in 2005. However, most certified forests have been managed by companies and progress in terms of promoting community-based forest management has remained slow. After 2006, when the export market increasingly shifted from North America and Europe to South America and Asia, government support decreased and concessions were reduced due to new reforms, coverage of certified forest areas dropped dramatically.

The interaction between public and private regulation has influenced the proper functioning of property rights and deforestation control in Bolivia.

First, unlike in Indonesia, comparatively clear forest tenure has existed on paper since the 1990s. Until then large-scale 'land occupation' movements led to many conflicting claims. From the 1990s onwards efforts have been made to regularize the land tenure and grant titles to private and indigenous people. The establishment of property rights was in line with certification requirements. However, the theoretical clear forest tenure has not had a great spin-off in practice, especially not in the case of private and communal rights, due to the weakening of traditional institutions (via adding an additional layer of institutions, the TCO, on top of traditional governing institutions, the villages), an expensive and slow land titling process, uncoordinated behaviour from different government agencies, and unbalanced agriculture and forestry policies (prioritizing agricultural use rather than forest use in granting land rights). Public property rights have historically been stronger. Public forests have often been exploited by timber companies in the form of concessions. Such concessions have also been actively seeking certification. The 2006 land reform, however, has cancelled the concession system, leading to a sharp decrease in certification coverage.

Second, even after this reform, public enforcement has remained weak, due to budget and staff constraints. Forestry law has remained largely unimplemented. It has therefore been a barrier to certification. Private parties have difficulties in enforcing their property rights, in a context where land conflicts are widespread and illegal logging abounds. For community forests, although they have formally been recognized under the new law and traditional governing institutions have not formally been abandoned, the introduction of an additional level of institutions has weakened traditional social control of forestry. Although certification has provided an additional layer of monitoring, its impact is limited due to the low coverage rate.

Third, traditional governing institutions involve community members in the decision-making and enforcement. They also provide a platform for collective decision-making. However, competition between the newly introduced TCOs and traditional institutions has weakened the traditional collective-choice arrangement. To obtain certification, communities need to either establish a community enterprise or resort to collaboration with companies. Either way, a more complicated governance system is established, which has rendered the coordination between actors more demanding.

5.1.3 North America (the US and Canada)

The case studies of the US and Canada have focused on how public and private regulation interact to address riparian buffer-zone protection in North America. The discussion focused on public and private forests in Canada and on private forests in the US. In Canada, it is the concessioners of public forests who conduct the daily management and therefore such forests have been treated as private property in the discussion. Comparatively clear forest tenure has been established in both countries, but the non-marketable values of the riparian forests have typically not been considered by private owners or managers. Therefore, regulation of riparian zone protection has been implemented.

When it comes to riparian forest protection, specifically the instalment of riparian buffer zones, research shows that certification schemes closely mimic public regulatory standards. On the one hand this overlap has made it easy to adopt certification. On the other hand it has also made it difficult to promote beyond-compliance protection.

Yet overall, many positive synergies have been shown to exist between public regulation and certification in addressing riparian zone protection. First, certification has made additional information available concerning the forest status and forestry activities. The coordination between public and private regulators in information sharing has also helped to reduce information costs. Second, there is some evidence showing that certification has been helpful in enhancing compliance with riparian buffer-zone rules. The coordination between public and private regulators has also helped to reduce redundancies in monitoring and hence to control monitoring costs. To incorporate considerations at the watershed-scale, however, has proven to be a challenge under both public and private regulation.

5.1.4 Sweden

The case study of Sweden has examined how public and private regulation interact to address biodiversity protection.

Different layers of rules exist under public regulation: forestry legislation and related regulations issued by the State Forestry Agency, other environmental legislation represented by the Environmental Code, as well as environmental/forestry objectives introduced by the Swedish parliament. To achieve the goals of public regulation, different types of instruments have been adopted, including command and control instruments, suasive instruments and incentive-based instruments, of which the first two types have been most important. The forest regulatory reform in the 1990s has led to a highly deregulated forest policy. The law has prescribed only a minimum level of performance requirements and has left the forest owners ample room to select the concrete measures with which to meet the requirements. Furthermore, the level of coercion has been low, leaving many command and control instruments not being complied with. Moreover, the use of suasive instruments has been reduced since the 1990s reform which has been accompanied by a reduction of the size and budget of the regulatory agency. Although these suasive instruments have enhanced environmental awareness among forest owners they have failed to change their inherent values and preferences.

Environmental campaigns and market demand have led to the quick establishment of forest certification in Sweden. Governmental support of certification schemes' development, implementation and enforcement has also played an important role in the wide adoption of certification. Two schemes are active there: FSC and PEFC, which cover more than half of the forests in Sweden. The competition between the two schemes and a bridging process have induced convergence of the standards of the two schemes.

The interaction between public and private regulation has influenced biodiversity protection via information generation, monitoring of forest activities and the incorporation of scale considerations. The standards of forest certification schemes and public regulation have been fairly coherent, with the former being more stringent on some issues. The general goals under binding laws and the 'Sustainable Forest' goal to a large extent have relied on certification schemes, which contain more specific guidance to operationalize. However, this does not imply that they have been fully compatible. The goals and standards under the Forestry Law, the Environmental Code and certification have not been fully coherent. The different requirements applicable to so-called set-aside areas in the two systems and the requirement to balance timber production and environmental protection with the more far-reaching requirements under certification schemes are examples of incompatible standards. The existence of multiple governance systems has also caused confusion and competing interpretations. Moreover, some scholars have argued that neither legal nor certification standards have sufficiently incorporated the latest scientific insights. In many situations, standards did not reflect the threshold recommended by scientific literature.

Similar to North America, public regulation and certification have contributed to information generation and strengthening of monitoring. Certification has played an even more important role in Sweden, not only because of its wider coverage but also because of its role to complement and specify the general and minimum legal requirements under the legal framework. Moreover, although certification has been an important complement to public regulation in Sweden it has not been its substitute. This is not only because their standards and ways of monitoring differ, but also because public regulation applies to all forest owners while certification applies only to certified forests. In recent years, criticism against forest certification has intensified, claiming that many certified forests have not been managed according to FSC standards and that the accountability mechanism has not been effective. Like the situation in public regulation where non-compliance is seldom followed by sanctions, the suspension of a certificate has rarely been used in practice. Another criticism has been that scientific uncertainties have been insufficiently accommodated in private standards. In addition, this case study has confirmed that certification schemes have difficulty in taking landscape issues sufficiently into consideration.

5.1.5 South Africa

South Africa faces serious overfishing. The government plays a central role in managing fish resources. Fisheries in South Africa are regulated under the TAC or TAE, which are then allocated to individual rights holders. The government also issues permits, develops management plans (OMPs) and requires compliance with technical standards. At each fishery level, two consultative forums (SWG and RMWG) are established to support the government in management. The existing fishing industry is the main participant of these forums. The TAC and TAE control the volume of catch and the fishing input (entry) respectively. The allocation of the TAC and TAE to individuals is accompanied by two types of fishing rights: individual quotas and limited entry.

During the apartheid period, fishing rights were concentrated in the hands of a few white-owned companies, excluding black and coloured people as well as traditional fishing communities. In the post-apartheid period, the government has tried to redistribute fishing rights to the historically excluded groups. The redistribution, however, has faced boycotts by the existing fishing industry. Critics have argued that the redistribution process has been window dressing, with too small quotas having been transferred to new entrants to support their survival. Concentration of fishing rights has prevailed in many sectors. Moreover, fishing rights have primarily been redistributed to the fishing companies with the best connections to politicians while traditional fishers have remained marginalized. The small fishers have continued to fish illegally.

As for certification, only one fishery has been certified by the MSC: the hake fishery. Hake is exploited by four industrial deep-sea trawlers. The certification is thought to have led to a reduction in seabird deaths. However, other environmental concerns have remained. Moreover, it has helped the deep-sea trawling industry's organization, the SADSTIA, to consolidate its fishing rights and to avoid further redistribution to historically marginalized groups and longliners.

The comparison between the inshore fishery governance in South Africa in general and the offshore hake fishery in particular has unravelled two striking differences: the first is related to the definition of property rights and the second to enforcement.

For inshore fishery a huge gap exists between the official fishing rights and the de facto property regime. The many traditional fishers, who have been marginalized during the apartheid period and who have not benefited from the post-apartheid redistribution, have formed a large informal sector. Combined with the general low level of enforcement, this has made inshore fisheries de facto open-access resources. Moreover, the late adoption of a limited entry system and the lenient awarding of fishing permits for linefishing have led to a misfit between property rights and the resources' features. The hake fishery presents a different story. Its most important species, the deepwater hake, has a much shorter commercial fishing history. It is more demanding to fish this species and hence fewer stakeholders are involved. Consequently, in this fishery the de jure fishing rights coincide with de facto ones. However, as mentioned before, in terms of distribution and fairness, this system is far from perfect. As for enforcement, due to the importance of the hake sector, it has been monitored better than inshore fisheries. The MSC has also contributed to further strengthening the enforcement of the hake fishery.

As far as coordination is concerned, the case study shows that the decision-making concerning fishery management has mainly involved the government and the existing fishing industry. New entrants and small-rights holders have hardly been involved. Neither has the MSC included these marginalized parties sufficiently. A general lack of information about fish stocks and catch has existed in South Africa's fisheries. However, increased governmental monitoring, the availability of a new scientific observer programme and the certification process have made more information available concerning the hake fishery. The certification has also played a role in promoting the internalization of externalities.

5.1.6 Mexico

Mexico is another country facing high fishing pressure and widespread overexploitation. Small-scale fishery plays a particularly important role in the Mexican fishing sector. It is on the fisheries exploited by small-scale fishers that this case study has focused.

Commercial fisheries have a comparatively short history in Mexico. For most of the nineteenth century, the fish industry was embedded in a state-led development paradigm. The state had ownership over fish resources and promoted fishery exploitation. The governmental promotion policies have led to a large-scale immigration to the fishery areas. Responding to low productivity and overexploitation, the government has shifted towards market-oriented development and privatization in the 1980s and 1990s. These developments have led to the elimination of cooperatives' exclusive access to many species and to a shift from territorial user rights (concessions of cooperatives) to a limited entry system (permits). Since then, both types of fishing rights have existed in Mexico, although the latter is dominant. Only a few cooperatives have survived the reform and have maintained their territorial user rights.

The institutions regulating fisheries have also changed several times since the 1960s, with the emphasis shifting between fish production and environmental protection. In the latest reform, environmental and fishery functions have been differentiated between different departments. The frequent institutional changes together with the existence of multiple agencies involved in fishery management have created challenges for coordination. Currently, the main instruments used to regulate fisheries are the National Fishery Chart (information provision) and NOMs (standards for different species). The new Fishery Law also requires the development of fishery management plans.

Mexico is active in fishery certification, with three certified fisheries and more under full assessments or pre-assessment. This case study has focused on two fisheries: the lobster fishery (certified) and the swimming crab fishery (pre-assessment).

A contrast has been found between the studied fisheries under territorial user rights and more common cases where fisheries are managed under a limited entry system. Such differences concern the definition of property rights, enforcement and coordination.

When it comes to property rights, the limited entry system has led permit holders to compete with each other in harvesting. The separation of permit holders and free fishers, together with a lack of public enforcement, has incentivized illegal fishing. This has resulted in a de facto open-access regime. Concessions have the potential to create better incentives, but have been limited to a few fisheries and have relied on the self-enforcement capacity of the concessioners. Successful examples have created opportunities for MSC certification. The case of the Red Rock Lobster Fishery and the Seris' fishery are such examples. In these cases, clear property rights have been established. The government has admitted exclusive fishing rights and self-governing institutions have developed their own rules in managing the resources. However, the granting of exclusive fishing rights has not automatically led to good performance. The self-governing capacity has only started to develop in response to external pressures. MSC certification has also played an important role in empowering the FEDECOOP and in consolidating the fishing rights. Pre-assessment and FIP have also been crucial in triggering the establishment of a public regulatory system for the crab fishery.

The lack of coordination between governmental agencies has created a gap in enforcement and has weakened enforcement capacity. This has been accompanied by widespread illegal fishing in Mexico. The lobster fishery and fishery of the Seri present two examples with stronger self-enforcement capacity, where the cooperatives' or Seris' governing institutions have adopted their own monitoring measures. MSC certification and pre-assessment have also contributed to the strengthening of enforcement capacity in these cases.

As regards coordination, there is a shortage of it among governmental agencies. For fisheries under the permit system, the separation between permit holders and actual fishers has complicated coordination. Only the former have formal legal status and have been involved in fishery management decision-making. Under the territorial user rights, self-governing institutions have provided opportunities for their members to coordinate. The certification has empowered the fishermen's organization, FEDECOOP, and has strengthened its capacity to further engage in policy-making.

5.1.7 New Zealand

Compared to many other countries, New Zealand presents one of the 'best practices' in fishery management. Three sectors are crucial in managing fish resources in New Zealand: the commercial sector, the recreational sector and the Maori.

The fishing rights system has shifted from traditional control by Maori communities to a regulated open-access system after the arrival of European settlers and further to a system represented by individual transferable quotas. Recently, three fishing sectors have been subject to different fishing rights arrangements. For the commercial sector, most economically valuable species have been introduced under the QMS system, in other words, the individual transferable quotas system. The Maori are engaged in both commercial and customary fishing. For commercial fishing, the Maori control more than one third of the fishing quotas. For the customary side, specific legislation defines their rights, but these rights are less clear and secure than commercial rights. Recreational fishing is also important in New Zealand, involving 20 per cent of its population. However, few regulations exist for this sector and people still have free access to fish for recreational use.

Parallel with a fishing rights reform in the 1980s, the New Zealand regulatory system also experienced a reform with the aim of deregulating many sectors, to make the state 'meaner and leaner' and to shift responsibilities to the industry. To achieve this goal, the government shifted from resources rental to costs recovery for the fishery exploitation activities and devolved some fishery management responsibilities to the industry (such as permit and trade registries). A co-management system started to develop, where the quota owners' associations collaborated with the government in deciding how fisheries were to be managed. To reconcile the interests between different fisheries sectors, plans were developed and supervised by the government.

The MSC first certified a fishery in 2011. So far, five fisheries have been certified and two more are under full assessment.

Although some limitations exist, such as spatial conflicts between customary, recreational and commercial fishing, compared to many other countries, the fishery governance, especially of and by the commercial sector is a success. This success is attributed to the clear definition of property rights, strong enforcement capacity and well-regulated coordination created by the co-management system.

First, the commercial rights have been well defined under the QMS system. The shift towards the cost-recovery system and the devolvement of management responsibility have promoted the establishment of self-management institutions. The co-management system, together with the influence of certification has contributed to an extensive definition of property rights. The long history of the entrenchment of the Maori in the market economy has accustomed them to the QMS system, which is a market-based instrument.

Second, the New Zealand government also has a strong capacity to enforce its fishery regulations. High reporting requirements, a sliding-scale system in enforcement, technological development, broad regulatory authority and coordination between different agencies have contributed to a strong enforcement capacity. Moreover, under the co-management system, many quota owners' associations also have developed their own monitoring programmes. The certification process has also helped to identify the insufficiencies of self-governance and governmental management, leading to further improvement.

Third, the co-management system has also promoted the coordination between various rights holders. However, conflicts between the different fishery sectors and between the fishing industry and environmental NGOs have not been reconciled yet.

5.1.8 Japan

Japan is a major fishery country, in terms of production, consumption and import. Japanese fisheries have five sectors: coastal fisheries, offshore, distant-water, marine aquaculture and inland aquaculture. Most fish products are for domestic consumption. The health of coastal fish stocks is better than that of the others.

Japan has the best established network of TUR fisheries: the fishermen's organizations, FCAs, have been granted exclusive fishing rights in coastal areas. Such fishing rights have evolved historically. Members of FCAs sometimes have divided into smaller groups and have formed FMOs to manage specific species. The offshore areas have been regulated under a limited entry system (via licences), though sometimes, they have functioned de facto as TUR fisheries.

Both central and local governments have played an important role in regulating fisheries. The government has taken different measures to protect fish resources, such as the introduction of resources protection plans, TACs, TAEs and so on. Coordination institutions have been established at different levels. However, cross-prefectural coordination has been limited.

Owing to distrust of NGOs, the peculiar fishery distribution system, the FCAs system and the lack of consumer awareness, MSCs have not rooted in Japan. A domestic, industrial and governmental-led certification scheme, MEL, has been established as a strategic response to the MSC. However, little information is available concerning the influence of MEL on fishery management. Therefore, this case study was limited to the assessment of the interaction between public regulation and property rights in Japan.

First, the species under the TURs system are usually not very migratory, so that clear geographical boundaries have been established. In the offshore areas, a limited entry system has been implemented and setting boundaries for specific migratory species is more difficult. However, sometimes, the fishermen in the same offshore areas are organized under the same FCAs, making self-management possible. The existence of fishermen's organizations also helps to define the user groups. Sometimes the TAC applies to fisheries managed by the limited entry system. The TAC defines the boundary or rights in terms of volume. In order to achieve the self-sufficiency rate of fishery products in Japan, the government, however, has incentives to set the TAC at too high a level, threatening the sustainability of the stocks.

Second, the co-management system with the involvement of both the government and the fishermen has contributed to information accumulation and has strengthened monitoring. The developed system has been more suitable to coastal areas concerning the definition of property rights and coordination (FCAs and FMOs contribute to the coordination in coastal areas and the less mobile characteristics of the species have also made coordination easier; whereas offshore resources require coordination between geographical areas, which usually has proven to be difficult) and has raised specific issues of scale. These factors probably have contributed to the better protection of coastal resources in comparison to offshore and distant-water fisheries.

5.2 Types of resources and corresponding property rights

5.2.1 Types of resources

Both forests and fish stocks are important common-pool resources, for which the exclusion of others is costly and consumption is rivalrous. The non-excludability and subtractability may lead to 'the tragedy of the commons' problem if access is not restricted. Forest comprises 31 per cent of the total land area, providing diversified environmental services.[1] Many people depend on forests for their livelihoods, with the estimate varying from a few hundred million up to 1.6 billion.[2] These include various groups of people (including indigenous people, communities living in or adjacent to forests, employees) involved in different activities (such as hunting, gathering, producing timber or energy).[3] The large number and diversity of user groups make it costly to exclude each other from exploiting the forests. Intensive use can go beyond ecological limits and can lead to deforestation and forest degradation.[4] Fishery is also a crucial resource, providing a significant part of animal protein and livelihood for 10 to 12 per cent of the world's population.[5] As these mobile resources are exploited by various types of actors, exclusion is also costly. Overexploitation also puts fishery resources under stress: 28.8 per cent of the resources are overexploited.[6]

One important difference between these two types of resources is the extent of mobility. Forests can be defined in many different ways, depending on the inclusion or exclusion of factors such as land use, tree height, tree density and so on.[7] In spite of the divergence, crucial elements of forests include the trees growing on specific land and land use.[8] The non-mobility of the trees and land makes it easy to demarcate their geographical

1 *Supra* Chapter 3, section 3.1.

2 FAO 2010, p. 121. For details see Chao 2012, p. 3, available at: www.forestpeoples.org/sites/fpp/files/publication/2012/05/forest-peoples-numbers-across-world-final_0.pdf.

3 Chao 2012, p. 3.

4 *Supra* Chapter 3, section 3.1.

5 *Supra* Chapter 4, section 4.1.

6 Ibid.

7 Schuck *et al.* 2002, pp. 15–16; Convention on Biological Diversity, 'Definitions: Indicative Definitions Taken from the Report of the ad hoc Technical Expert Group on Forest Biological Diversity', available at: www.cbd.int/forest/definitions.shtml; United Nations Environment Programme, 'Forest Definition and Extent', available at: www.unep.org/vitalforest/Report/VFG-01-Forest-definition-and-extent.pdf.

8 For example, a broadly accepted definition is provided by the FAO, which defines a forest as 'land spanning more than 0.5 hectares with trees higher than 5 meters and a canopy cover of more than 10 percent, or trees able to reach these thresholds in situ. It does not include land that is predominantly under agricultural or urban land use'. The explanatory note 1 further clarifies: 'Forest is determined both by the presence of trees and the absence of other predominant land uses. The trees should be able to reach a minimum height of 5 meters'. Page 3, www.fao.org/docrep/017/ap862e/ap862e00.pdf.

boundaries and they are hence suitable for local level management. As mentioned earlier, forests can have multiple functions[9] such as providing a habitat for species, controlling climate and conserving water. Species and water are mobile and their protection requires governance on a large scale.

Fish is a mobile resource living in inland, coastal areas and high seas. Different species have various extents of mobility. Some are highly migratory or straddling and can migrate long distances across national borders. These species usually live in Exclusive Economic Zones or in the high seas.[10] There are also more sedentary species for which the demarcation is comparatively easy. Different extents of mobility influence the nature of the desired property rights.

5.2.2 Property rights

The extent of resources' mobility influences the possibility to demarcate the boundary of the resources and the users' rights and the costs to exclude other parties. Therefore, establishment of property rights for forests and fish stocks differs. More complete property rights can be established for forests, for which clear geographical boundaries can be determined. In the five case studies for forests, ownership was established, either of forests (e.g. in the US, Canada, Sweden, and Indonesia) or of the land where forests grow (e.g. Bolivia). Owners have broad rights over their property, including access, harvest, management, exclusion, and alienation. The owners can either be the public/state, private actors or communities. Sometimes, less complete rights are established for forests. For example, private parties can gain rights to access, harvest and manage a specific part of publicly owned forests via permits or concessions.[11] In these cases, ownership and less complete users' rights co-exist in the same forests: ownership is a public property right, while user's rights are private. This research treats the forests as public, private or communal according to the party in which the management right is vested.[12]

The migratory nature of fish makes a clear demarcation and the exclusion of other parties from exploiting the same stocks difficult. It creates challenges to the establishment of ownership over fish resources. The state may claim ownership/sovereignty over the resources within its territory

9 Notice that often a rather anthropocentric perspective with respect to forests is followed, whereby the value of the forest is mainly considered as a resource for use by humans rather than for its own sake. See Hey 2016, p. 2.

10 www.fao.org/docrep/009/a0653e/a0653e04.htm.

11 This is the main approach to managing public forests in Canada. Permits/concessions have also been established in public forests in Indonesia and Bolivia.

12 For example, in Indonesia, community-based forest management policies provide a different extent of rights to harvest and manage forests for communities in public forests. These types of forests are discussed under the heading 'Community forests'.

(e.g. Mexico). However, it is usually the private parties/communities who harvest the resources. These user rights for fish are less complete than for forests, including only parts of the elements of ownership.[13] In the strict legal sense, these rights are not always admitted as property rights.[14] However, it does imply some control of the parties over fish resources, such as access, harvest and management. This research does not discuss the legal nature of property rights, but uses the term inclusively, referring to the rights/interests/privileges of the fishing actors.

Even for less complete property rights, their boundaries need to be established. The boundary of forests is decided geographically. As for fishing rights, the boundary can be decided according to geographical scope, species or volume. The territorial user rights are established geographically. Under the territorial user rights, the rights to harvest specific species (and exceptionally all species) in demarcated areas are exclusively vested in individuals or communities. The territorial user rights are more suitable for coastal areas and less migratory species, such as the rights of FCAs in Japan, the rights over Baja California lobster and the Seris' rights over crab in Mexico. The boundaries of property rights of migratory species cannot be properly established geographically. In these cases, volume and species are often used for defining property rights. For example, under the quotas schemes, the rights to harvest a specific species up to a certain volume are granted to individuals or communities. Individual quotas have been used broadly and individual transferable quotas have been considered an important tool in promoting sustainable fishery management by many scholars.[15] In the studied cases, individual quotas were used broadly in South Africa while New Zealand has one of the most comprehensive and best known ITQ systems. In addition to TURs and quotas, a limited entry system also exists broadly in, for example, South Africa and Mexico. In this situation, no quotas are established individually. Each rights holder has the right to harvest specific species. The access is limited in terms of the number of permits, fleets, size of engine and so on.

Property rights for fish are less complete than for forests. This implies that the different elements of ownership, such as the rights to access,

13 For example, Schlager and Ostrom differentiate four types of fishing rights holder: owners, proprietors, claimants and authorized users. Owners enjoy comprehensive bundles of rights, including access and withdrawal, management, exclusion and alienation. Proprietors have most bundles except for alienation rights. Claimants have rights of access, withdrawal and management. The rights of authorized users are limited only to access and withdrawal. See Schlager & Ostrom 1999, p. 90.

14 See the discussion in Chapter 4, section 2.2.2. For example, fishing rights for coastal fisheries are treated as property rights under fishery law while the rights of licence holders for offshore fisheries are not. In this book, they are all treated as 'fishing rights' for our purpose is to focus on the management of fisheries.

15 Grafton 1996; Squires *et al.* 1998; McCay *et al.* 1995; Branch 2009; Gibbs 2009.

harvest, manage, exclude and alienate are vested in different parties, either the fishing rights holders or the owners. The property rights systems examined combine characteristics of public, private and communal property rights. Property rights under limited entry systems are the most vaguely defined. The rights holders can access and harvest specific resources in specific areas. But such rights are not exclusive and many actors operate in the same areas. To manage the fishery, multiple actors need to cooperate. However, since the volume of fish which can be harvested is not restricted individually, actors tend to compete with each other and have little incentive to coordinate in managing the resources. Therefore, in this example, it is often the government which prepares management plans and determines how the resources are to be protected.[16] The rights of exclusion and alienation are also vested in the government.[17] Therefore, the limited entry system is closest to a public rights system.

Under the individual quotas systems, individuals have rights to access and harvest the resources. Individual quotas take away the incentives for rights holders to 'race for fish' and provide incentives to protect the targeted species. However, individuals alone cannot decide how the fishery is managed. In some cases, the government plays a dominant role in deciding how a fishery is managed.[18] The fishermen usually do not participate in the decision-making and do not decide the operational rules concerning the use of the fisheries.[19] Sometimes, the fishermen self-organize to achieve common goals and the self-governing institutions can share some rights of management with the government.[20] Usually, the rights holders have no authority to decide who can access the resources and to alienate their rights. The rights to exclude and to alienate still reside with the government. This system therefore has combined characteristics of different property rights. The individual transferable quota system in New Zealand is an exception. The quotas are granted to individuals permanently and the rights holders can transfer the quotas to other parties; they then also have rights to alienation.

The territorial user rights (TURs) can be granted to both individuals and communities but the latter is more common.[21] In this case, the

16 Such as in South Africa and Mexico. An exception is Japan, where fishery management is highly decentralized. Even for fisheries in offshore areas, which are formally operating under the licensing system, fishermen's organizations play an important role in coordinating the activities of their members and deciding how fisheries are managed.

17 The government determines who has access to the resources and who is excluded from using them.

18 Such as in South Africa.

19 This means direct decision-making, and does not exclude potential public consultation in public decision-making.

20 New Zealand is such an example.

21 The establishment of a territorial user rights system in abalone fisheries in South Africa is such an example. See Chapter 4, section 4.4.

communities have the right to access, harvest and manage fisheries. Moreover, the self-governing institutions have the right to determine who can access their resources and hence have the right to exclude as well. Only the right to alienate is reserved for the government.

5.3 Public and private regulation

This part compares public and private regulatory instruments adopted to address forest and fishery problems. For public regulation, this overview will focus on the national level and examine the specific regulatory instruments used in each country studied. This section compares several commonly used regulatory instruments, such as planning, permits and concessions as well as performance standards. More particularly, the question of what specific instruments are being used to control forest degradation and depletion of fish stocks will be addressed. As far as private regulation is concerned, this research focuses on certification, specifically the FSC, the PEFC and the MSC. The next section will focus on the interaction of public and private regulation (5.3) as well as on the necessary preconditions for the proper functioning of property rights, discussed in the theoretical framework (5.4).

5.3.1 Public regulation

The case studies show that different regulatory instruments are being used to regulate forests and fisheries in different countries. The analysis of North America and Sweden subsequently focuses on riparian forest protection and biodiversity protection. Therefore a limited number of instruments will be discussed: spatial planning of riparian buffer zones in North America and command and control/suasive instruments targeting biodiversity in Sweden. Instruments discussed in other case studies can either influence the establishment of property rights, such as spatial planning and permits, or influence directly how the forests/fisheries are managed, such as more specific performance standards (limits to tree felling, equipment restriction and so on). This part compares these instruments which have been used for forest and fishery regulation.

5.3.1.1 Spatial planning

Both forestry and fishery concern the exploitation of spatial areas. Different actors may use the same or different resources in the same areas, such as timber harvest (either by large companies, private owners or indigenous communities), food collection, plantations, or agriculture in the same forests, or fish capture (either by commercial, recreational or customary users), aquaculture, or marine biodiversity conservation in the same marine areas. The co-existence of multiple activities potentially leads to spatial conflicts. For example, in Indonesia conflicts between commercial timber companies and indigenous people are widespread.[22] In addition, the fast expansion of palm oil planting, mining and agriculture led to the conversion of forests.[23] The competition between timber harvest, agriculture and cattle farming in Bolivia has also led to rapid deforestation.[24] Spatial conflicts may also exist in marine areas. The cases of South Africa and Mexico exemplify that unclear fishing rights lead to conflicts between commercial fishermen and small-scale fishermen operating in the same marine areas. Even in New Zealand where clear commercial fishing rights have been established, conflicts have arisen between different fishing sectors (commercial, recreational and customary), and between fishermen and farmers (who use the marine areas for aquaculture) and environmental groups (who prefer to establish environmental protection areas).[25]

Spatial planning is an important tool to decide which activities are allowed in specific areas. The permission of actors to harvest forests or to capture fish needs to comply with the spatial planning. Therefore spatial planning and the establishment of property rights are closely linked. However, the coordination of different interests in spatial planning is not easy. The case of Indonesia illustrates this. Both central and regional governments have the authority to plan space at respective levels, often resulting in conflicting plans.[26] In addition to the Ministry of Forests which is responsible for deciding the boundary of forested areas, non-forestry agencies also determine the areas for their related activities according to sector-based legislation, such as the areas for agriculture, mining and the establishment of eco-regions. Though the spatial planning law requires coordination among the actors involved in planning land use, it is not actually implemented.[27] In summary, the conflicts between different levels of government and different departments have led to incoherent spatial planning. Marine spatial planning is also argued to be an instrument to integrate different marine interests, such as fishery, aquaculture, offshore oil and gas exploitation, ecosystem protection and so on.[28] To establish comprehensive marine spatial planning is, however, a difficult task. The area-based management is often conducted from a sectoral perspective. For example, in New Zealand fisheries under QMS systems are managed at the level of quota management areas, determined by the fisheries agency. The establishment of marine protection areas is carried out according to

22 See *supra* Chapter 3, section 3.4.3.1.
23 Indrarto *et al.* 2012, pp. 4–9; Palo & Vanhanen 2012; Gupta, Van der Grijp & Kuik 2012, pp. 122–124.
24 Müller, Pacheco & Montero 2014, p. 10; Müller *et al.* 2013.
25 Bess & Rallapudi 2007.
26 Gupta, Van der Grijp & Kuik 2012, p. 128.
27 Indrato *et al.* 2012, p. 21.
28 E.g. Douvere 2008; Douvere & Ehler 2009; Agardy, Notarbartolo di Sciara & Christie 2011.

the Marine Protection Act by the Department of Conservation.[29] Efforts have been made trying to integrate diverse marine usages, such as developing fisheries' plans.[30] However, the process of establishing fisheries plans has been cumbersome. It concerns only fishery sectors and includes mainly broad management goals rather than strict spatial planning. A comprehensive marine spatial plan is still yet to be developed in New Zealand.[31]

5.3.1.2 Permits and concessions

Permits and concessions are commonly used by the government to restrict access to forest and fish resources. In Indonesia, the Ministry of Forests is responsible for issuing most logging permits. Local governments can issue some 'lesser' permits and other authorities can issue permits for mining and estate crops plantation. These different agencies sometimes issue conflicting permits, thereby accelerating deforestation.[32] In Bolivia, concessions have been used to authorize private actors and indigenous people to access public forests since 1996. Authorization is also needed to access and harvest private and communal forests. Under the concessions and authorizations, five types of forest tenure have been clearly distinguished on paper, though the institutional arrangement may make some forest tenure difficult to establish and function in practice. In fishery governance, permits/licences/concessions are also broadly used to authorize companies/individuals/communities to harvest specific species in determined areas. For example, in South Africa, one needs to obtain a permit to execute fishing rights. Under the total allowable catch (TAC) system, permit holders are allocated quotas for specific species. Under the total allowable effort (TAE) system, the number of permits or number of vessels, or size of engines is limited. In Japan, access to fish resources is granted in two ways, fishery rights or fishery licences. The fishery rights systems are area-based. With the licences, the fisheries are managed under a limited entry system.

5.3.1.3 Technical standards

Spatial planning and permits are related to the establishment of property rights. In addition, public regulation can also determine how forests/fisheries are to be managed directly, and hence influence the execution of property rights. These technical standards include, for example the

29 Bess & Rallapudi 2007, p. 723.
30 Ibid., pp. 726–727.
31 See *supra* Chapter 4, section 4.6.3.3.5. See also Environmental Defence Society 2011, p. 43, available at: www.pepanz.com/assets/Uploads/EEZ-Policy-Document-EDS-4.pdf.
32 See *supra* Chapter 3, section 3.4.3.2.

minimum logging cycle, a minimum cut diameter by tree species, the type of fishing gear, fisheries closures and so on. The biodiversity standards in Sweden show both the advantages and the limits of what can be achieved with performance standards. However, the design of such standards is usually based on western science which is more suitable for resource management by commercial parties on a large scale than for small-scale forestry and fishery which are often artisanal, low impact, not capital intensive and often lack information. Using the same standards for commercial and small-scale forestry/fishing has created challenges for the latter, weakening their property rights, as has happened, for example, in Bolivia[33] and South Africa.[34]

5.3.2 Private regulation

The examination of private regulation focuses on certification schemes, specifically FSC and PEFC in forest governance and the MSC in fishery governance. In forest governance, both FSC and PEFC are well-known transnational schemes with broad coverage. The FSC was initiated by environmental NGOs in 1993. Its stringent standards provoked caution in the forest industry, which responded by establishing its own industry-led national certification schemes. These national schemes were later endorsed by the PEFC. The competition between the FSC and PEFC has led to the ratcheting up and convergence of standards.[35] The development of national FSC and PEFC standards in Sweden is an example of this.[36]

A different picture exists for fishery certification: so far the MSC is the most acknowledged and broadly used transnational fishery certification scheme for capturing fish. The FOS, though it has broad coverage as well, is not well accepted because of vague standards and relying only on existing documentation. Some countries have developed their own national schemes as alternatives for the dominant MSC, such as the national scheme in Japan. However, in the international arena, competition between certification schemes, like in forest governance, is not so obvious. Unlike forest certification, which developed in response to an international regulatory gap, the MSC is established in a heavily regulated field. Therefore, it is strongly influenced by international law. The original MSC principles and criteria are based on a 1995 FAO Code of Conduct. The UN General

33 The regulation of small private properties and TCOs is modelled on the regulation of large concessions. This creates challenges for local communities who are more used to subsistence use rather than large-scale commercial use. See Chapter 3, section 3.5.4.2.3.

34 In South Africa, some small-scale fishers are classified under the category 'limited commercial sector'. They are subject to similar regulation to the industrial sector. See Chapter 4, section 4.3.2.

35 Cashore *et al.* 2007; Overdevest 2010.

36 See Chapter 3, section 3.7.3.3.

Assembly Resolution 61/105 led to better protection of vulnerable marine ecosystems in the recent revision of MSC standards.[37]

The MSC was initiated by the WWF, which was also an initiator of the FSC. The MSC also learnt from the experience with the FSC. However, important differences exist between the two schemes. First, the FSC has its own global standards, the FSC Principles and Criteria. To tailor the general standards to local situations, many national or regional standards have been developed by its national affiliates.[38] Such forests are evaluated against the local standards. But if local standards do not exist, forests are evaluated by certifiers against global standards.[39] The MSC, however, has only global standards. In the beginning, the evaluation team appointed by certifiers developed 'an assessment tree' comprising detailed indicators and scoring guides for the assessment of specific fisheries according to global standards. This led to inconsistencies among certifications for different fisheries. Therefore, since 2008, a 'default assessment tree' was introduced by the MSC to apply the same methodology for assessment of different fisheries.[40]

Second, the governance structure is also different. FSC is a 'devolved, chamber-based, membership organization'.[41] It has three chambers: economic, environmental and social chambers, with equal voting rights. It is open to individuals and organizations and can determine FSC policy through General Assembly meetings or electing international and national board members.[42] The membership-based structure allows broad participation in FSC decision-making, but on the other hand, also leads to long procedures. The MSC has chosen deliberately a more efficient, but less participatory approach. It has adopted a streamlined governance structure without members. Its decision power has been placed with the Board of Trustees.[43] This structure, however, has been criticized for its lack of democratic accountability.[44]

Third, they also focus differently on standard setting. The FSC standards require sustainability in terms of environmental, economic and social issues. However, the MSC emphasizes only environmental issues. The protection of economic and social interests does not go beyond legal compliance.[45]

Another issue concerns the accountability mechanisms of the certification schemes. Public regulation bases its legitimacy and accountability on parliamentary democratic principles. However, 'how, to whom and against

37 See Chapter 4, section 4.4.2.3.2.
38 Gulbrandsen 2010, p. 55.
39 Ibid., p. 56.
40 www.msc.org/documents/scheme-documents/msc-scheme-requirements/methodologies/Fisheries_Assessment_Methodology.pdf/view.
41 Gale & Haward 2004, p. 15.
42 Ibid.
43 Ibid., p. 24.
44 Ibid., pp. 26–29.
45 Chapter 4, section 4.1.

what norms certification programmes are held accountable' causes more concern.[46] Accountability includes both external and internal accountability. External accountability requires the programme to respond to the demands of external parties, such as environmental groups, business and governments, which can be accommodated by the above-mentioned governance structure. Internal accountability needs to ensure compliance with certification standards, which can be achieved via assessment and objection procedures.[47] The above analysis shows that the FSC can better address external accountability via more participatory governance structures. As for internal accountability, both schemes have adopted specific objection procedures which allow for the challenging of certified forestry and fisheries. However, both procedures have been criticized in the literature. In addition to the high costs and long procedures which impede the initiation of formal objection procedures, such procedures rarely lead to the revocation of certification in practice.[48] The case studies of Sweden and New Zealand provide examples. Dissatisfaction with certification has led to the withdrawal of support of the major ENGOs in Sweden from FSC Sweden. The RFBPS in New Zealand objected to the first two certifications of the hoki fishery. It felt that its concerns were not sufficiently considered. Therefore, it did not participate in the third assessment, but initiated its own buyers' guide programme.[49]

5.4 The interaction of public and private regulation

The case studies show that public and private regulation often interact in governing forest and fishery problems. This part examines how such interaction plays out similarly or differently in the two examined domains. First, whether certification standards include requirements directly linked to property rights influences interaction in forest and fishery governance. Second, the different characteristics of the resources also influence the role of government in certification. The third section shows that although public and private regulation are complementary, the latter cannot be replaced by the former. The differences between the two regulatory approaches are explored in section 5.5.4.

5.4.1 Standards

As mentioned earlier, forest certification includes standards regarding environmental, economic and social issues. Clear forest tenure and respect for

46 Gulbrandsen & Auld 2016, p. 43.

47 Ibid., pp. 43–46.

48 E.g. until 2015, only two out of 31 objection procedures for MSC certification lead to the upholding or denying of certification, ibid., p. 55.

49 Weeber & Wallace 2008.

the rights of indigenous people are usually preconditions for a forest to obtain certification. Property rights are also regulated under domestic law. The case studies of Indonesia and Bolivia show that many public regulatory instruments, such as spatial planning, permits and CBFM promotion measures, can influence property rights. Here it will be analysed how public and private regulation interact on paper and in practice. The stringency and compatibility of certification standards and domestic law regarding property will be compared. In the case studies of North America and Sweden, the focus will be on the limitation of property rights. The analysis will be on how public and private regulation interact to address subsequently riparian zone protection and biodiversity protection issues. These environmental issues are regulated directly both under certification schemes and public regulation.

MSC standards, however, are deliberately narrowed down to environmental issues. Social issues like property rights are also reflected in terms of legal compliance requirements. Therefore, a direct comparison of the standards regarding property rights under certification and domestic law does not make much sense. The analysis of interaction hence focuses on the practical level: to what extent the interaction between public and private regulation has led to a strengthening or refining of property rights.

5.4.2 The role of the government

Although the government can issue general regulations regarding forest management, it are usually the private parties, such as forest companies, private owners or communities who conduct daily management and develop management plans. These private parties can apply for certification and improve their performance according to the requirements of certification. The government usually does not participate in the certification procedures themselves, but often plays an indirect role in certification, such as providing expertise and financial support in agenda setting. This can be done by providing resources to clients of certification and enacting public procurement.[50] Fish resources are migratory and are often exploited by many users in a broad geographical scope. Sometimes, local fishermen or their organizations conduct management themselves. However, it is not uncommon that the government decides how a fishery is to be managed and develops the management plan itself.[51] In this case, the government acts as the client of certification.[52] Even if private parties are the client, as in the case studies, the government can influence the whole certification process, from nominating the evaluation team, providing financial

50 Gulbrandsen 2014.

51 South Africa is such an example.

52 Gulbrandsen 2010, p. 125. E.g. Alaska Department of Fish and Game is the holder of the certificate for the Alaska salmon fisheries.

support, to providing information for assessment and conducting measures to satisfy the conditions required for certification.[53] These more direct roles of the government in certification are shown in the examples of certified fisheries in the three case studies.[54] Given the fact that the government usually does not participate directly in certification processes of forestry, its influence in this domain is best understood in general terms. On the contrary, the analysis of interaction is based on specific examples when it comes to fishery.

5.4.3 Complementarities between public and private regulation

In the studied countries, forestry and fishery are governed by both public and private regulation. The case studies show that the countries scoring high on good governance and public regulation have the most forests and fisheries certified.[55] Moreover, certified forests and fisheries in developing countries are usually already better regulated before certification than most other users in the same country. For example, the case study of Indonesia shows that the certified community-managed forests were already managed well due to prior public regulation.[56] In South Africa, the certified hake fishery was prioritized in public regulation and hence a higher level of public enforcement already existed in hake fishery before certification compared to other fisheries in that country. The pre-existence of good public regulation made it easier for the hake fishery to obtain certification.[57] However, since good public regulation is missing in developing countries, it is more difficult for the majority of forests and fisheries to obtain certification in these countries. However, the case studies also show that sometimes the efforts to obtain certification provide incentives to improve public regulation. The management of swimming crab in Mexico is such an example. Public regulation concerning swimming crab in most areas of the Gulf of California was lacking. These areas were managed under a limited entry system, where resource overexploitation was widespread. The Seri, however, effectively controlled the fisheries in the Infiernillo Channel through self-governance. The early attempts of the Seris' fishery to be pre-assessed for certification led to the first public management tool, the crab NOM. The increased demand for crab triggered the effort to certify the whole fishery in the Gulf of California in the late 2000s. A fishery improvement project

53 See Foley 2013, p. 301.

54 South Africa, Mexico and New Zealand.

55 For example, forest certification is more widespread in North America and Sweden than in Indonesia and Bolivia. Japan is an exception. Though it ranked high in terms of good governance, the MSC is not active there. An explanation for the absence of the MSC in Japan is provided in Chapter 7, section 4.7.3.3.

56 *Supra* Chapter 3, section 3.4.4.2.

57 *Supra* Chapter 4, section 4.6.2.

and a pre-assessment were conducted. These efforts have led to the development of many public management measures.[58]

When public regulation is lacking, certification is most likely to bring public regulation about when financial incentives are present, such as increased demand or a price premium. The example of the crab fishery in Mexico, just presented, showed that there was a desire to have certification, but public regulation was lacking. In that case the demand for certification pushed public regulation. In addition, when public regulation is already in place, certification to some extent accredits public regulation.[59] The evaluation and certification process help to expose insufficiencies in public regulation and the triggers to solve them. Sometimes certification leads to changes in legislation. For example, the certification of the Mexico Baja California Red Rock Lobster Fishery identified the insufficiency of the current regime in reducing ecosystem risks. This led to the changes in the lobster NOM and also the inclusion of ecological risk considerations in the General Law on Sustainable Fisheries and Aquaculture.[60] In Sweden, the certification criterion regarding retention of trees was also later incorporated into government regulation.[61]

5.4.4 Differences between public and private regulation

Although public and private regulation can be complementary, important differences exist. First, the case studies show that standards under public and private regulation are not always coherent. In North America and Bolivia, certification standards closely mimic public regulation. However, Indonesia provides an example of contradictory norms in public and private regulation regarding indigenous people's property rights. Public and private regulation of fisheries in Sweden also shows differences in standards.[62] As far as fishery governance is concerned, the MSC does not have specific requirements concerning compliance with domestically determined property rights.

Second, and more important, private and public regulation operate in different ways. This can again be illustrated by the case of Sweden: whereas public regulation applies at the national or regional level, certification applies to certified forests/fisheries and operates at the management unit level.[63]

58 *Supra* Chapter 4, section 4.5.5.2.

59 The existence of a good management system is a criterion in both forest and fishery certification. Public regulations are a crucial part of the management system, especially in fishery management. Therefore the evaluation process inevitably also involves the assessment of public regulation.

60 *Supra* Chapter 4, section 4.5.5.1.2.

61 *Supra* Chapter 3, section 3.7.4.1.

62 *Supra* Chapter 3, section 3.7.4.1.

63 *Supra* Chapter 3, section 3.7.4.2.

Third, both public and private regulations are subject to their own limitations. Public regulation requires enforcement capacity. This is a problem for many developing countries, which are subject to staff, financial and expertise scarcity and vulnerable to corruption and elite capture.[64] Even in developed countries with stronger governance capacity, limits exist. For example, reduced budgets undermined the usage of suasive instruments in Sweden.[65] As far as certification is concerned, insufficiency of accountability mechanisms has been noted. The cases of Sweden and New Zealand show that certifiers are reluctant to revoke certifications when challenged by third parties.[66]

5.5 Preconditions for a proper functioning of property rights

In Chapter 2 we have indicated that the literature pointed to the fact that the establishment and maintenance of property rights create various transaction costs. After identifying those particular costs (related to the clear definition of property rights, enforcement, coordination, information creation and sharing as well as scaling) we argued that those costs in fact are preconditions for a proper functioning of the various types of property rights. We will now verify in what respects public and private regulation affect those preconditions for a proper functioning of property rights and are able to overcome the limitations of property rights in the cases of forestry and fisheries.

5.5.1 Definition of property rights

This book examined three types of property rights over common-pool resources: public property rights, private property rights and communal property rights. For forests, these are usually formulated as ownership over forests/the land or less complete forest tenure. Three types of fishing rights were discussed: the limited entry system, individual quotas and territorial user rights. These fishing rights often have combined characteristics of public, private or communal property rights. Some literature argues that either public property rights or private property rights are the only solution to the common-pool resources problem.[67] The case studies we conducted do not support these arguments. For example, private forests are more popular in America while public property rights dominate in Canada.[68] However, the case studies show that both countries perform relatively well in spite of the different types of property rights adopted. For fishery

64 See the case studies of South Africa, Mexico, Indonesia and Bolivia.

65 *Supra* Chapter 3, section 3.7.3.2.

66 *Supra* Chapter 3, section 3.7.4.2 ; Chapter 4, section 4.6.5.1.

67 Ophuls 1973, p. 228; Demsetz 1967; Johnson 1972.

68 *Supra* Chapter 3, section 3.6.3.1.

governance, many have recommended individual quotas, especially individual transferable quotas systems.[69] However, the case studies show that territorial user rights (Mexico and Japan) can also induce good performance.[70] Moreover, conversely, the individual quotas in South Africa have not been able to guarantee sustainable management.[71]

The case studies also show that shifts of communal property rights to public or private property rights have been based on unsubstantiated confidence in the effectiveness of the private or public property rights. Nationalization may transform 'previously limited access common-pool resources into open access resources' when the government does not have the capacity to enforce its public property rights over vast resources.[72] Examples are constituted by the cases of Indonesia and Bolivia. 'Likewise, the substitution of common property for private ownership has been environmentally and socially deleterious, where privatization has meant the loss of access to critical resources by local populations and private entrepreneurs are motivated solely by short-term profit, discounting the future value of the resource'.[73] For example, in Indonesia communal property rights used to be the dominant approach to regulating forests, based on customary law. However, since colonization customary law and communal property rights have been replaced by state ownership of forests.[74] Traditional control over forests has hence been dismantled. The government, however, has no capacity to control the vast forests in Indonesia, leading to rampant illegal logging.[75] The 1917 Constitution of Mexico established state ownership over natural resources and enshrined state intervention in allocating and safeguarding resources.[76] The state-led development policies encouraged outside encroachment and resource poaching. Especially in the 1970s and 1980s, the export-oriented fisheries policies led to large-scale immigration to the fishery areas. Responding to low productivity and overexploitation, the 1980s and 1990s witnessed a shift towards market-oriented development and privatization. It eliminated the cooperatives' exclusive access to many species, leading to a shift from territorial user rights (concessions of cooperatives) to a limited entry system (permits). This arrangement further aggravated destructive resource use.[77]

69 Chu 2009; Griffith 2008; Symes & Crean 1995.
70 See the example of Baja California Rock Lobster Fishery in Mexico and the coastal fisheries in Japan.
71 South African fisheries are subject to serious illegal fishing and overexploitation. The hake fishery is an exception with prioritized public enforcement.
72 Young 2001, p. 286.
73 Ibid.
74 *Supra* Chapter 3, section 3.4.3.1.
75 *Supra* Chapter 3, section 3.4.5.1.
76 Mexican Constitution Law, Article 27. See also Young 2001, p. 286.
77 Young 2001.

Rather than the type of property rights, it seems that the clear definition of property rights and their content are more important. A comparatively complete property right is crucial to incentivize sustainable behaviour. This is in line with the theoretical literature, more particularly Schlager and Ostrom who argue that at the least, communities need to have rights to access, harvest and manage resources to induce effective communal management.[78] The CBFM policies in Indonesia constitute an example. Although aiming at promoting community management of forests, these policies often provided limited rights to the communities, failing to incentivize responsible behaviour.[79]

In addition to the types of rights awarded, the institutional arrangement is also crucial. For example, in Bolivia, the rights of indigenous people over forests are recognized in law. However, the law also required the establishment of a new institution, the TCO, on top of traditional self-governing institutions (usually the villages) as a precondition of confirming the communities' property rights. Geographically separated and culturally different villages have been joined by the TCO and thus have weakened the basis of self-governance. The co-existence of different levels of governance has also weakened the capacity to conduct effective control in Indonesia and Bolivia. Therefore, unsuitable institutions can impede the definition of clear property rights.

The case studies have also shown that the match between property rights and the cultural and economic characteristics of the user groups is important. For example, replacing communal property rights with private property rights for indigenous people or local communities is problematic because they are often exploiting natural resources for subsistence use and are not used to large-scale operations, intensive use and commercialized activities. Private forests or ITQ system fisheries are usually regulated by instruments most suitable for large-scale intensive use and for commercial entrepreneurs. Small-scale fisheries that have been defined as a 'limited commercial sector' have found it difficult to comply with the same property rights rules and regulations that apply to commercial fishers.[80] Another example is that communities in Indonesia and Bolivia have hardly succeeded in obtaining certificates in forest management. The case of New Zealand provides an example where communities have successfully adapted to a private property regime. The Maori have two types of fishing rights: the ITQ as commercial fishing rights and their customary rights. The ITQ is market-based and most suitable for capitalist market economies. The Maori have been well integrated into the market economy and have accommodated the ITQs.

78 Schlager & Ostrom 1999, p. 105.
79 *Supra* Chapter 3, section 3.4.4.2.
80 *Supra* Chapter 4, section 4.3.2.

They possess more than one third of the quotas and organize their own companies to manage these quotas.[81]

The establishment of property rights includes some parties and excludes others. How to address the excluded parties, especially those who have historically exploited the resources, is crucial to ensure that de jure property rights are respected. If the historical users are excluded without proper compensation and alternative employment, their exclusion leaves them little choice but to continue exploiting the resources illegally, as shown in the cases of Indonesia, Bolivia and South Africa. The establishment of ITQs in New Zealand has also led many fishermen to lose quotas. However, the existence of alternative employment and allocation based on historical production together with government buy backs has provided fishermen with opportunities to leave the industry.[82]

Both public and private regulation can influence the definition of property rights. The influence of public regulation can be found in various ways. For example, CBFM policies define the scope of rights granted to communities; the institutional arrangements for communal property rights (establishing new ones like TCO in Bolivia, or restructuring local communities according to the Java model in Indonesia)[83] depend on public regulation. Permits/concessions are crucial for establishing property rights. The limited entry system and ITQ system are also products of public regulation, such as restricting the number of vessels/permits and TAC and allocating them to individuals. Forest certification also has standards regarding property rights. To get certified, clear forest tenure needs to be established and the rights of indigenous people need to be respected. This may act as a barrier for certification in many developing countries where such conditions are not satisfied, such as Bolivia and Indonesia. However, this also suggests that in the certified forests, property rights arrangements may be fine-tuned for the forests to obtain certification. The influence of the MSC on property rights is more indirect given the limited direct relevance of property rights in the standards. Case studies, however, do show that certification has the potential to refine the content of property rights, such as further dividing the quotas for hoki into different stocks in New Zealand.[84] Certification also has the potential to consolidate the rights of certified parties and to marginalize their competitors, as shown in the case of South Africa.[85]

81 *Supra* Chapter 4, section 4.6.3.2.1.
82 *Supra* Chapter 4, section 4.6.3.1.
83 *Supra* Chapter 3, section 3.4.3.1 and section 3.5.3.1.
84 *Supra* Chapter 4, section 4.6.5.2.
85 *Supra* Chapter 4, section 4.5.2.

5.5.2 Enforcement

The safeguarding of property rights can be monitored and assured by public and private regulation. As far as communal property rights are concerned, self-governing institutions can also oversee the compliance with property rights.

The good governance indicators provide an indication of the enforcement capacity of public regulators. This capacity is related to the availability of staff, budget, expertise, vulnerability to corruption, the authority provided to a public agency and the relationship between different regulatory bodies. For example, the regulatory fishery agencies in New Zealand have far-reaching authority in enforcement, such as random entry, search and questioning. These far-reaching powers and the cooperation between agencies contribute to strong enforcement capacity.[86] Mexico provides an opposite example. The main competent authority CONAPESCA has little enforcement power. The lack of coordination between CONAPESCA and other agencies has even created an enforcement gap for several years.[87] Indonesia provides another example showing that conflicts between different departments and levels of government can lead to conflicting enforcement activities.[88] In addition, the levels of enforcement may be unequally distributed across a country. For example, in South Africa, generally speaking, many fisheries have been subject to illegal fishing and overexploitation. However, since the enforcement of the economically most valuable fishery in this country, hake has been prioritized, this fishery is better protected than the other fisheries. This means that the unequal allocation of enforcement capacity may lead to a sharp contrast between forests/fisheries.

Certification can provide additional monitoring and assurance to public enforcement capacity. The cases of certified fisheries show that insufficiencies of public enforcement are sometimes identified in the certification process and that the repair of these insufficiencies is a requirement for obtaining certification.[89] In America, anecdotal evidence shows that certification has contributed to the improvement of compliance with public regulation regarding riparian buffer-zone protection.[90] However, certification does not automatically improve monitoring and oversight. Certifiers may be captured by their customers in order to secure income. The case studies of Sweden and New Zealand have shown that objections to certification seldom have serious consequences.[91]

86 *Supra* Chapter 4, section 4.6.6.2.
87 *Supra* Chapter 4, section 4.6.2.
88 *Supra* Chapter 3, section 3.4.4.2.
89 See the example of South Africa, Mexico and New Zealand.
90 *Supra* Chapter 3, section 3.6.4.2.
91 *Supra* Chapter 3, section 3.7.4; Chapter 4, section 4.6.5.1.

Public regulation, self-governing institutions and private regulation rely on different ways of monitoring: self-governing institutions rely on local users who conduct daily management activities and have easiest access to the information regarding the resources and exploitation activities. Public regulators rely on the monitoring of government officials and the certification schemes require the expertise of forest/fishery professionals. The latter actors may have advantages in terms of economies of scale or expertise, but are further away from the daily activities in forests/fisheries and monitoring can therefore be more costly. It is thus crucial to figure out whether these different approaches of monitoring complement or replace/weaken each other. Indonesia provides an example where traditional self-governing institutions have been dismantled by law and self-enforcement is replaced by weak public enforcement.[92] In Bolivia, the traditional governing institutions, the villages have not been abolished but an additional level of governing institution has been added on top of the villages, de facto weakening their functioning.[93] Certification may also require the replacement of traditional governing institutions by new governing ones such as cooperatives or enterprises.[94] There are also examples where public regulators recognize traditional property rights and their governing institutions. Such recognition increases the self-governing capacity of local communities. The lobster and crab fisheries in Mexico and the development of fishing rights in Japan are examples where the recognition by public authorities has strengthened the capacity to exert social control by communities. In North America, the coordination between public and private regulators has helped to reduce enforcement costs. In New Zealand, public regulatory bodies and fishermen's organizations have collaborated to strengthen monitoring and assurance.

5.5.3 Coordination

Many different actors are involved in exploiting forests or fish stocks. Coordination between the various parties, both among rights holders or between forest/fishing rights holders and other parties is crucial for the proper functioning of property rights.

The different types of property rights vary in their capacity to coordinate the behaviour of the different actors involved. Under public property rights, the state or other level of government is the rights holder, who also decides the allocation of user rights and acts as coordinator. As far as communal property rights are concerned, self-governing institutions play an important role in coordinating the behaviour of its members. These

92 *Supra* Chapter 3, section 3.4.5.3.
93 *Supra* Chapter 3, section 3.5.5.3.
94 See the examples of Indonesia and Bolivia.

self-governing institutions decide who has rights to access and to harvest the resources and how. Private actors are rights holders under private property rights. Coordination problems arise especially when multiple parties are allowed to exploit the same resources, as in the fisheries cases. The individual quotas system grants the right to harvest specific species up to a certain volume. This means individual quotas diminish the incentives to move first to catch fish. However, multiple actors may exploit the same stocks in the same or adjacent areas. Moreover, individual parties may not have incentives to coordinate their actions by sharing information or adopting more efficient and environmentally friendly methods.[95] In New Zealand, a co-management system has been implemented, encouraging the establishment of quota owners' organizations. In this system, quotas are still granted to individuals but the organizations play an important role in pooling resources and coordinating the behaviour of individual quota owners.[96] In this particular case, a hybrid system has evolved, combining private and communal rights.

Both public and private regulation can influence the coordination of forest and fishery management. Spatial planning is an instrument to demarcate the areas used for forestry and other forest-related activities, such as mining, plantation and agriculture. Marine spatial planning is also promoted to coordinate fishing activities with other activities using marine areas, though its usage in the examined cases is more limited. New Zealand provides an example with little coordination between the commercial, communal and recreational sectors. The fishery planning only concerns coordination among different fishery sectors, but not coordination between fishery sectors and other sectors which have an interest in using marine areas.[97] Sometimes, specific governmental institutions are established to promote coordination. Japan provides such an example, where coordination committees have been established at prefectural, cross-prefectural and national levels.[98] The case study shows that prefectural coordination committees have worked together with fishermen's organizations in developing management plans. However, cross-prefectural coordination has rarely occurred in practice.[99] Coordination among governmental agencies can also be problematic. Departmentalism and a contested decentralization process have led to conflicting decisions and regulatory gaps among differential sectors or different levels of governments, as shown in the case studies of Indonesia, Bolivia and Mexico.

Certification can also provide a platform for multiple stakeholders to collaborate. The standards of development and governance of the

95 *Supra* Chapter 4, section 4.2.2.2.
96 *Supra* Chapter 4, section 4.6.3.
97 *Supra* Chapter 4, section 4.6.6.3.
98 *Supra* Chapter 4, section 4.7.3.2.
99 *Supra* Chapter 4, section 4.7.4.3.

certification schemes usually involve industry, rights holders, social groups and environmental interests. As discussed earlier, the governance structure of forest certification is more inclusive than in fishery certification, resulting in more public involvement in the former.[100] In the individual certification process, stakeholders are identified and their opinions are considered in the evaluation.[101] However, sometimes public involvement is criticized as insufficient.[102] The insufficient involvement of environmental groups has been a cause of concern. As shown in the cases of Sweden and New Zealand, dissatisfaction with certification has led some environmental groups to withdraw their support from certification schemes. This misalignment is problematic since environmental groups are usually an important impetus for the increasing demand for certified products.[103]

The co-existence of different regulatory schemes may also add complexity to the governance of forests and fisheries and a cause of confusion among regulatees. For example, multiple institutions have been created to manage community forests in Indonesia and Bolivia: traditional self-governing institutions, institutions reorganized or created by the government (e.g. the reformed villages in Indonesia and the TCO in Bolivia) as well as cooperatives/enterprises created to obtain certification. These complexities have added to coordination costs. The example of Sweden also shows that the multiple layers of regulation often lead to confusion among forest owners in deciding their forest management activities. The community-managed fisheries in Mexico (lobster) and Japan provide examples of better coordination. In the lobster fishery in Mexico, the cooperatives are rights holders, which are organized under the federation of FEDECOOP. The FEDECOOP applied for certification. Therefore, the same institutions conduct self-management and operate under public and private regulation. In Japan, the autonomous management bodies, FCAs, originated in the Edo era. These rights holders are acknowledged by law and conduct self-management in practice. Path-dependent institutional arrangements maintain consistency in decision-making and align decision-making with local culture and customs.

5.5.4 Information

The availability of information regarding resources status, forest/fishing activities and their interaction is crucial for resources management. Local users, government institutions and certification assessment can all act as information generators. These different sources of information all have their advantages and disadvantages. Local actors are well informed about

100 *Supra* Chapter 5, section 5.3.2.
101 As seen in the case study of fishery governance.
102 E.g. the hake fishery in South Africa.
103 Gulbrandsen & Auld 2016.

local issues, such as the status of the resources, local livelihood demands and behaviours. The indigenous exploitation of forests in Indonesia is such an example. The use of fallow periods, fire and cleared land for agriculture are based on local knowledge and respect ecological limits. However, the replacement of customs by commercial exploitation and agriculture by outsiders, has had devastating effects on the tropical forests.[104]

Compared to local knowledge, 'the state has a regional and national vantage point and a repertoire of tools and techniques not normally available to local institutions'.[105] However, to obtain information about vast forest or marine areas remains a challenge, especially when the capacity of regulatory agencies is weak, such as in Indonesia, Bolivia, South Africa and Mexico. Even in countries with comparatively strong governmental capacity, information insufficiency is not uncommon. New Zealand is such an example. This country became a marine superpower overnight because of the declaration of the EEZ. Many measures have been adopted to overcome information problems. For example, New Zealand has imposed high reporting requirements on quota holders, vessel holders and fish receivers. Cross-checks ensure the accuracy of information. Some new technologies, such as VMS, are also widely used in Japan, New Zealand and some important fisheries in South Africa to make available real-time information. In Sweden, many informational instruments have been used to guide the behaviour of forest owners.

Certification can also play an important role in information generation and spreading. Certification schemes have rules concerning the documentation of processes related to forest and fishery management. Information is made available during the certification process. Assessment teams gather information from existing documentation and literature produced by clients, scientists, government authorities and other stakeholders. The publication of final reports and of surveillance reports also increases the transparency of information. Assessment reports also identify gaps in knowledge and issue conditions or recommendations accordingly, which provide drivers for information production.[106]

For resources management it is beneficial to build a robust bridge between those complementary types of knowledge. For example, in North America public and private regulators coordinate in sharing of information and rely on the information provided by each other.[107] In South Africa, the government prioritizes its activities in the hake fishery. The adoption of VMS provides real-time information. The fishing industry's organization SADSTIA also has its own scientific observer programme generating information. In New Zealand and Japan, co-management is

104 *Supra* Chapter 3, section 3.4.4.2.
105 Berkes 2009, p. 1694.
106 Cano Chacón 2013.
107 *Supra* Chapter 3, section 3.6.4.2.

promoted to take advantage of the information of both governmental and local users.[108]

5.5.5 Scale

Forest and fishery management requires considerations at different scales. Issues such as the protection of sedentary fish species and the retention of trees can be addressed at a local scale. But issues like biodiversity protection and riparian forest protection require landscape considerations. Regulatory approaches vary in their capacity to address different scales. Certification is conducted at management unit level and is therefore particularly tailored to local situations. Coordination between different units is needed in order to incorporate higher-level considerations. Public regulation can address issues at both local level (such as the management of fisheries in South Africa) or at larger scale (e.g. fishing plans in New Zealand). However, the case study of North America and Sweden shows that even public regulation is often insufficient in incorporating landscape considerations.

Decentralization is often recommended to regulate natural resources. It is believed to enable decision-making close to local users and hence to contribute to resource management. However, the case studies show that when the system is captured and corrupt and lack of coordination among government agencies and different levels of government is aggravated by obscure and conflicting legislation, the decentralization process itself can be a driver of environmental problems. For example, in Indonesia, conflicting legislation concerning decentralization and recentralization reflects a tug of war between different levels of government. Multiple agencies often conduct conflicting spatial planning and issue overlapping permits over the same area. The self-interested motives of bureaucrats behind these conflicting behaviours have not been overcome yet. In Bolivia, conflicting interests among different levels of government and different departments also exist, such as the reluctance of municipalities to support TCOs' application and the nesting of the land regularization process under the agricultural model. Such conflicts are also reflected in the legislation. One example is the requirement of an FES in titling private land and TCOs are more favourable to agriculture than to forestry use.

5.5.6 Externalities

An important goal of property rights in addressing common-pool resources problems is to internalize externalities and to limit the incentives of free-riding. Property rights, however, have their own limitations. Establishing

108 *Supra* Chapter 4, section 4.6.6.4; section 4.7.4.4.

property rights over forest and fish resources can help to internalize the economic values of the resources, and hence contribute to solving part of the environmental problems, such as deforestation and the overexploitation of targeted fish species. However, property rights are less useful in internalizing the non-market values of resources. For example, the supporting value of forests for other species, and their functions in soil and water conservation will not be internalized by establishing property rights. Indeed, as the case studies of North America and Sweden have shown, forest degradation may still occur even when clear forest tenure has been established. This has been one of the reasons for creating riparian zone protection by regulation. The fishery chapter also shows the limitations of three types of fishing rights. The limited entry system does not eradicate the incentives to move first in catching fish. Under the individual quotas system, fishermen need to coordinate their activities for resource heterogeneity to be taken into account, which is not always easily achievable. Moreover, property rights holders have not sufficiently considered by-catch and the protection of habitats and ecosystems. Even the TURs leave non-targeted species and ecosystem protection as externalities. As was argued in Chapter 2, property rights are crucial in overcoming common-pool resources problems. However, they do not prevent rights holders from using property rights in ways that can harm third parties. It is for this reason that public and private regulation do not only have to intervene in the establishment and functioning of the property rights, but also in guaranteeing that the use of the property rights will not result in externalities.

5.6 General trends

In the next chapter conclusions will be drawn about the comparative case studies. This chapter ends by pointing out general trends that can be derived from the case studies. In many of the examined countries, shifts could be observed towards decentralization to local authorities and privatization, whereby increasingly control over natural resources was assigned to private actors or self-governing communities.[109]

These phenomena also have had an important influence on establishing property rights. Although historically many natural resources were exploited by local communities, as a result of the influence of colonization and decolonization, state ownership and control over natural resources became common in many countries. An example is constituted by the case of Indonesia. Forests used to be managed by communities. During the colonial period the colonial government claimed ownership over resources and controlled land in Java. The later post-colonial Indonesian government

109 For an overview of those developments see Saito 2008, p. 5 and see Webb & Shivakoti 2008; De Vries 2000.

retained this state ownership. To some extent similar developments took place in other countries examined in the case studies, such as Bolivia and New Zealand. Indonesia and Bolivia started to recognize private forest tenure in the 1990s. Though purely private property rights over fish resources are difficult to establish, many countries also conferred more control over fish resources on private parties, for example by expanding fishing permits to individuals/companies and by supporting individual quotas systems in New Zealand and South Africa. In addition, the long-ignored communal property rights have also started to regain importance. The rights of local communities, especially indigenous people, over natural resources have started to become recognized in Indonesia, Bolivia and New Zealand, at least on paper.

Also in resources management, decentralization and privatization processes can be observed to varying degrees. The US and Canada are federal countries, where a division of authority over forest regulation exists between central and local (state/provincial) governments. In Sweden decentralization resulted from the new Forestry Act which was implemented in the 1990s. This Act consists of a framework law adopting only general and minimum standards. Under the guidance of the principle 'freedom with responsibility', the industry has been provided with discretion to choose the measures to achieve legal goals. The coercive level of command and control instruments has not been high and informational instruments have widely been used to guide the behaviour of the industry/forest owners. In New Zealand, the government has also tried to transfer some fishery regulatory power to the industry, such as quota registry and fishery research. In Japan, fishermen's organizations have always played an important role in fishery management. In developing countries, decentralization became popular in the 1980s. In some countries, like Mexico, decentralization occurs 'at a very cautious rate and lacks a cohesive long-term plan'.[110] Some developing countries have gone through more comprehensive decentralization, in terms of both finance and administration, such as Bolivia and Indonesia.[111]

In some developed countries decentralization has improved the management of common-pool resources problems. This success depended on several conditions, such as societal pressure, information and accounting systems, parallel decentralization in administrative functions and finance as well as technical and administrative capacity.[112] These conditions, however, are not always satisfied in developing countries. For in Bolivia and Mexico, local government usually has limited capacity (in terms of staff and budget) in forest/fishery regulation. Conflicts exist between different

110 OECD 2011a, p. 54.

111 *Supra* Chapter 3, section 3.4.3.2; section 3.5.3.2.

112 Ibid.

levels of agencies in decision-making.[113] Indonesia provides an example of lack of accountability, capture and the absence of legal clarity leading to conflicting decisions among different levels of government. These conflicting decisions lead to unclear and overlapping boundaries between properties.

The delegation process shifts more responsibility for resources management and protection from public authorities to resource users themselves. In other words, rights holders (either private actors or local communities) have more discretion in executing their own property rights. However, to what extent delegation leads to clearer and more secure property rights, and to the promotion of sustainable resources management activities, depends on the incentives, information and capacity of rights holders. For example, in Sweden, delegation has led to the active involvement of private forest owners in forest management and protection activities. However, it has also led to the reduction of the coercive level of command and control instruments, leading to a consistent fairly high level of non-compliance. Decreased budgets of the Swedish Forest Agency have also restricted the use of informational instruments. Although some informational instruments have positively affected forestry management, they have failed to change the underlying values and preferences of forest owners. In New Zealand, the fishing industry holds permanent quotas over specific fish resources. With strong property rights (commercial fishing rights) and the partial delegation of responsibilities by the government, fishermen have formed associations, which have imposed management rules on their members. However, conflicts still exist among commercial, recreational and customary fishing rights holders. Environmental groups and non-commercial fishers have opposed the further devolution of responsibilities out of fear that this would further strengthen the position of the commercial sector and marginalize other sectors.

The decentralization processes have allowed more involvement of private actors in environmental governance, thereby providing opportunities for private regulation to develop. Forest and fishery certifications have spread widely across the world. Although property rights are mainly a domestic law issue, certification schemes (which are often transboundary) can have influence on property rights as well. Forest certification requires clear property rights as a precondition for certification. The MSC influences property rights indirectly. It can empower the certified parties, as shown by the examples of South Africa and the Baja California Rock Lobster fishery. Another side of the same coin is that certification can also be used strategically by clients to marginalize competitors who exploit the same resources by denying them market access. In addition, the content of

113 The different attitudes of municipalities and central agencies in supporting TCO application are such an example. See Chapter 3, section 3.5.4.2.1.

property rights can also be fine-tuned under the influence of certification. For example, in the New Zealand case the evaluation of certification has promoted the separation of quotas for different hoki stocks. Hence, certification does have an important influence on the way in which property rights are managed at the domestic level as well.

In sum, the case studies show that important shifts took place in many countries towards the delegation and privatization of managing common-pool resources. Delegation and privatization are no guarantee to the successful management of common-pool resources problems. In some cases, mostly developed countries, these shifts have brought the places of decision-making closer to local resource users, thereby increasing local participation and reducing decision-making and enforcement costs and tailoring policies to local circumstances. In other cases, mostly in developing countries, they have (also) led to regulatory capture, legal uncertainty for property rights holders and conflicting decisions among levels of government.

6 Concluding remarks

6.1 Starting points

We started our analysis of smart mixes in governing common-pool resources in the introduction by stating that the two domains which were central to our analysis, forestry and fishery, are classic examples of Hardin's tragedy of the commons. The starting point is that in the absence of any type of legal framework there may be incentives for a 'race for trees or fish' since no individual would receive any benefit from taking protective measures. That starting point has been the basis for advocating the allocation of property rights (either public, private or communal) to remedy the tragedy of the commons scenario. In order for those property rights to function effectively a variety of public and private regulations has been created which in turn also interact with the way in which the specific property rights are able to protect the specific resources.

That (some type of public or private) regulation is necessary to establish those property rights and to let them function in their protection of the common-pool resources has been well established. However, less is known of the way in which public and private regulation interact to address common-pool resources problems in relation to the establishment and functioning of property rights.

6.2 Goal

Precisely because there are few studies that have addressed specifically how, in particular legal systems public and private regulation do interact in their protection of common-pool resources, the goal of our study was to examine how this interaction is shaped in specific legal systems. The crucial question for our study was whether this interaction between public and private regulation can be qualified as a 'smart' mix in protecting this common-pool resource. To be clear: with our search for 'smart' interactions between public and private regulation we never implied that there would only be one optimal model of interaction which would constitute necessarily the ideal smart mix. The goal of this study has not been to propose just one specific mix of regulatory approaches which would fit all different types of scenarios. In fact, the case studies we undertook in this research show a wide diversity of governance structures for common-pool resources. It is precisely this wide variety, combined with the differences in the characteristics of resources in the national contexts, which have often shaped the particular institutions and the mixes of instruments in particular countries. Whereas the goal of our study was therefore not to determine one ideal smart mix (as this can be highly context-dependent and various combinations could under specific circumstances work well in a particular context and therefore be smart), we did notice as a result of the case studies that interactions between the different instruments do matter and that more particularly the relationship between public and private regulation may affect the way in which the property rights are able to provide an adequate protection to the common-pool resources. Whereas defining one single smart mix is therefore neither possible nor a goal to be achieved, the case studies have indicated that it is possible to show that some interactions may, under specific conditions, work better than others.

The goal of our study was to analyse this relationship, more particularly the interaction between public and private regulation and the influence of that interaction on the establishment and functioning of property rights, both theoretically as well as at a practical level. Based on the literature that addressed the way in which property rights, public and private regulation can protect common-pool resources, Chapter 2 started by providing a theoretical framework which guided the further research. Although commentators have indicated under which particular conditions property rights, public and private regulation can be expected to protect common-pool resources, the more challenging question is obviously how this interaction between public and private regulation and their influence on the functioning of property rights takes place in reality. In order to analyse the theoretical framework presented in Chapter 2, the two chapters which followed included case studies on forestry (Chapter 3) and case studies on fishery (Chapter 4). The obvious reason for limiting the analysis to these two domains was that on the one hand it kept the project manageable (addressing all potential common-pool resources problems that could emerge would of course be impossible), but on the other it focused on two important, yet different, typical common-pool resources problems. Given the differing natures of forests and fisheries it is equally interesting to examine how the structure of property rights and the interaction between public and private regulation in relation to the functioning of the property rights differ between those two areas.

In addition to providing a general sketch of how international law, domestic law and private regulation generally deal with forest governance on the one hand (Chapter 3) and fishery on the other hand (Chapter 4). Detailed case studies were provided focusing on the institutional regimes in specific legal systems. It is only through this more detailed analysis of the nature of the property rights, but more particularly the interaction between public and private regulation and the relationship to the functioning of the property rights that a better understanding can be obtained on how the different institutional regimes in specific countries function and affect the protection of the specific common-pool resources.

As explained in the specific chapters we purposely selected different countries for the cases of forest governance and fishery, not only because some countries may be more relevant for forest governance (like Sweden) whereas others may be more interesting for fishery (like New Zealand), but also because it allowed us to obtain a broad, yet relatively specific and detailed overview of the institutional regime (both public and private regulation as well as of the nature of the property rights) in nine legal systems. We deliberately selected in each case countries from the northern hemisphere, the developed world and countries from the southern hemisphere, representing the developing world.

Whereas the specific chapters on forest governance and fisheries provided comparative conclusions in which the governance structures in the legal regimes examined were compared, Chapter 5 also provided a comparative chapter in which the results of the case studies (presented in Chapters 3 and 4) were critically reviewed, more particularly in the light of the theoretical framework presented in Chapter 2.

6.3 Theoretical framework

The background for the analysis in the case studies was the theoretical framework that we presented in Chapter 2. We basically introduced the various instruments that may affect the effectiveness of various institutions in addressing the tragedy of the commons. We explained that the tragedy of the commons can, according to commentators, be remedied via different types of property rights, the most important distinction being made between public, private or communal property. However, in order to create those property rights the literature has indicated that specific preconditions (sometimes referred to as transaction costs) affect the way in which private and public regulation may enable property rights to protect the common-pool resources. We used those preconditions, identified in Chapter 2 (more particularly the definition of the property rights, enforcement, coordination, information and scale) to examine how the different institutional arrangements affect the property rights function. In that respect we stressed that property rights are of course de jure created through public regulation, but in many legal systems one can now increasingly notice that the functioning of the property rights (more particularly in their protection of common-pool resources) is no longer only affected by public regulation, but increasingly by private regulation (such as certification schemes) as well. We, however, hypothesized that public and private regulation do not function independently of each other, but interact and that, moreover, this interaction between public and private regulation will affect the functioning of the various property rights in the way they protect the common-pool resources.

Those were the starting points which constituted the basis for the subsequent case studies presented in Chapters 3 and 4.

One issue guiding our analysis was, as our title suggests, that we wished to analyse to what extent specific interactions could be considered as 'smart mixes'. Although we already indicated that there is not one single smart mix, but that this study should rather be considered as a search for (various possible) smart mixes, the preconditions for a proper functioning of the property rights were used as criteria to judge the 'smartness' of the interaction between public and private regulation in the way that they affected the functioning of the specific property rights.

In order to answer the question whether one particular instrument mix can be considered smarter than another, we have used the five preconditions for the proper functioning of property rights as benchmarks. Broadly this means that we have examined to what extent a specific mix of legal and policy instruments (in our specific case more particularly the interaction between public and private regulation in their relationship to the functioning of property rights) would be able to increase the protection of particular common-pool resources. It is more particularly these preconditions (definition, enforcement, coordination, information and scale) that allow to indicate to what extent a particular interaction between public and private regulation is better or less able to provide an adequate protection of the specific common-pool resource. Of course we have on purpose refrained from referring to this analysis explicitly as an effectiveness analysis. The reason for that caution is that within the scope of our study it was not possible (and also not the goal of our study) to analyse for example whether a particular mix directly led to a specific outcome in terms of improving the quality of the protected common-pool resources. That would require research into causal relationships in which we have not engaged. That does not, however, mean that we have not, within the scope of the case studies, provided an assessment of the desirability of particular mixes. We have, however, done so on the basis of a review of literature and policy documents indicating for example whether as a result of changes in the interaction between public and private regulation in relation to the functioning of property rights the situation of a particular resource (like a specific fish stock) improved or not. Within that restricted framework it was therefore possible, within the scope of the case studies, to indicate under which particular conditions specific interactions between private and public regulation were better or less able to protect specific common-pool resources. It is therefore possible, on the basis of the study of this documentation, as we did in Chapters 3, 4 and 5 to provide some indications of which interactions may have worked better and which combinations or interactions could certainly not be qualified as 'smart mixes'. An important conclusion from those case studies is that the effectiveness of either public or private regulation in its protection of common-pool resources cannot be examined in isolation, precisely because public and private regulation interact and are therefore mutually dependent.

In addition we will now formulate a few more general conclusions based both on the theoretical framework, but mostly on the case study research.

6.4 Concluding observations

6.4.1 Relevance of the nature of the resource

A first, seemingly obvious but still relevant conclusion is that the nature of the resource to be protected will to an important extent affect the optimal design of the specific property rights. That became clear, as was also indicated in the comparative chapter[1] when regarding the difference between the way in which forests can be protected in comparison to the governance of fish stocks. The mere fact that forests and more particularly trees do not have the tendency to move shows that the type of property rights and also the regulations affecting the functioning of the property rights will be different from fisheries where mobility is obviously substantially larger. Note, moreover, that some rights, such as for example fishery rights, are not pure property rights in the sense that the fishers do not 'own' the fish in the ocean; they are, as was extensively discussed, user rights. That played an important role as was established for example in the way in which fishery territorial user rights were considered suitable in the case of the Baja California lobster and the Seris' rights over crab in Mexico, whereas the same territorial user rights, based on a specific geographical scope, would be less significant when regulating migratory species. It is the first important point to be taken into account, meaning that the institutional arrangement should obviously reflect the nature of the specific common-pool resource to be regulated.

6.4.2 The choice of property rights

Early literature often regarded public property rights or private property rights as 'the' solution for common-pool resources problems.[2] During the post WWII period, state ownership of natural resources was dominant in many countries. Recent decades have shown more support for a shift towards private property rights. The case studies show, however, that none of the three types of property rights has absolute advantages over the other.

1 See Chapter 5, section 5.2.1.

2 For a discussion of such literature, see Ostrom 1990, pp. 8–13.

Canada is dominated by public forest ownership while private ownership is more popular in the US and Sweden. In all these three countries, forests have been protected comparatively well. The lobster fishery in Mexico and the coastal fisheries in Japan provide examples of well-functioning communal property rights regimes. At the same time, the case studies of fisheries in South Africa and forestry in Bolivia and Indonesia showed suboptimal management of common-pool resources based subsequently on a private and public property rights regime. This means we have found examples of comparatively good and poor management of common-pool resources under all three types of property rights regimes. The case studies thus show that none of these types of property rights dominates the others in resource management.

At the same time, the absence of priority among the three types of property rights does not mean that all types of property rights can apply in all contexts. The choice of property rights needs to be based on the fit between them and the characteristics of the resources and the resource users. The blind shift from communal property rights towards public property rights (as in Indonesia and Bolivia) or privatization (e.g. Mexico) has had devastating effects. The policy of changing property rights and replacing the existing way of control needs to be made cautiously. The fishing rights in coastal fisheries in Japan, for instance, have their roots in the Edo era. The self-governing institutions are developed based on the earlier fishermen's associations. This does not mean that historical arrangements should always be kept and that property rights shifts cannot be successful. New Zealand provides an example where open-access regimes have been successfully replaced by individual transferable quota systems. The compensation for fishing-industry leavers and the exposure of Maori communities to commercial fishing have prepared them for this market-based instrument.[3]

What these shifts in property rights regimes also show is that the creation or existence of property rights as such does not necessarily guarantee the proper management of common-pool resources. The establishment of communal property rights in the Seris' fishery and the Baja California Rock Lobster Fishery provide examples. The reason for this was that under the previous property rights regime, there was a deteriorating environmental status and external pressure created by poaching. That precisely

3 These are problems often influencing the functioning of ITQ systems in other countries. The transferability leads to the concentration of quotas and the loss of fishing rights of some small-scale and less efficient fishermen. Without property compensation to help them leave and chose an alternative livelihood, the excluded parties may still engage in fishing activities, forming a large informal sector. Indigenous people usually have a long history of exploiting natural resources for subsistence use. But they are usually not prepared for market-based instruments. These problems have been solved in New Zealand, therefore promoting a successful shift of property rights.

provided incentives for local communities to organize, develop and implement self-governing rules. So our case studies suggest that neither the presence, nor the type of property rights is decisive for the management of common-pool resources problems. However, this does not imply that we draw the conclusion that property rights' role in resource management all depends on the circumstances. Rather, we suggest that the way in which these property rights function depends on the specific earlier mentioned preconditions (definition, enforcement, coordination, information and scale).

6.4.3 Definition of property rights

Our study suggests that certainty of tenure is important for the quality of the protection awarded to common-pool resources, irrespective of whether this is allocated via public, private or common property rights. It is not surprising that more particularly in the developing countries that were examined in the case studies it was shown that the certainty of tenure was often lacking as a result of which environmental degradation took place. The case of Indonesia is illustrative: formally the state has control over the natural resources, including the forests, but de facto there is considerable uncertainty concerning tenure, with devastating effects on forest governance. The case of fishing rights in Mexico also underscores this point: the shift from the permit system to exclusive fishing rights for the Seris worked well since the access rules developed by the Seris defined relatively clear property rights.

In addition to the establishment/recognition of property rights, the content of property rights is also important. Case studies show that many regulatory instruments influence the content of property rights. For example, spatial planning and logging/fishing permits help to define the boundaries of property rights. In other words, the plan determines who has rights to exploit which resources in which geographical areas under which conditions. The coherence of those instruments is crucial to demarcate property rights. Indonesia provides an example that conflicting permits lead to overlapping forest rights and so many land conflicts. The coherence between public and private regulation is also crucial in defining property rights. In Indonesia, public and private regulation include conflicting standards concerning the protection of communal forest rights (customary rights). The conflicts make certification standards concerning property rights difficult to satisfy, and become a barrier for forest certification. In addition, public and private regulation also include standards regarding fishery and forest management. The standards often tend to be more agreeable to the large-scale commercial forestry/fishing industry, but more difficult for small-scale loggers/fishermen to comply with. Such performance standards hence weaken the status of smaller actors and their property rights.

6.4.4 Enforcement

A point which clearly came out of the case studies is that one important aspect of the clear definition of property rights obviously also consists of the enforcement of those rights. It does not suffice to allocate particular property rights (such as, for example, exclusive fishing rights); the de jure property rights should also correspond with the de facto situation. If that is not the case a lack of enforcement will de facto lead to open access. This could, for example, be noticed in Mexico where a lack of public enforcement created incentives for illegal fishing, but also in South Africa where the loss of opportunities for small-scale fishers resulted in a large informal sector and again a de facto open access, unavoidably leading to a deterioration of the resource. Those examples also show that it may not suffice to remove property rights from existing groups without giving them the opportunity to secure alternative sources of income; that may unavoidably lead to open access and hence to resource depletion.

In enforcement, a positive interaction between public and private regulation in relation to the protection of the property rights of the common-pool resource can be identified. The fact that private certification agencies equally monitor compliance with public regulations can in principle lead to additional enforcement and also save on enforcement costs. There is evidence in some of the case studies, for example concerning forest governance in the US, that private certification institutions step into monitor compliance, especially in cases where the public monitoring system fails. Also in Canada there is evidence of synergies in the monitoring and enforcement of compliance, thus leading to a reduction in the duplication of monitoring efforts. It is precisely that, given the limited resources to monitor public authorities, private certification is held as an instrument to complement public enforcement.

Notwithstanding these positive interactions (smart mixes), as far as enforcement is concerned, it should also be recalled that private certification can, also as far as enforcement is concerned, merely complement public enforcement, but not replace it; even in the instance where private certification institutions discover non-compliance with the standards, a revocation or suspension of the certification will rarely be applied.[4]

This leads to the conclusion that in the domain of enforcement, private certification may play an important complementary role in addition to public enforcement, but it will not be able to substitute it.

6.4.5 Coordination

Since many different actors are involved in exploiting the forests or fish

4 Leading for example in Sweden to the withdrawal of an NGO from the private certification scheme.

resources, coordination between the various parties is important for resource management. Particularly when multiple parties are allowed to exploit the same resources it can be important to install incentives that help overcome the reluctance of resource users to coordinate their actions by sharing information or adopting more efficient and environmental friendly methods. In New Zealand, for example, a co-management system has been implemented wherein quotas are still granted to individuals but organizations play an important role in pooling resources and coordinating the behaviour of individual quota owners.

Both public and private regulation can influence the coordination of forest and fishery management. Governmental institutions are sometimes established to promote coordination, such as the coordination committees that have been established in Japan at prefectural, cross-prefectural and national levels. However, departmentalism and contested decentralization processes can impede coordination among governmental agencies as the case studies of Indonesia, Bolivia and Mexico have shown. Moreover, spatial planning is often used as an instrument to demarcate the areas used for forestry and fishery and to coordinate activities. Besides, certification can provide a platform for multiple stakeholders to collaborate in standard setting and learning and improvement activities, although public involvement in certification is sometimes perceived as insufficient. However, the co-existence of public and private regulation may also complicate coordination by adding complexity. The example of Sweden shows that the multiple layers of regulation often lead to confusion among forest owners in deciding their forest management activities.

6.4.6 *Information*

The availability of information regarding the resources status, forest/fishing activities and their interaction is crucial for resources management. Local users, government institutions and certification assessment can all act as information generators. Whereas local actors may be well informed about local issues, such as the status of the resources, local livelihood demands and behaviours, state actors can have the capacity to generate oversight at the system level. However, obtaining information about vast forest or marine areas remains a challenge, sometimes even in countries with comparatively strong governmental capacity such as New Zealand. Certification can also play an important role in information generation and spreading. Certification schemes have rules concerning the documentation of processes related to forest and fishery management. For resources management it can be beneficial to build a robust bridge between those complementary types of knowledge. For example, in North America public and private regulators coordinate in sharing information and rely on the information provided by each other.

6.4.7 Scale

Forest and fishery management require consideration on different scales. Whereas issues such as the protection of sedentary fish species and the retention of trees can be addressed at local level, issues like biodiversity protection and riparian forest protection require landscape considerations. Public regulation can address issues at both local level (such as the management of fishery in South Africa) or on a larger scale (e.g. fishing plans in New Zealand). However, the case studies of North America and Sweden show that even public regulation is often insufficient in incorporating landscape considerations. Certification is conducted at management-unit level and is therefore less suitable for upscaling.

Decentralization is often recommended to regulate natural resources on a proper scale. It is believed to enable decision-making close to local users and hence to contribute to resource management. However, the case studies show that when the system is captured and corrupt and lack of coordination among government agencies and different levels of government is aggravated by obscure and conflicting legislation, the decentralization process itself can be a driver of environmental problems. In Bolivia, for example, conflicting interests among different levels of governments and different departments exist, such as the reluctance of municipalities to support TCO application and the nesting of the land regularization process under the agricultural model.

6.4.8 The need for government regulation

It also became clear from the case studies that, differently from what may sometimes be considered, when private property rights are used, such as for example individual quota systems, there still is an important role for the government.[5] Hence, the mere creation of a private property right (for example an individual quota system for fish) does not imply that government regulation would not be needed anymore. To the contrary, as both the forestry and the fishery chapters showed, there still is an important role for the government in deciding how the fishery would be managed. Also spatial planning, both with respect to forest governance as well as in fisheries, may be an important tool of public regulation that could accompany the establishment of a (private) property rights regime.

That is yet a second important conclusion: the case studies show that also in cases where (both in forestry and in fishery) property rights of a more private nature are created (for example an individual quota system), this does not eliminate the need for (public) regulation. The experience

5 As discussed earlier, some literature views private property rights as the only solution for 'the tragedy of the commons' problem. See Demsetz 1967; Johnson 1972.

from the case studies rather underscores a point already made in Chapter 2, that the creation of (private) property rights requires a regulatory framework regarding the establishment and functioning of those property rights.

6.4.9 The role of private certification

In many of the countries that were studied we could notice that private governance systems, more particularly private certification, are playing an important role. In fact, both with respect to fishery and forest governance, private certification played a role in the functioning of property rights in almost all of the legal systems that were examined.[6]

Although private actors therefore play an increasingly important role also in the governance of common-pool resources, the case studies equally showed a few limits of private governance systems which should also be taken into account:

- First, it should be remembered that the case studies showed that private certification systems play an important but at the same time a relatively limited role. For example as far as fisheries are concerned it appeared that the Marine Stewardship Council (MSC) only covers roughly 10 per cent of the world capture harvest, and, moreover, most of those certified fisheries are located in developed countries. In fact only 8 per cent of the certified fisheries are located in developing countries.
- Notwithstanding the advantages of private certification, these systems can at most play a supportive role in interaction with public regulation, but not constitute a substitute for public regulation. One reason is that enforcement of private certification systems may be weak. It appeared that notwithstanding problems with certification, non-compliance rarely leads to a revocation of the certification. However, one should be careful about arguing that this is only a weakness of private governance systems. After all, the enforcement of public regulation is often also weak.

6.4.10 Discrete design of decentralization and deregulation

Decentralization and privatization are processes which are assumed to increase public participation, strengthen private and communal property rights and hence to bring about desirable environmental outcomes. However, the case studies show that those theoretical advantages do not always materialize. A few conditions need to be satisfied for successful decentralization, such as population mobility, information and accounting systems, parallel decentralization in administrative functions and finance as

6 With the exception of Japan as far as fisheries are concerned.

well as technical and administrative capacity. This can be challenging in many developing countries. If institutions cannot be designed in a way to make such conditions available, decentralization may not lead to desirable outcomes. Privatization promotes the adoption of more flexible, diverse instruments and involves multiple actors in environmental governance. When more discretion is devolved to regulated actors, it is crucial to ensure sufficient protection incentives and to strengthen the capacity of private actors.

6.4.11 Towards a smart interaction between public and private regulation

With the proliferation of private regulation, forest and fishery protection increasingly relies on the interaction between public and private regulation. Complementary interaction can strengthen their capacity in realizing protection goals and conflicting forms of regulation may counteract the effect of each other. North America provides examples with coherent public and private standards. The coherence promotes the spread of private certification and allows the public and private regulators to rely on each other's work. The coordination between them helps to reduce enforcement costs. Indonesia, however, possesses conflicting standards regarding forest rights under public regulation and private regulation. It acts as a barrier to the spread of certification.

Good public regulation often serves as a basis for private regulation to develop. The private certification process can also help to identify the insufficiencies in public regulation and lead to improvements in both public standard setting and implementation. Sometimes, when strong certification incentives exist (such as the demand to broaden access to the market), private certification can also help to bring public regulation into place.[7]

In addition, even when public and private regulation are operating in a complementary way, they cannot replace each other. Public regulation can apply at the national or regional level where certification only applies to certified forests/fisheries and operates at the management-unit level. In addition, certification control is based on available public and company documentation. Only selected site visits are conducted. Hence the inspection does not cover all certified areas. Moreover, both public and private regulation are subject to their own limitations. Many case studies have provided examples of positive interactions between public regulation and private certification. The comparative chapter already provided quite a few examples of complementarities between public and private regulation. In some cases compliance with public regulation is a condition for certification under the private scheme. In other cases the emergence of a private

7 See the example of the swimming crab fishery in Mexico. *Supra* Chapter 4, section 4.5.5.2.

scheme stimulates the government to either issue new public regulations or to increase the level of existing standards.

Many examples of those types of positive interactions were provided in the case studies (and in Chapter 5). For example, in the case study on forest governance in North America it was found that certification made additional information available concerning forestry activities and reduced information costs. Certification also induced compliance with the riparian buffer-zone rules. Reference is also made to interactions whereby private certification mimics government policy approaches and in most cases requires at least compliance with public regulation and in many cases to go beyond the requirements under public regulation. The fishery example in Mexico moreover nicely showed how MSC certification induced the renewal of a concession, forcing the operator (FEDECOOP) also to comply with legal requirements. Many examples are thus provided where a positive interaction between public regulation and private certification takes place. However, the case studies equally provide examples of unsuccessful interactions. For example, the Indonesian case study showed contradictory norms regarding indigenous people's property rights in public and private regulation. The case of Indonesia is a particularly challenging one since the existing customary law and communal property were replaced by state ownership of the forests, which was, however, not accompanied by effective enforcement, which may have contributed to a continuing deforestation in Indonesia. A similar problem of contradictory norms could also be observed in Sweden.

Finally, although the case studies mostly show positive interactions between public and private regulation and there is some evidence (like in the case of Indonesia) of negative interactions, it equally appears from the case studies that private certification has an important role to play to supplement public regulation, but it is not able to completely replace public regulation. Private certification cannot be expected to address all the limitations and shortcomings of public regulation. For example, in a case where public regulation would fail to determine clear forest tenure it cannot be expected that this would be solved via private certification. The reason is that the definition of property rights is always a matter of public regulation.

6.4.12 Tragedy of the commons and externalities

An important theoretical, as well as practical, point to recall, and already referred to in Chapter 2, is that the public and private regulation in their interaction and relationship with property rights may be able to solve Hardin's tragedy of the commons. The smart interaction between public and private regulation may reduce the transaction costs in creating property rights and thus reduce the perverse incentives resulting from open access, the race for trees or the race for fish stocks. However, even in a system wherein private and public regulation optimally interact in their relationship with property rights to protect natural resources, externalities, namely costs for third parties, could still be created. As we indicated in Chapter 5, both the case studies concerning forest governance as well as fisheries indicate that the allocation of (public, private or common) property rights does not necessarily guarantee a sustainable use of the particular resource. To give an illustration from the domain of forest governance, a forest owner could still decide to use his forest in an unsustainable manner (for example via monoculture), thus reducing the ecosystem services provided by the forest. That shows that the interaction between public and private regulation in its relationship to property rights may be relevant to remedy the tragedy of the commons, but that in addition further instruments[8] are needed in order to provide incentives to internalize external costs as well.

6.5 Context specificity

From the above conclusions it follows that it is difficult to provide generalizations. The adequacy of the protection of common-pool resources seems to depend on the certainty of the property rights (not so much on their nature) and on the way in which a smart interaction between public and private regulation is able to allow those property rights to be established and to provide adequate protection to the common-pool resource. A point not explicitly addressed, but probably the proverbial 'elephant in the room', is that the effectiveness of the interactions as well as the certainty of land tenure may to an important extent be related to good governance in a particular legal system. It is a point that we have not yet explored in detail, either in the comparative analysis, or in these concluding remarks. But it is precisely for that reason that in every case study, parameters of good governance were also provided. It is therefore equally not surprising that a rough generalization as far as good governance is concerned leads to a North–South divide, which is to some extent also paralleled with problems in land tenure. Of course the precise relationship between good governance, certainty of land tenure and the adequacy of the protection of the specific common-pool resources goes beyond the scope of this book and should be the subject of further study.

We have identified in the case studies some good practices, namely interactions between public and private regulation which, according to the literature studied, seemed to have positive effects on the way in which the property rights were able to protect the specific common-pool resource. We equally found in some cases examples of where the interactions were less successful but cannot lead to any generalizations beyond some of the

8 Not necessarily only public or private regulation, but eventually also financial incentives, subsidies, taxes or emission trading.

tentative conclusions we formulated in the previous section. One has to be very careful with generalizations precisely since the case studies showed that the effectiveness of the institutional design with respect to the interaction between public and private regulation, affecting the way in which property rights protect the common-pool resource, is highly context specific. There is therefore not just one 'smart mix', but some interactions that, under the specific conditions depending upon the country context, may work better than others. It is important to take this context specificity into account also when drawing normative conclusions from the results of our study.

6.6 Limits

We already indicated ourselves a few obvious limits to our study. One limit inherent in our study is that we analysed public regulation in its interaction with private regulation and property rights. Many other instruments aiming at the protection of common-pool resources are available and some of those were discussed in the theoretical chapter. One could, for example, think of market-based instruments, such as financial incentives like taxes or subsidies but one could equally imagine the role of private law instruments, such as contracts or liability rules. It would have needlessly complicated our study, but might be interesting to be included in a further search for other 'smart mixes'.

For the same reasons we also limited our study to two resources (forests and fishery) and nine legal systems. Although we do think that this already limited study leads to a few interesting observations, the studies undoubtedly have their limits as well and one should, as we already mentioned, therefore be careful with generalizations.

6.7 Further research

The indicated limits of our research at the same time also indicate that there is substantial scope for further research into this fascinating domain, examining smart mixes between legal and policy instruments in the fight against (transboundary) environmental harm.

Beyond a further examination of other common-pool resources and different legal systems it would be worthwhile to include other instruments in the analysis of smart mixes as well. Given the changing role of the state, whereby tasks that traditionally belonged to the regulator are increasingly allocated to private actors, the search for smart mixes in policy instruments could be enlarged by examining how smart public–private partnerships could be developed, both as far as standard setting, but also enforcement are concerned.

Mixing of policy instruments seems to be an important new development and will certainly be expanded to other policy areas as well. In this book we mostly adopted a positive analysis, describing the mixes between public and private regulation as well as the relationship with property rights. Another question would be how one can judge the desirability of particular interactions or mixes. This could for example ask the question of who designs specific mixes, in other words who is the orchestrator of specific interactions. But the question could also be asked whether criteria or conditions for specific interactions could also be developed. Adding more instruments to the policy agenda will as such not necessarily guarantee that social policy goals can be achieved better. Finally, the increasing intervention of private actors can equally lead to questions of legitimacy and accountability as public participation could be strongly reduced, more particularly in comparison to the design of public regulation.

The smart mixes will therefore undoubtedly still lead to an interesting research and policy agenda beyond the domains of forest and fishery governance which were the main subjects of this study.

References

Aalders *et al.* 2008
Aalders, E., Akroyd, J., Ward, T. & Short, K., 'Case Study 4: The New Zealand Hoki, A: The Fishery', in Phillips, B., Ward, T. & Chaffee, C. (eds.), *Eco-Labelling in Fisheries: What Is It All About?*, Oxford: John Wiley & Sons, 2008, pp. 145–161.

Agardy, Notarbartolo di Sciara & Christie 2011
Agardy, T., Notarbartolo di Sciara, G. & Christie, P., 'Mind the Gap: Addressing the Shortcomings of Marine Protected Areas through Large Scale Marine Spatial Planning', *Marine Policy*, 2011, Vol. 35(2), pp. 226–232.

Agblede 2010
Agblede, H., Forest Stewardship Council's (FSC) Certification Dispute in Sweden: A Case Study Involving the Swedish Society for Nature Conservation (SSNC) and Svenska Cellulosa Aktiebolaget (SCA, 35–37), Second cycle, A2E. Uppsala: SLU, Dept. of Urban and Rural Development, 2010, available at: http://stud.epsilon.slu.se/1837/.

Agrawal & Ribot 1999
Agrawal, A. & Ribot, J., 'Accountability in Decentralization: A Framework with South Asian and West African Cases', *The Journal of Developing Areas*, 1999, Vol. 33(4), pp. 473–502.

Alchian 1965
Alchian, A.A., 'Some Economics of Property Rights', *Il Politico*, 1965, Vol. 30(4), pp. 816–829.

Alchian 1979
Alchian, A.A., 'Some Implications of Recognition of Property Right Transaction Costs', in Brunner, K. (ed.), *Economics and Social Institutions. Insights from the Conferences on Analysis & Ideology*, Boston: Springer, 1979, pp. 233–252.

Alchian & Woodward 1988
Alchian, A.A. & Woodward, S.L., 'The Firm is Dead; Long Live the Firm: A Review of Oliver E. Williamson's "The Economic Institutions of Capitalism: Firms, Markets, Relational Contracting"', *Journal of Economic Literature*, 1988, Vol. 26, pp. 65–79.

Alexander & Peñalver 2012
Alexander, G.S. & Peñalver, E.M., *An Introduction to Property Theory*, Cambridge: Cambridge University Press, 2012.

Allen 1991
Allen, D.W., 'What Are Transaction Costs?', *Research in Law and Economics*, 1991, Vol. 14, pp. 1–18.

Allen 2000

Allen, D.W., 'Transaction Costs', in Bouckaert, B. & De Geest, G. (eds.), *Encyclopedia of Law and Economics*, Cheltenham: Edward Elgar, 2000, pp. 893–926.

Allison 2001

Allison, E.H., 'Big Laws, Small Catches: Global Ocean Governance and the Fisheries Crisis', *Journal of International Development*, 2001, Vol. 13(7), pp. 933–950.

Andersen 1983

Andersen, P., '"On Rent of Fishing Grounds": A Translation of Jens Warming's 1911 Article, with an Introduction', *History of Political Economy*, 1983, Vol. 15(3), pp. 391–396.

Anderson & Leal 1991

Anderson, T.L. & Leal, D.R., *Free Market Environmentalism*, London: Palgrave Macmillan, 1991.

Angelsen & Rudel 2013

Angelsen, A. & Rudel, T.K., 'Designing and Implementing Effective REDD+ Policies: A Forest Transition Approach', *Review of Environmental Economics and Policy*, 2013, Vol. 7(1), pp. 91–113.

Angelstam *et al.* 2011

Angelstam, P., Andersson, K., Axelsson, R., Elbakidze, M., Jonsson, B.G. & Roberge, J.-M., 'Protecting Forest Areas for Biodiversity in Sweden 1991–2010: Policy Implementation Process and Outcomes on the Ground', *Silva Fennica*, 2011, Vol. 45(5), pp. 1111–1133.

Annala 1996

Annala, J.H., 'New Zealand's ITQ System: Have the First Eight Years Been a Success or a Failure?', *Reviews in Fish Biology and Fisheries*, 1996, Vol. 6(1), pp. 43–62.

Arnason 2012

Arnason, R., 'Property Rights in Fisheries: How Much Can Individual Transferable Quotas Accomplish?', *Review of Environmental Economics and Policy*, 2012, Vol. 6(2), pp. 217–236.

Arnold 2008

Arnold, L.L., 'Deforestation in Decentralised Indonesia: What's Law Got to Do with It', *Law, Environment & Development Journal*, 2008, Vol. 4(2), pp. 77–101.

Attwood & Farquhar 1999

Attwood, C.G. & Farquhar, M., 'Collapse of Linefish Stocks between Cape Hangklip and Walker Bay, South Africa', *South African Journal of Marine Science*, 1999, Vol. 21(1), pp. 415–432.

Auld 2007

Auld, G., The Origins and Growth of Social and Environmental Certification Programs in the Fisheries' Sector. Paper read at 11th Annual Conference of the International Society for New Institutional Economics, June 21–23, at Reykjavik, Iceland, 2007.

Baird & Quastel 2011

Baird, I.G. & Quastel, N., 'Dolphin-Safe Tuna from California to Thailand: Localisms in Environmental Certification of Global Commodity Networks', *Annals of the Association of American Geographers*, 2011, Vol. 101(2), pp. 337–355.

Barr *et al.* 2006

Barr, C., Resasudarmo, I.A.P., Dermawan, A. & McCarthy, J. (eds.), *Decentralization of Forest Administration in Indonesia: Implications for Forest Sustainability, Economic Development and Community Livelihoods*, Bogor: CIFOR, 2006.

Barrera 2011

Barrera, A., Turning Legal Pluralism into State-Sanctioned Law: Assessing the Implications of the New Constitutions and Laws in Bolivia and Ecuador, 2011, German Institute of Global and Area Studies Working Papers. No. 176, available at: http://papers.ssrn.com/sol3/papers.cfm?abstract_id=1920297.

Bartley 2003

Bartley, T., 'Certifying Forests and Factories: States, Social Movements, and the Rise of Private Regulation in the Apparel and Forest Products Fields', *Politics & Society*, 2003, Vol. 31(3), pp. 433–464.

Bartley 2011

Bartley, T., 'Transnational Governance as the Layering of Rules: Intersections of Public and Private Standards', *Theoretical Inquiries in Law*, 2011, Vol. 12(2), pp. 517–542.

Bartley 2014

Bartley, T., 'Transnational Governance and the Re-Centered State: Sustainability or Legality?', *Regulation & Governance*, 2014, Vol. 8(1), pp. 93–109.

Barzel 1985

Barzel, Y., 'Transaction Costs: Are They Just Costs?', *Journal of Institutional and Theoretical Economics*, 1985, Vol. 141(1), pp. 4–16.

Barzel 1989

Barzel, Y., *Economic Analysis of Property Rights*, Cambridge: Cambridge University Press, 1989.

Basurto 2005

Basurto, X., 'How Locally Designed Access and Use Controls Can Prevent the Tragedy of the Commons in a Mexican Small-Scale Fishing Community', *Society and Natural Resources*, 2005, Vol. 18(7), pp. 643–659.

Basurto *et al.* 2012

Basurto, X., Cinti, A., Bourillón, L., Rojo, M., Torre, J. & Hudson Weaver, A., 'The Emergence of Access Controls in Small-Scale Fishing Commons: A Comparative Analysis of Individual Licenses and Common Property-Rights in Two Mexican Communities', *Human Ecology*, 2012, Vol. 40, pp. 597–609.

Becker & León 2000

Becker, C.D. & León, R., 'Indigenous Forest Management in the Bolivian Amazon: Lessons from the Yuracaré People', in Gibson, C.C., McKean, M.A. & Ostrom, E. (eds.), *People and Forests: Communities, Institutions, and Governance*, Cambridge, MA: MIT Press, 2000, pp. 164–189.

Benneker 2008

Benneker, C., *Dealing with the State, the Market and NGOs. The Impact of Institutions on the Constitution and Performance of Community Forest Enterprises in the Lowlands of Bolivia*, Doctoral thesis, Wageningen University, 2008.

Berkes 2009

Berkes, F., 'Evolution of Co-management: Role of Knowledge Generation, Bridging Organizations and Social Learning', *Journal of Environmental Management*, 2009, Vol. 90(5), pp. 1692–1702.

Berry 2006

Berry, A., *Branching Out: Case Studies in Canadian Forest Management*, Bozeman, MT: PERC, 2006.

Bess 2005

Bess, R., 'Expanding New Zealand's Quota Management System', *Marine Policy*, 2005, Vol. 29(4), pp. 339–347.

Bess & Rallapudi 2007
Bess, R. & Rallapudi, R., 'Spatial Conflicts in New Zealand Fisheries: The Rights of Fishers and Protection of the Marine Environment', *Marine Policy*, 2007, Vol. 31(6), pp. 719–729.
Blomley 2016
Blomley, N. 'The Boundaries of Property: Complexity, Relationality, and Spatiality', *Law & Society Review*, 2016, Vol. 50(1), pp. 224–255.
Borch 2010
Borch, T., 'Tangled Lines in New Zealand's Quota Management System: The Process of Including Recreational Fisheries', *Marine Policy*, 2010, Vol. 34(3), pp. 655–662.
Boscolo & Vargas Rios 2007
Boscolo, M. & Vargas Rios, M.T., 'Forest Law Enforcement and Rural Livelihoods in Bolivia', in Tacconi, L. (ed.), *Illegal Logging: Law Enforcement, Livelihoods and the Timber Trade*, London: Earthscan, 2007, pp. 191–217.
Boström 2003
Boström, M., 'How State-Dependent is a Non-State-Driven Rule-Making Project? The Case of Forest Certification in Sweden', *Journal of Environmental Policy & Planning*, 2003, Vol. 5(2), pp. 165–180.
Bouckaert 2010
Bouckaert, B., *Property Law and Economics*, Cheltenham: Edward Elgar, 2010.
Bourillón-Moreno 2002
Bourillón-Moreno, L., *Exclusive Fishing Zone as a Strategy for Managing Fishery Resources by the Seri Indians, Gulf of California, Mexico*, Diss., Tucson: University of Arizona, 2002.
Branch 2009
Branch, T.A., 'How Do Individual Transferable Quotas Affect Marine Ecosystems?', *Fish and Fisheries*, 2009, Vol. 10(1), pp. 39–57.
Bromley 1991
Bromley, D.W., *Environment and Economy: Property Rights and Public Policy*, Oxford/New York: Blackwell Publishing, 1991.
Brown, Brown & Binkley 1993
Brown, T.C., Brown, D. & Binkley, D., 'Laws and Programs for Controlling Nonpoint Source Pollution in Forest Areas', *Journal of the American Water Resources Association*, 1993, Vol. 29(1), pp. 1–13.
Bryant, Nielsen & Tangley 1997
Bryant, D., Nielsen, D. & Tangley, L., *The Last Frontier Forests: Ecosystems and Economies on the Edge. What is the Status of the World's Remaining Large, Natural Forest Ecosystems?* Report, Washington, DC: World Resources Institute, Forest Frontiers Initiative, 1997.
Buchanan 1975
Buchanan, J., *The Limits of Liberty: Between Anarchy and Leviathan*, Chicago, IL: University of Chicago Press, 1975.
Buchanan & Tollison 1984
Buchanan, J.M. & Tollison, R.D., *The Memory of Public Choice*, Vol. II, Ann Arbor, MI: University of Michigan Press, 1984.
Bush *et al.* 2013
Bush, S.R., Toonen, H., Oosterveer, P. & Mol, A.P.J., 'The "Devil's Triangle" of MSC Certification: Balancing Credibility, Accessibility and Continuous Improvement', *Marine Policy*, 2013, Vol. 37, pp. 288–293.

Caddy & Seijo 2005
Caddy, J.F. & Seijo, J.C., 'This is More Difficult than We Thought! The Responsibility of Scientists, Managers and Stakeholders to Mitigate the Unsustainability of Marine Fisheries', *Philosophical Transactions of the Royal Society of London B: Biological Sciences,* 2005, Vol. 360(1453), pp. 59–75.
Cafaggi & Renda 2012
Cafaggi, F. & Renda, A., Public and Private Regulation: Mapping the Labyrinth, CEPS Working Document, 2012.
Cafferata *et al.* 2003
Cafferata, B., Dahl, D., Hughes, S. & Waldorf, J., Role of an Independent Forestry Watchdog in an Era of Certification, submitted to the XII World Forestry Congress by the Forest Practices Board of British Columbia, November 14, 2002.
Campbell 2002
Campbell, J.Y., 'Differing Perspectives on Community Forestry in Indonesia', in Colfer, C.J. & Resosudarmo, I.A.P. (eds.), *Which Way Forward? People, Forests and Policymaking in Indonesia,* Washington, DC: Resources for the Future, 2002, pp. 10–125.
Canadian Council of Forest Ministers 2000
Canadian Council of Forest Ministers, *Criteria and Indicators of Sustainable Forest Management in Canada: National Status 2000,* Ottawa: Natural Resources Canada, 2000.
Cancino, Uchida & Wilen 2007
Cancino, J.P., Uchida, H. & Wilen, J.E., 'TURFs and ITQs: Collective vs. Individual Decision Making', *Marine Resource Economics,* 2007, Vol. 22(4), pp. 391–406.
Cano Chacón 2013
Cano Chacón, M., *The Role of Information of the Marine Stewardship Council Certification Process in Developing Countries: A Case Study of Two MSC Fisheries Certified in Mexico,* Diss., Halifax: Dalhousie University, 2013.
Cardona *et al.* 2014
Cardona, W.C., De Jong, W., Zuidema, P.A. & Boot, R., 'Diverse Local Regulatory Responses to a New Forestry Regime in Forest Communities in the Bolivian Amazon', *Land Use Policy,* 2014, Vol. 39, pp. 224–232.
Carlson & Palmer 2016
Carlson, A. & Palmer, C., 'A Qualitative Meta-Synthesis of the Benefits of Eco-Labeling in Developing Countries', *Ecological Economics,* 2016, Vol. 127, pp. 129–145.
Carraway *et al.* 2002
Carraway, B., Donellan, J., Simpson H. & Work, D., Voluntary Implementation of Forestry Best Management Practices in East Texas, Texas Forest Service, 2002.
Cashore & Stone 2012
Cashore, B. & Stone, M.W., 'Can Legality Verification Rescue Global Forest Governance?: Analyzing the Potential of Public and Private Policy Intersection to Ameliorate Forest Challenges in Southeast Asia', *Forest Policy and Economics,* 2012, Vol. 18, pp. 13–22.
Cashore & Stone 2014
Cashore, B. & Stone, M.W., 'Does California Need Delaware? Explaining Indonesian, Chinese, and United States Support for Legality Compliance of Internationally Traded Products', *Regulation & Governance,* 2014, Vol. 8, pp. 49–73.

Cashore *et al.* 2007

Cashore, B., Auld, G., Bernstein, S. & McDermott, C., 'Can Non-State Governance "Ratchet Up" Global Environmental Standards? Lessons from the Forest Sector', *Review of European Community & International Environmental Law*, 2007, Vol. 16(2), pp. 158–172.

Chaffee 2008

Chaffee, C., Annual Surveillance Report as Required under the Marine Stewardship Council Program, Red Lobster, Baja California, Mexico, 2008.

Chandra & Tisdell 1998

Chandra, R. & Tisdell, C., 'Good Governance in Sustainable Development: The Impact of Institutions', *International Journal of Social Economics*, 1998, Vol. 25(6/7/8), pp. 1310–1325.

Chao 2012

Chao, S., Forest Peoples: Numbers across the World, Moreton-in-Marsh: Forest Peoples Programme, 2012, available at: www.forestpeoples.org/sites/fpp/files/publication/2012/05/forest-peoples-numbers-across-world-final_0.pdf.

Chu 2009

Chu, C., 'Thirty Years Later: The Global Growth of ITQs and their Influence on Stock Status in Marine Fisheries', *Fish and Fisheries*, 2009, Vol. 10(2), pp. 217–230.

Cinti *et al.* 2010

Cinti, A., Shaw, W., Cudney-Bueno, R. & Rojo, M., 'The Unintended Consequences of Formal Fisheries Policies: Social Disparities and Resource Overuse in a Major Fishing Community in the Gulf of California, Mexico', *Marine Policy*, 2010, Vol. 34(2), pp. 328–339.

Cisneros-Montemayor *et al.* 2013

Cisneros-Montemayor, A.M., Cisneros-Mata, M.A., Harper, S. & Pauly, D., 'Extent and Implications of IUU Catch in Mexico's Marine Fisheries', *Marine Policy*, 2013, Vol. 39, pp. 283–288.

Clement, Wells & Gallagher 2008

Clement, G., Wells, R. & Gallagher, C.M., 'Industry Management within the New Zealand Quota Management System: The Orange Roughy Management Company', in Townsend, R.E., Shotton, R. & Uchida, H. (eds.), *Case Studies in Fisheries Self-Governance*, Rome: FAO, 2008, pp. 277–290.

Cole 2002

Cole, D.H., *Pollution and Property: Comparing Ownership Institutions for Environmental Protection*, Cambridge: Cambridge University Press, 2002.

Cole 2010

Cole, D.H., 'New Forms of Private Property: Property Rights in Environmental Goods', in Bouckaert, B. (ed.), *Property Law and Economics*, Cheltenham: Edward Elgar, 2010, pp. 225–269.

Connor & Shallard 2010

Connor, R. & Shallard, B., 'Evolving Governance in New Zealand Fisheries', in Grafton, R.Q., Hilborn, R., Squires, D., Tait, M. & Williams, M. (eds.), *Handbook of Marine Fisheries, Conservation and Management*, Oxford: Oxford University Press, 2010, pp. 347–359.

Contreras-Hermosilla & Vargas Rios 2002

Contreras-Hermosilla, A. & Vargas Rios, M.T., *Social, Environmental and Economic Dimensions of Forest Policy Reforms in Bolivia*, Washington DC: Forest Trends and CIFOR, 2002.

Cooper 2004–2005
Cooper, T., 'Picture This: Promoting Sustainable Fisheries through Eco-Labeling and Product Certification', *Ocean & Coastal Law Journal*, 2004–2005, Vol. 10(1–2), pp. 1–49.
Cooter & Ulen 2008
Cooter, R. & Ulen, T., *Law and Economics*, 5th edn., Harlow: Addison-Wesley, 2008.
Cooter & Ulen 2010
Cooter, R. & Ulen, T., *Law and Economics*, 6th edn., Harlow: Addison-Wesley, 2010.
Costello & Deacon 2007
Costello, C. & Deacon, R., 'The Efficiency Gains from Fully Delineating Rights in an ITQ Fishery', *Marine Resource Economics*, 2007, Vol. 22, pp. 346–361.
Cudney-Bueno *et al.* 2009
Cudney-Bueno, R., Bourillón, L., Sáenz-Arroyo, A., Torre-Cosio, J., Turk-Boyer, P. & Shaw, W.W., 'Governance and Effects of Marine Reserves in the Gulf of California, Mexico', *Ocean & Coastal Management*, 2009, Vol. 52(3–4), pp. 207–218.
Cunningham 2013
Cunningham, E., Catch Shares in Action: Mexican Baja California FEDECOOP Benthic Species Territorial Use Rights for Fishing System, Environmental Defense Fund, available at: http://fisherysolutionscenter.edf.org/design-stories/mexican-baja-california-fedecoop-benthic-species-territorial-use-rights-fishing, 2013.
Department Agriculture, Forestry and Fisheries Republic of South Africa 2012
Department Agriculture, Forestry and Fisheries Republic of South Africa, 'Policy for the Small Scale Fisheries Sector in South Africa', *Government Gazette*, No. 35455, 20 June 2012.
Dasgupta 1982
Dasgupta, P., *The Control of Resources*, Cambridge, MA: Harvard University Press, 1982.
De Alessi 2012
De Alessi, M., 'The Political Economy of Fishing Rights and Claims: The Maori Experience in New Zealand', *Journal of Agrarian Change*, 2012, Vol. 12(2–3), pp. 390–412.
De Pourcq, Thomas & Van Damme 2009
De Pourcq, K., Thomas, E. & Van Damme, P., 'Indigenous Community-based Forestry in the Bolivian Lowlands: Some Basic Challenges for Certification', *International Forestry Review*, 2009, Vol. 11(1), pp. 12–26.
De Vries 2000
De Vries, M.S., 'The Rise and Fall of Decentralization: A Comparative Analysis of Arguments and Practices in European Countries', *European Journal of Political Research*, 2000, Vol. 38(2), pp. 193–224.
Deepwater Group Limited 2011
Deepwater Group Limited, Sustainable Management of New Zealand's Hoki Fisheries, 2011, available at: www.cport.net/assets/uploads/files/Sustainable%20Management%20of%20New%20Zealand%20Hoki%20120711.pdf.
Delaney 2015
Delaney, A., 'Japanese Fishing Cooperative Associations: Governance in an Era of Consolidation', in Jentoft, S. & Chuenpagdee, R. (eds.), *Interactive Governance for Small-Scale Fisheries: Global Reflections*, Berlin: Springer, pp. 263–280.

Delgado *et al.* 2003
Delgado, C.L., Wada, N., Rosegrant, M.W., Meyer, S. & Ahmed M., *Outlook for Fish to 2020: Meeting Global Demand*, Washington, DC: International Food Policy Resources Institute, 2003.
Demsetz 1967
Demsetz, H., 'Toward a Theory of Property Rights', *American Economic Review*, 1967, Vol. 57(2), pp. 347–359.
Denman 1978
Denman, D.R., *The Place of Property: A New Recognition of the Function and Form of Property Rights in Land*, Berkhamsted: Geographical Publications, 1978.
Dobbin, Simmons & Garrett 2007
Dobbin, F., Simmons, B. & Garrett, G., 'The Global Diffusion of Public Policies. Social Construction, Coercion, Competition, or Learning?', *Annual Review of Sociology*, 2007, Vol. 33, pp. 449–472.
Douvere 2008
Douvere, F. 'The Importance of Marine Spatial Planning in Advancing Ecosystem-Based Sea Use Management', *Marine Policy*, 2008, Vol. 32(5), pp. 762–771.
Douvere & Ehler 2009
Douvere, F. & Ehler, C.N., 'New Perspectives on Sea Use Management: Initial Findings from European Experience with Marine Spatial Planning', *Journal of Environmental Management*, 2009, Vol. 90(1), pp. 77–88.
Duery & Vlosky 2005
Duery, S. & Vlosky, R., 'Bolivia: A Global Leader in Certification', *Forest Products Journal*, 2005, Vol. 55, pp. 8–19.
Ebeling & Yasué 2009
Ebeling, J. & Yasué, M., 'The Effectiveness of Market-Based Conservation in the Tropics: Forest Certification in Ecuador and Bolivia', *Journal of Environmental Management*, 2009, Vol. 90(2), pp. 1145–1153.
Eberlein *et al.* 2014
Eberlein, B., Abbott, K.W., Black, J., Meidinger, E. & Wood, S., 'Transnational Business Governance Interactions: Conceptualization and Framework for Analysis', *Regulation & Governance*, 2014, Vol. 8(1), pp. 1–21.
Edvardsson 2004
Edvardsson, K., 'Using Goals in Environmental Management: The Swedish System of Environmental Objectives', *Environmental Management*, 2004, Vol. 34(2), pp. 170–180.
Elbakidze, Angelstam, Andersson, Nordberg & Pautov 2011
Elbakidze, P., Angelstam, P., Andersson, K., Nordberg, M. & Pautov, Y., 'How Does Forest Certification Contribute to Boreal Biodiversity Conservation? Standards and Outcomes in Sweden and NW Russia', *Forest Ecology and Management*, 2011, Vol. 262, pp. 1983–1995.
Ellefson *et al.* 2001
Ellefson, P. Kilgore, M. & Philips, M., Monitoring Compliance with BMPs: The Experience of State Forestry Agencies, *Journal of Forestry*, 2001, Vol. 99(1), pp. 11–17.
Ellefson *et al.* 2006
Ellefson, P. Kilgore, M. & Granskog, J., 'State Government Regulation of Forestry Practices Applied to Nonfederal Forests: Extent and Intensity of Agency Involvement', *Journal of Forestry*, 2006, Vol. 104(8), pp. 401–406.

Environmental Defence Society 2011
Environmental Defence Society, Governing our Oceans: Environmental Reform for the Exclusive Economic Zone, 2011, available at: www.pepanz.com/assets/Uploads/EEZ-Policy-Document-EDS-4.pdf.
Espinosa-Romero *et al.* 2014
Espinosa-Romero, M.J., Rodriguez, L.F., Hudson Weaver, A., Villanueva-Aznar, C. & Torre, J., 'The Changing Role of NGOs in Mexican Small-Scale Fisheries: From Environmental Conservation to Multi-Scale Governance', *Marine Policy*, 2014, Vol. 50, pp. 290–299.
Espinoza & Dockry 2014
Espinoza, O. & Dockry, M.J., 'Forest Certification in Bolivia: A Status Report and Analysis of Stakeholder Perspectives', *Forest Products Journal*, 2014, Vol. 64(3–4), pp. 80–89.
Espinoza-Tenorio, Espejel & Wolff 2011
Espinoza-Tenorio, A., Espejel, I. & Wolff, M., 'Capacity Building to Achieve Sustainable Fisheries Management in Mexico', *Ocean & Coastal Management*, 2011, Vol. 54(10), pp. 731–741.
Espinoza-Tenorio *et al.* 2011
Espinoza-Tenorio, A., Espejel, I., Wolff, M. & Zepeda-Domínguez, J.A., 'Contextual Factors Influencing Sustainable Fisheries in Mexico', *Marine Policy*, 2011, Vol. 35(3), pp. 343–350.
Esty 2006
Esty, D., 'Good Governance at the Supranational Scale: Globalizing Administrative Law', *The Yale Law Journal*, 2006, Vol. 115(7), pp. 1490–1562.
FAO 2010
FAO, *Global Forest Resources Assessment 2010*, Rome: FAO, 2010.
FAO 2012a
FAO, *The State of World Fisheries and Aquaculture*, Rome: FAO, 2012.
FAO 2012b
FAO, FAO Global Capture Production Database updated to 2012, available at: ftp://ftp.fao.org/fi/stat/Overviews/CaptureStatistics2012.pdf.
FAO
FAO, Fishery and Aquaculture Country Profiles, the Republic of South Africa, available at: www.fao.org/fishery/facp/ZAF/en.
FAO 2014
FAO, *The State of World Fisheries and Aquaculture*, Rome: FAO, 2014.
Faure & Skogh 2003
Faure, M. & Skogh, G., *The Economic Analysis of Environmental Policy and Law: An Introduction*, Cheltenham: Edward Elgar, 2003.
Feeny *et al.* 1990
Feeny, D., Berkes, F., McCay, B. & Acheson, J.M., 'The Tragedy of the Commons: Twenty-Two Years Later', *Human Ecology*, 1990, Vol. 18(1), pp. 1–19.
FERN 2001
FERN, *Behind the Logo: An Environmental and Social Assessment of Forest Certification Schemes*, Moreton-in-Marsh: FERN, 2001.
Finkbeiner & Basurto 2015
Finkbeiner, E.M. & Basurto, X., 'Re-defining Co-management to Facilitate Small-Scale Fisheries Reform: An Illustration from Northwest Mexico', *Marine Policy*, 2015, pp. 433–441.

Foley 2013
Foley, P., 'National Government Responses to Marine Stewardship Council (MSC) Fisheries Certification: Insights from Atlantic Canada', *New Political Economy*, 2013, Vol. 18(2), pp. 284–307.

Foley & Havice 2016
Foley, P. & Havice, E., 'The Rise of Territorial Eco-Certifications: New Politics of Transnational Sustainability Governance in the Fishery Sector', *Geoforum*, 2016, Vol. 69, pp. 24–33.

Foley & McCay 2014
Foley, P. & McCay, B., 'Certifying the Commons: Eco-Certification, Privatization, and Collective Action', *Ecology and Society*, 2014. Vol. 19(2), pp. 28–40.

Gale & Haward 2004
Gale, F. & Haward, M., Public Accountability in Private Regulation: Contrasting Models of the Forest Stewardship Council (FSC) and Marine Stewardship Council (MSC), Refereed paper presented to the Australasian Political Studies Association Conference University of Adelaide 29 September–1 October 2004.

Garcia, Rice & Charles 2014a
Garcia, S.M., Rice, J. & Charles, A., 'Governance of Marine Fisheries and Biodiversity Conservation: A History', in Garcia, S.M., Rice, J. & Charles, A. (eds.), *Governance of Marine Fisheries and Biodiversity Conservation: Interaction and Coevolution*, Chichester: John Wiley & Sons, 2014, pp. 3–17.

Garcia, Rice & Charles 2014b
Garcia, S.M., Rice, J. & Charles, A., 'Governance of Marine Fisheries and Biodiversity Conservation: Convergence or Coevolution', in Garcia, S.M., Rice, J. & Charles, A. (eds.), *Governance of Marine Fisheries and Biodiversity Conservation: Interaction and Coevolution*, Chichester: John Wiley & Sons, 2014, pp. 18–36.

Gezelius & Hauck 2011
Gezelius, S.S. & Hauck, M., 'Toward a Theory of Compliance in State Regulated Livelihoods: A Comparative Study of Compliance Motivations', in Developed and Developing World Fisheries', *Law & Society Review*, 2011, Vol. 45(2), pp. 435–470.

Gibbs 2008
Gibbs, M.T., 'The Historical Development of Fisheries in New Zealand with respect to Sustainable Development Principles', *The Electronic Journal of Sustainable Development*, 2008, Vol. 1(2), pp. 23–33.

Gibbs 2009
Gibbs, M.T., 'Individual Transferable Quotas and Ecosystem-Based Fisheries Management: It's All in the T', *Fish and Fisheries*, 2009, Vol. 10(4), pp. 470–474.

Gold & Zuckerman 2014
Gold, M.E. & Zuckerman, R.B., 'Indonesian Land Rights and Development', *Columbia Journal of Asian Law*, 2014, Vol. 28(1), pp. 41–70.

Gondor & Morimoto 2011
Gondor, D., & Morimoto, H., 'Role of World Wildlife Fund (WWF) and Marine Stewardship Council (MSC) in Seafood Eco-Labelling Policy in Japan', *Sustainability Accounting, Management and Policy Journal*, 2011, Vol. 2(2), pp. 214–230.

Gordon 1954
Gordon, H.S., 'The Economic Theory of a Common-Property Resource: The Fishery', *Journal of Political Economy*, 1954, Vol. 62(2), pp. 124–142.

Grafton 1996
Grafton, R.Q., 'Individual Transferable Quotas: Theory and Practice', *Reviews in Fish Biology and Fisheries*, 1996, Vol. 6(1), pp. 5–20.
Grafton *et al.* 2006
Grafton, R.Q., Arnason, R., Bjørndal, T., Campbell, D., Campbell, H.F., Clark, C.W., Connor, R., Dupont, D.P., Hannesson, R., Hilborn, R., Kirkley, J.E., Kompas, T., Lane, D.E., Munro, G.R., Pascoe, S., Squires, D., Steinshamn, S.I., Turris, B.R. & Weninger, A., 'Incentive-Based Approaches to Sustainable Fisheries', *Canadian Journal of Fisheries and Aquatic Sciences*, 2006, Vol. 63(3), pp. 699–710.
Griendle 2007
Griendle, M., 'Good Enough Governance Revisited', *Development Policy Review*, 2007, Vol. 25(5), pp. 533–574.
Griffith 2008
Griffith, D.R., 'The Ecological Implications of Individual Fishing Quotas and Harvest Cooperatives', *Frontiers in Ecology and the Environment*, 2008, Vol. 6(4), pp. 191–198.
Griffiths 2000
Griffiths, M.H., 'Long-Term Trends in Catch and Effort of Commercial Linefish off South Africa's Cape Province: Snapshots of the 20th Century', *South African Journal of Marine Science*, 2000, Vol. 22(1), pp. 81–110.
Gulbrandsen 2004
Gulbrandsen, L.H., 'Overlapping Public and Private Governance: Can Forest Certification Fill the Gaps in the Global Forest Regime?', *Gobal Environmental Politics*, 2004, Vol. 4(2), pp. 75–99.
Gulbrandsen 2005
Gulbrandsen, L.H., 'Sustainable Forestry in Sweden: The Effect of Competition among Private Certification Schemes', *The Journal of Environment & Development*, 2005, Vol. 14(3), pp. 338–355.
Gulbrandsen 2010
Gulbrandsen, L.H., *Transnational Environmental Governance: The Emergence and Effects of the Certification of Forest and Fisheries*, Cheltenham: Edward Elgar, 2010.
Gulbrandsen 2014
Gulbrandsen, L.H., 'Dynamic Governance Interactions: Evolutionary Effects of State Responses to Non-State Certification Programs', *Regulation & Governance*, 2014, Vol. 8(1), pp. 74–92.
Gulbrandsen & Auld 2016
Gulbrandsen, L.H. & Auld, G., 'Contested Accountability Logics in Evolving Nonstate Certification for Fisheries Sustainability', *Global Environmental Politics*, 2016, Vol. 16(2), pp. 42–60.
Gupta, Van der Grijp & Kuik 2012
Gupta, J., Van der Grijp, N. & Kuik, O., *Climate Change, Forests and REDD: Lessons for Institutional Design*, London/New York: Routledge, 2012.
Hansen *et al.* 2013
Hansen, M.C., Potapov, P.V., Moore, R., Hancher, M., Turubanova, S.A., Tyukavina, A., Thau, D., Stehman, S.V., Goetz, S.J., Loveland, T.R., Kommareddy, A., Egorov, A., Chini, L., Justice, C.O. & Townshend, J.R.G., 'High-Resolution Global Maps of 21st-Century Forest Cover Change', *Science*, 2013, Vol. 341(6160), pp. 850–853.

Hara *et al.* 2014

Hara, M.M., Rogerson, J., De Goede, J. & Raakjaer, J., 'Fragmented Participation in Management of the Fishery for Small Pelagic Fish in South Africa – Inclusion of Small-Rights Holders is a Complex Matter', *African Journal of Marine Science*, 2014, Vol. 36(2), pp. 185–196.

Harada & Wiyono 2014

Harada, K. & Wiyono, 'Certification of a Community-Based Forest Enterprise for Improving Institutional Management and Household Income: A Case from Southeast Sulawesi, Indonesia', *Small-Scale Forestry*, 2014, Vol. 13(1), pp. 47–64.

Hardin 1968

Hardin, G., 'The Tragedy of the Commons', *Science*, 1968, Vol. 162(3859), pp. 1243–1248.

Harris 2010

Harris, P., Beneficial Management Practices for Riparian Zones in Atlantic Canada, Island Nature Trust, 2010, available at: www.islandnaturetrust.ca/wp-content/uploads/2010/04/Beneficial-Management-Practices-for-Riparian-Zones-in-Atlantic-Canada1.pdf

Harte 2008

Harte M., 'Assessing the Road Towards Self-Governance in New Zealand's Commercial Fisheries', in Townsend, R.E., Shotton, R. & Uchida, H. (eds.), *Case Studies in Fisheries Self-Governance*, Rome: FAO, 2008, pp. 323–334.

Hauck 2009

Hauck, M., 'Crime, Environment and Power: Revisiting the Abalone Fishery', *South African Journal of Criminal Justice*, 2009, Vol. 22(2), pp. 229–245.

Hauck & Gallardo-Fernández 2013

Hauck, M. & Gallardo-Fernández, G.L., 'Crises in the South African Abalone and Chilean Loco Fisheries: Shared Challenges and Prospects', *Maritime Studies*, 2013, Vol. 12(3), pp. 1–20.

Hauck & Kroese 2006

Hauck, M. & Kroese, M., 'Fisheries Compliance in South Africa: A Decade of Challenges and Reform 1994–2004', *Marine Policy*, 2006, Vol. 30(1), pp. 74–83.

Helms 1998

Helms, J.A., *The Dictionary of Forestry*, Bethesda, MD: Society of American Foresters, 1998.

Hernandez & Kempton 2003

Hernandez, A. & Kempton, W., 'Changes in Fisheries Management in Mexico: Effects of Increasing Scientific Input and Public Participation', *Ocean & Coastal Management*, 2003, Vol. 46(6–7), pp. 507–526.

Hersoug 2002

Hersoug, B., *New Zealand Experience with Rights-Based Fisheries Management – Unfinished Business*, Delft: Eburon, 2002.

Hey 2016

Hey, E., *Advanced Introduction to International Environmental Law*, Cheltenham, UK/Northampton, MA: Edward Elgar, 2016.

Hinrichs, Muhtaman & Irianto 2008

Hinrichs, A., Muhtaman, D.R. & Irianto, N., *Forest Certification on Community Land in Indonesia*, Eschborn: Deutsche Gesellschaft für Technische Zusammenarbeit, 2008.

Hokimoto 2009
Hokimoto, K., 'The Present Situation and Problems of the FCA in Japan', *Journal of National Fisheries University*, 2009, Vol. 58(1), pp. 53–58.
Honoré 1961
Honoré, A.M., 'Ownership', in Guest, A.G. (ed.), *Oxford Essays in Jurisprudence*, London/New York: Oxford University Press, 1961, pp. 107–147.
Hosonuma *et al.* 2012
Hosonuma, N., Herold, M., De Sy, V., De Fries, R.S., Brockhaus, M., Verchot, L., Angelsen, A. & Romijn, E., 'An Assessment of Deforestation and Forest Degradation Drivers in Developing Countries', *Environmental Research Letters*, 2012, Vol. 7(4), pp. 1–13.
Huppert 2005
Huppert, D.D., 'An Overview of Fishing Rights', *Reviews in Fish Biology and Fisheries*, 2005, Vol. 15(3), pp. 201–215.
Husak *et al.* 2005
Husak, A.L., Grado, S.C., Bullard, S.H. & Moffat, S.O., 'Silvicultural Best Management Practice Compliance Monitoring Programs in the Southern United States', *Southern Journal of Applied Forestry*, 2005, Vol. 29(1), pp. 48–52.
Hutchings 2000
Hutchings, J.A., 'Collapse and Recovery of Marine Fishes', *Nature*, 2000, Vol. 406, pp. 882–885.
Hysing 2009a
Hysing, E., 'Governing without Government? The Private Governance of Forest Certification in Sweden', *Public Administration*, 2009, Vol. 87(2), pp. 312–326.
Hysing 2009b
Hysing, E., 'From Government to Governance? A Comparison of Environmental Governing in Swedish Forestry and Transport', *Governance*, 2009, Vol 22(4), pp. 647–672.
Hysing & Olsson 2005
Hysing, E. & Olsson, J., 'Sustainability through Good Advice? Assessing the Governance of Swedish Forest Biodiversity', *Environmental Politics*, 2005, Vol. 14(4), pp. 510–526.
Ice, Schilling & Vowell 2010
Ice, G.G., Schilling, E. & Vowell, J., 'Trends for Forestry Best Management Practices Implementation', *Journal of Forestry*, 2010, Vol. 108(6), pp. 267–273.
Ice *et al.* 2004
Ice, G., Dent, L., Robben, J., Cafferata, P., Light, J., Sugden, B. & Cundy, T., 'Programs Assessing Implementation and Effectiveness of State Forest Practice Rules and BMPs in the West', *Water, Air and Soil Pollution: Focus*, 2004, Vol. 4(1), pp. 143–169.
Imperial & Yandle 2005
Imperial, M.T. & Yandle, T., 'Taking Institutions Seriously: Using the IAD Framework to Analyze Fisheries Policy', *Society & Natural Resources*, 2005, Vol. 18(6), pp. 493–509.
Indrarto *et al.* 2012
Indrarto, G.B., Murharjanti, P., Khatarina, J., Pulungan, I., Ivalerina, F., Rahman, J., Prana, M.M., Resosudarmo, I.A.P. & Muharrom, E., The Context of REDD+ in Indonesia: Drivers, Agents and Institutions, Cifor Working Paper No. 92, 2012.

Intertek Moody Marine 2012
Intertek Moody Marine, Second Re-assessment for New Zealand Hoki Fisheries, 2012.
Isaacs, Hara & Raakjaer 2007
Isaacs, M., Hara, M. & Raakjaer, J., 'Has Reforming South African Fisheries Contributed to Wealth Redistribution and Poverty Alleviation?', *Ocean & Coastal Management*, 2007, Vol. 50(5–6), pp. 301–313.
ITS Global 2011
ITS Global, The Economic Contribution of Indonesia's Forest-Based Industries, 2011, available at: www.itsglobal.net/sites/default/files/itsglobal/ITS_ Indo forest_Economic_Report.pdf.
Izawa & Makino 2005
Izawa, A. & Makino, M., 'MSC Certification and its Implementation for Japan's Fisheries – Its Role and Issues', *Global Environmental Research*, 2005, Vol. 9(2), pp. 151–156.
Jack 1998
Jack, D., Of Markets and Forests: Certification and Sustainable Forestry in Bolivia, Deming, WA: Watson Foundation, 1998, available at: http://citeseerx.ist.psu.edu/viewdoc/download?doi=10.1.1.203.1244& rep=rep1&type=pdf.
Jackson *et al.* 2001
Jackson, J.B.C., Kirby, M.X., Berger, W.H., Bjorndal, K.A., Botsford, L.W., Bourque, B.J., Bradbury, R.H., Cooke, R., Erlandson, J., Estes, J.A., Hughes, T.P., Kidwell, S., Lange, C.B., Lenihan, H.S., Pandolfi, J.M., Peterson, C.H., Stenecke, R.S., Tegner, M.J. & Warner, R.R., 'Historical Overfishing and the Recent Collapse of Coastal Ecosystems', *Science*, 2001, Vol. 293(5530), pp. 629–637.
Jessop 1994
Jessop, B., 'Post Fordism and the State', in Ash, A. (ed.), *Post-Fordism: A Reader*, Oxford: Blackwell Publishing Ltd, 1994, pp. 251–279.
Johansson 2012
Johansson, J., 'Challenges to the Legitimacy of Private Forest Governance – The Development of Forest Certification in Sweden', *Environmental Policy and Governance*, 2012, Vol. 22(6), pp. 424–436.
Johansson *et al.* 2013
Johansson, T., Hjältén, J., De Jong, J. & Von Stedingk, H., 'Environmental Considerations from Legislation and Certification in Managed Forest Stands: A Review of their Importance for Biodiversity', *Forest Ecology and Management*, 2013, Vol. 303, pp. 98–112.
Johansson 2014
Johansson, J., 'Towards Democratic and Effective Forest Governance? The Discursive Legitimation of Forest Certification in Northern Sweden', *Local Environment*, 2014, Vol. 19(7), pp. 803–819.
Johansson & Keskitalo 2014
Johansson, J. & Keskitalo, E.C.H., 'Coordinating and Implementing Multiple Systems for Forest Management: Implications of the Regulatory Framework for Sustainable Forestry in Sweden', *Journal of Natural Resources Policy Research*, 2014, Vol. 6(3), pp. 117–133.
Johnson 1972
Johnson, O.E.G., 'Economic Analysis, the Legal Framework and Land Tenure Systems', *Journal of Law and Economics*, 1972, Vol. 15(1), pp. 259–276.

Jordan 2008

Jordan, A., 'The Governance of Sustainable Development: Taking Stock and Looking Forwards', *Environment and Planning C: Government and Policy*, 2008, Vol. 26(1), pp. 17–33.

Kalfagianni & Pattberg 2013

Kalfagianni, A. & Pattberg, O., 'Fishing in Muddy Waters: Exploring the Conditions for Effective Governance of Fisheries and Aquaculture', *Marine Policy*, 2013, Vol. 38, pp. 124–132.

Kamieniecki 2000

Kamieniecki, S., 'Testing Alternative Theories of Agenda Setting: Forest Policy Change in British Columbia, Canada', *Policy Studies Journal*, 2000, Vol. 28(1), pp. 176–189.

Karlsen, Hermansen & Dreyer 2012

Karlsen, K.M., Hermansen, O. & Dreyer, B.M., 'Eco-Labeling of Seafood: Does It Affect the Harvesting Patterns of Norwegian Fishermen?', *Marine Policy*, 2012, Vol. 36(5), pp. 1123–1130.

Keskitalo *et al.* 2009

Keskitalo, E.C.H., Sandstrom, C., Tysiachniouk, M.S. & Johansson, J., 'Local Consequences of Applying International Norms: Differences in the Application of Forest Certification in Northern Sweden, Northern Finland, and Northwest Russia', *Ecology and Society*, 2009, Vol. 14(2), online at: www.ecologyandsociety.org/vol14/iss2/art1/.

Kilgore & Blinn 2004

Kilgore, M.A. & Blinn, C.R., 'Policy Tools to Encourage the Application of Sustainable Timber Harvesting Practices in the United States and Canada', *Forest Policy and Economics*, 2004, Vol. 6(2), pp. 111–127.

Kilgore, Ellefson & Phillips 2003

Kilgore, M.A., Ellefson, P.V. & Phillips, M.J., 'Ensuring the Application of Sound Forest Practices on Private Forests: Challenges Facing the Design and Implementation of State Compliance Monitoring Programmes', in Teeter, L.D., Cashore, B.W. & Zhang, D. (eds.), *Forest Policy for Private Forestry: Global and Regional Challenges*, Wallingford/New York: CABI Publishing, 2003, pp. 117–128.

Kilgore, Ellefson & Phillips 2004

Kilgore, M.A., Ellefson, P.V. & Phillips, M.J., 'BMP Compliance Monitoring Programs in the Eastern United States', *Water, Air and Soil Pollution: Focus*, 2004, Vol. 4(1), pp. 119–130.

Klassen, Romero & Putz 2014

Klassen, A., Romero, C. & Putz, F.E. (eds.), *Forest Stewardship Council Certification of Natural Forest Management in Indonesia: Required Improvements, Costs, Incentives, and Barriers*, Vienna: IUFRO, 2014.

Kolding, Bener & Bavinck 2014

Kolding, J., Bener, C. & Bavinck, M., 'Small-scale Fisheries: Importance, Vulnerability and Deficient Knowledge', in Garcia, S.M., Rice, J. & Charles, A. (eds.), *Governance of Marine Fisheries and Biodiversity Conservation: Interaction and Coevolution*, Chichester: John Wiley & Sons, 2014, pp. 317–332.

Lallemand *et al.* 2014

Lallemand, P., Berg, M., Hansen, M., Venter, C. & Purves, M., 'An Analysis of the Economic Benefits of MSC Certification for the South African Hake Fishery', 2014, available at: http://ir.library.oregonstate.edu/xmlui/bitstream/handle/1957/55162/Purves%20392%20IIFET.pdf?sequence=1.

Lantto & Mörkenstam 2008

Lantto, P. & Mörkenstam, U., 'Sami Rights and Sami Challenges: The Modernization Process and the Swedish Sami Movement, 1886–2006', *Scandinavian Journal of History,* 2008, Vol. 33, pp. 26–51.

Lee, Smyth & Boutin 2004

Lee, P., Smyth, C. & Boutin, S., 'Quantitative Review of Riparian Buffer Width Guidelines from Canada and the United States', *Journal of Environmental Management,* 2004, Vol. 70(2), pp. 165–180.

León *et al.* 2012

León, R., Uberhuaga, P., Benavides, J.P. & Andersson, K., 'Public Policy Reforms and Indigenous Forest Governance: The Case of the Yuracaré People in Bolivia', *Conservation and Society,* 2012, Vol. 10(2), pp. 195–207.

Lindayati 2002

Lindayati, R., 'Ideas and Institutions in Social Forestry Policy', in Colfer, C.J.P. & Resosudarmo, I.A.P. (eds.), *Which Way Forward? People, Forests, and Policymaking in Indonesia,* Washington DC: Resources for the Future, 2002, pp. 36–59.

Lister 2012

Lister, J., *Corporate Social Responsibility and the State: International Approaches to Forest Co-regulation,* Vancouver: UBC Press, 2012.

Liu 2014

Liu, J., 'The Government's Roles in Transnational Forest Governance', *Recht der Werkelijkheid,* 2014, Vol. 35(3), pp. 54–78.

Lock & Leslie 2007

Lock, K. & Leslie, S. New Zealand's Quota Management System: A History of the First 20 Years, Motu Working Paper 07-02, 2007, available at: http://papers.ssrn.com/sol3/papers.cfm?abstract_id=978115.

Lopuch 2008

Lopuch, M., 'Benefits of Certification for Small-Scale Fisheries', in Ward, T. & Phillips, B. (eds.), *Seafood Ecolabelling: Principles and Practice,* Oxford: Wiley-Blackwell, 2008, pp. 307–321.

Lucier & Shepard 1997

Lucier, A.A. & Shepard, J.P., 'Certification and Regulation of Forestry Practices in the United States: Implications for Intensively Managed Plantations', *Biomass and Bioenergy,* 1997, Vol. 13(4–5), pp. 193–199.

Lueck 1989

Lueck, D., 'The Economic Nature of Wildlife Law', *Journal of Legal Studies,* 1989, Vol. 18(2), pp. 291–324.

Makino 2010

Makino, M., 'Japanese Coastal Fisheries', in Grafton, R.Q., Hilborn, R., Squires, D., Tait, M. & Williams, M. (eds.), *Handbook of Marine Fisheries, Conservation and Management,* Oxford: Oxford University Press, 2010, pp. 287–298.

Makino 2011

Makino, M., *Fisheries Management in Japan: Its Institutional Features and Case Studies,* Dordrecht/Heidelberg/London/New York: Springer Science & Business Media, 2011.

Makino & Matsuda 2005

Makino, M. & Matsuda, H., 'Co-management in Japanese Coastal Fisheries: Institutional Features and Transaction Costs', *Marine Policy,* 2005, Vol. 29(5), pp. 441–450.

Marine Stewardship Council 2008

Marine Stewardship Council and Governance Review Commission, 'Simpler Faster and More Consistent: MSC Launches New Fishery Assessment Methodology', 2008.

Markopoulos 2002

Markopoulos, M., 'The Role of Certification in Community-Based Forest Enterprise', in Meidinger, E., Elliott, C. & Oesten, G. (eds.), *Social and Political Dimensions of Forest Certification*, Eifelweg: Verlag, 2002, pp. 105–131.

Matsuda *et al.* 2010

Matsuda, H., Makino, M., Tomiyama, M., Gelcich, S. & Castilla, J.C., 'Fishery Management in Japan', *Ecological Research*, 2010, Vol. 25(5), pp. 899–907.

McCarthy 2000

McCarthy, J., 'The Changing Regime: Forest Property and Reformasi in Indonesia', *Development and Change*, 2000, Vol. 31(1), pp. 91–129.

McCarthy 2012

McCarthy, J.F., 'Certifying in Contested Spaces: Private Regulation in Indonesian Forestry and Palm Oil', *Third World Quarterly*, 2012, Vol. 33(10), pp. 1871–1888.

McCarthy *et al.* 2006

McCarthy, J., Barr, C., Resosudarmo, I. & Dermawan, A. 'Origins and Scope of Indonesia's Decentralization Laws', in Barr, C., Resosudarno, I., Dermawan, A., McCarthy, J.F., Moeliniono, M. & Setiono, B. (eds.), *Decentralization of Forest Administration in Indonesia: Implications for Forest Sustainability, Economic Development and Community Livelihoods*, Bogo: Center for International Forestry Research (CIFOR), 2006, pp. 31–57.

McCay *et al.* 1995

McCay, B.J., Creed, C.F., Finlayson, A.C., Apostle, R., & Mikalsen, K., 'Individual Transferable Quotas (ITQs) in Canadian and US Fisheries', *Ocean & Coastal Management*, 1995, Vol. 28(1–3), pp. 85–115.

McCay *et al.* 2014

McCay, B.J., Micheli, F., Ponce-Díaz, G., Murray, G., Shester, G., Ramírez-Sánchez, S. & Weisman, W., 'Cooperatives, Concessions, and Co-management on the Pacific Coast of Mexico', *Marine Policy*, 2014, Vol. 44, pp. 49–59.

McDermott, Cashore & Kanowski 2010

McDermott, C., Cashore, B. & Kanowski, P., *Global Environmental Forest Policies: An International Comparison*, London/Washington, DC: Earthscan, 2010.

McDermott, Noah & Cashore 2008

McDermott, C.L., Noah, E. & Cashore, B., 'Differences that "Matter"? A Framework for Comparing Environmental Certification Standards and Government Policies', *Journal of Environmental Policy and Planning*, 2008, Vol. 10(1), pp. 47–70.

McGinnis & Ostrom 2008

McGinnis, M. & Ostrom, E., 'Will Lessons from Small-Scale Social Dilemmas Scale Up?', in Biel, A., Eek, D., Gärling, T. & Gustafsson, M. (eds.), *New Issues and Paradigms in Research on Social Dilemmas*, New York: Springer, 2008, pp. 189–211.

Medina, Pokorny & Campbell 2009

Medina, G., Pokorny, B. & Campbell, B.M., 'Community Forest Management for Timber Extraction in the Amazon Frontier', *International Forestry Review*, 2009, Vol. 11(3), pp. 408–420.

Meidinger 2006

Meidinger, E., 'The Administrative Law of Global Private–Public Regulation: The Case of Forestry', *European Journal of International Law*, 2006, Vol. 17(1), pp. 47–87.

Memon & Kirk 2010

Memon, P.A. & Kirk, N.A., 'Barriers to Collaborative Governance in New Zealand Fisheries: Pt II', *Geography Compass*, 2010, Vol. 4(7), pp. 789–802.

Memon & Kirk 2011

Memon, P.A. & Kirk, N.A., 'Institutional Reforms in New Zealand Fisheries as an Ecological Modernization Project', *Society & Natural Resources*, 2011, Vol. 24(1), pp. 995–1010.

Millennium Ecosystem Assessment 2005

Millennium Ecosystem Assessment, *Ecosystem and Human Well-Being: Synthesis*, Washington, DC: Island Press, 2005.

Mincher 2008

Mincher, R., 'New Zealand's Challenger Scallop Enhancement Company: From Reseeding to Self-Governance', in Townsend, R.E., Shotton, R. & Uchida, H. (eds.), *Case Studies in Fisheries Self-Governance*, Rome: FAO, 2008, pp. 307–321.

Ministry for Primary Industries (New Zealand) 2015

Ministry for Primary Industries, The Status of New Zealand's Fisheries, available at: http://fs.fish.govt.nz/Doc/24000/status-of-nz-fisheries-2015.pdf.ashx, 2015.

Moody Marine 2008

Moody Marine Ltd, First Surveillance Report 1 for Reassessment of New Zealand Hoki Fishery, 2008.

Moye 2010

Moye, P.A., 'Private Certification Versus Public Certification in the International Environmental Arena: The Marine Stewardship Council and Marine Eco-Label Japan Fisheries Certification Schemes as Case Studies', *Vanderbilt Journal of Transnational Law*, 2010, Vol. 43(2), pp. 533–551.

MSC 2004

MSC, Final Certification Report for South African Hake Trawl Fishery, 2004.

MSC 2015

MSC, South Africa Hake Trawl Fishery, Public Certification Report, 2015.

Mueller 2004

Mueller, D.C., 'Public Choice: An Introduction', in Rowley, C. & Schneider, F. (eds.), *The Encyclopedia of Public Choice*, New York/Boston/Dordrecht/London/Moscow: Kluwer Academic Publishers, 2004.

Muhtaman & Prasetyo 2006

Muhtaman, D.R. & Prasetyo, F.A., 'Forest Certification in Indonesia', in Cashore, B., Gale, F., Meidinger, E. & Newsom, D. (eds.), *Confronting Sustainability: Forest Certification in Developing and Transitioning Countries*, New Haven, CT: Yale School of Forestry and Environmental Studies, 2006, pp. 33–68.

Müller, Pacheco, & Montero, 2014

Müller, R., Pacheco, P. & Montero, J., The Context of Deforestation and Forest Degradation in Bolivia: Drivers, Agents and Institutions, 2014, Center for International Forestry Research, Occasional Paper, No. 108, available at: http://www.cifor.org/publications/pdf_files/OccPapers/OP-108.pdf.

Müller *et al.* 2013

Müller, R., Pistorius, T., Rohde, S., Gerold, G. & Pacheco, P., 'Policy Options to Reduce Deforestation Based on a Systematic Analysis of Drivers and Agents in Lowland Bolivia', *Land Use Policy*, 2013, Vol. 30(1), pp. 895–907.

Murota 2011

Murota, T., Fishery Commons in Japan: Their Legal Framework and Recent Crises,

Conference paper at the Conference Sustaining Commons: Sustaining our Future, the Thirteenth Biennial Conference of the International Association for the Study of the Commons, 10–14 January 2011, Hyderabad, India.

Nanda 2006

Nanda, P., 'The "Good Governance" Concept Revisited', *The ANNALS of the American Academy of Political and Social Science*, 2006, Vol. 603(1), pp. 269–283.

Nebel *et al.* 2005

Nebel, G., Quevedo, L., Bredahl Jacobsen, J. & Helles, F., 'Development and Economic Significance of Forest Certification: The Case of FSC in Bolivia', *Forest Policy and Economics*, 2005, Vol. 7(2), pp. 175–186.

Newsom, Bahn & Cashore 2006

Newsom, D., Bahn, V. & Cashore, B., 'Does Forest Certification Matter? An Analysis of Operation-Level Changes Required during the Smart Wood Certification Process in the United States', *Forest Policy and Economics*, 2006, Vol. 9(3), pp. 197–208.

Nielsen & Hara 2006

Nielsen, J.R. & Hara, M., 'Transformation of South African Industrial Fisheries', *Marine Policy*, 2006, Vol. 30(1), pp. 43–50.

Nittler & Nash 1999

Nittler, J.B. & Nash, D.W., 'The Certification Model for Forestry in Bolivia', *Journal of Forestry*, 1999, Vol. 97(3), pp. 32–36.

Novy-Hildesley & Short 2003

Novy-Hildesley, J. & Short, K., 'Community-Based Certification: A Route to Sustainable Fisheries', in Phillips, B., Ward, T. & Chaffee, C. (eds.), *Eco-Labeling in Fisheries: What Is It About*, Oxford: Wiley-Blackwell, 2003, pp. 162–175.

Nylund 2010

Nylund, J.-E., *Swedish Forest Policy since 1990: Reforms and Consequences*, Uppsala: Swedish University of Agricultural Sciences, Institutionen för skogens produkter, 2010.

Nurrochmat *et al.* 2014

Nurrochmat, D.R., Dharmawan, A.H., Obidzinski, K., Dermawan, A. & Erbaugh, J.T., 'Contesting National and International Forest Regimes: Case of Timber Legality Certification for Community Forests in Central Java, Indonesia', *Forest Policy and Economics*, 2014, Vol. 68, pp. 54–64.

Obidzinski *et al.* 2014

Obidzinski, K., Dermawan, A., Andrianto, A., Komarudin, H. & Hernawan, D., 'The Timber Legality Verification System and the Voluntary Partnership Agreement (VPA) in Indonesia: Challenges for the Small-Scale Forestry Sector', *Forest Policy and Economics*, 2014, Vol. 48, pp. 24–32.

O'Carroll 2004

O'Carroll, A., *Where Land and Waters Meet: An Assessment of Canada's Riparian Forest Management Standards*, Edmonton, Alberta: Global Forest Watch Canada, 2004.

OECD 2011a

OECD, *Fisheries Policy Reform: National Experiences*, Paris: OECD, 2011.

OECD 2011b

OECD, 'Introducing Market-Based Reforms to Ensure Sustainability in New Zealand Fisheries', in *Fisheries Policy Reform: National Experiences*, Paris: OECD, 2011, pp. 31–44.

OECD 2013

OECD, *OECD Review of Fisheries: Policy and Summary Statistics*, Paris: OECD, 2013.

Olson 1973

Olson, M., 'Evaluating Performance in the Public Sector', in Moss, M. (ed.), *The Measurement of Economic and Social Performance*, Cambridge: NBER, 1973, pp. 355–409.

Olson 1986

Olson, M., 'Toward a More General Theory of Governmental Structure', *The American Economic Review*, 1986, Vol. 76(2), pp. 120–125.

Oosterveer 2010

Oosterveer, P., Ecolabelling and Seafood Certification in Equitable Benefit Sharing of Tuna Fisheries, 2010, pp. 1–15, available at: http://awsassets.panda.org/downloads/background_paper__ecolabelling_and_certification_schemes.pdf.

Ophuls 1973

Ophuls, W., 'Leviathan or Oblivion', in Daly, H.E. (ed.), *Toward a Steady State Economy*, San Francisco, CA: Freeman, 1973.

Orzechowski 1977

Orzechowski, W., 'Economic Models of Bureaucracy: Survey, Extensions, and Evidence', in Borcherding, T.E. (ed.), *Budgets and Bureaucrats: The Sources of Government Growth*, Durham, NC: Duke University Press, 1977, pp. 199–259.

Ostrom 1990

Ostrom, E., *Governing the Commons: The Evolution of Institutions for Collective Actions*, Cambridge: Cambridge University Press, 1990.

Ostrom 2010

Ostrom, E., 'Beyond Markets and States: Polycentric Governance of Complex Economic Systems', *American Economic Review*, 2010, Vol. 100(3), pp. 641–672.

Overdevest 2010

Overdevest, C., 'Comparing Forest Certification Schemes: The Case of Ratcheting Standards in the Forest Sector', *Socio-Economic Review*, 2010, Vol. 8(1), pp. 47–76.

Overdevest & Zeitlin 2014

Overdevest, C. & Zeitlin, J., 'Constructing a Transnational Timber Legality Assurance Regime: Architecture, Accomplishments, Challenges', *Forest Policy and Economics*, 2014, Vol. 48, pp. 6–15.

Pacheco 2004a

Pacheco, P., Law Compliance: Bolivia Case Study, Rapport à la FAO, Rome. Meilleures pratiques pour l'application des lois dans le secteur forestier, 2004, available at: www.fao.org/forestry/19583-06647b501b992f13fa80edbbb8942712c.pdf.

Pacheco 2004b

Pacheco, P., 'What Lies behind Decentralisation? Forest, Powers and Actors in Lowland Bolivia', *The European Journal of Development Research*, 2004, Vol. 16(1), pp. 90–109.

Pacheco 2005

Pacheco, P., Decentralization of Forest Management in Bolivia: Who Benefits and Why, 2005, available at: www.cifor.org/publications/pdf_files/interlaken/pablo_pacheco.pdf.

Pacheco 2011

Pacheco, P., Land Tenure, Forest and Political Reforms: A Look at their Implications for Common-Property Forests in Lowland Bolivia, Conference Paper: Sustaining Commons: Sustaining Our Future, the Thirteenth Biennial Conference of the

International Association for the Study of the Commons, 2011.

Pacheco & Benatti 2015

Pacheco, P. & Benatti, J.H., 'Tenure Security and Land Appropriation under Changing Environmental Governance in Lowland Bolivia and Pará', *Forests*, 2015, Vol. 6(2), pp. 464–491.

Pacheco, De Jong & Johnson 2010

Pacheco, P., De Jong, W. & Johnson, J., 'The Evolution of the Timber Sector in Lowland Bolivia: Examining the Influence of Three Disparate Policy Approaches', *Forest Policy and Economics*, 2010, Vol. 12(4), pp. 271–276.

Pacheco *et al.* 2008

Pacheco, P., Barry, D., Cronkleton, P. & Larson, A.M., *The Role of Informal Institutions in the Use of Forest Resources in Latin America*, Bogor: Center for International Forestry Research (CIFOR), 2008.

Paillet *et al.* 2010

Paillet, Y., Bergès, L., Hjältén, J., Ódor, P., Avon, C., Römermann, M.B., Bijlsma, R.-J., De Bruyn, L., Fuhr, M., Grandin, U., Kanka, R., Lundin, L., Luque, S., Magura, T., Matesanz, S., Mészáros, I., Sebastià, M.-T., Schmidt, W., Standovár, T., Tóthmérész, B., Uotila, A., Valladares, F., Vellak, K. & Virtanen, R., 'Biodiversity Differences between Managed and Unmanaged Forests: Meta Analysis of Species Richness in Europe', *Conservation Biology*, 2010, Vol. 24(1), pp. 101–112.

Palo & Vanhanen 2012

Palo, M. & Vanhanen, H., *World Forests from Deforestation to Transition?*, Dordrecht/London/Boston: Kluwer Academic Publishers, 2012.

Pauly *et al.* 1998

Pauly, D., Christensen, V., Dalsgaard, J., Froese, R. & Torres Jr., F., 'Fishing down Marine Food Webs', *Science*, 1998, Vol. 279(5352), pp. 860–863.

Pauly *et al.* 2002

Pauly, D., Christensen, V., Guénette, S., Pitcher, T.J., Sumaila, U.R., Walters, C.J., Watson, R. & Zeller, D., 'Towards Sustainability in World Fisheries', *Nature*, 2002, Vol. 418, pp. 689–695.

Pellegrini 2009

Pellegrini, L., Forest Management in Bolivia, Honduras and Nicaragua: Reform Failures?, ISS Working Paper Series/General Series, 2009.

Pellegrini & Dasgupta 2011

Pellegrini, L. & Dasgupta, A., 'Land Reform in Bolivia: The Forestry Question', *Conservation and Society*, 2011, Vol. 9(4), pp. 274–285.

Peña-Claros & Dockry 2010

Peña-Claros, M. & Dockry, M.J., Bolivia, 2010, available at: www.tropenbos.org/file.php/645/11-m.pe%C3%B1a-claros-r.guzman-m.dockry.pdf.

Pérez-Ramírez, Ponce-Díaz & Lluch-Cota 2012

Pérez-Ramírez, M., Ponce-Díaz, G. & Lluch-Cota, S., 'The Role of MSC Certification in the Empowerment of Fishing Cooperatives in Mexico: The Case of Red Rock Lobster Co-managed Fishery', *Ocean & Coastal Management*, 2012, Vol. 63, pp. 24–29.

Pérez-Ramírez *et al.* 2012

Pérez-Ramírez, M., Phillips, B., Lluch-Belda, D. & Lluch-Cota, S., 'Perspectives for Implementing Fisheries Certification in Developing Countries', *Marine Policy*, 2012, Vol. 36(1), pp. 297–302.

Peters 1994
Peters, B.G., 'Managing the Hollow State', *International Journal of Public Administration*, 1994, Vol. 17(3–4), pp. 739–756.
Phillips, Bourillón & Ramade 2008
Phillips, B., Bourillón, L. & Ramade, M., 'Case Study 2: The Baja California, Mexico, Lobster Fishery', in Ward, T. & Phillips, B. (eds.), *Seafood Ecolabelling, Principles and Practice*, Chichester: John Wiley & Sons, 2008, pp. 259–268.
Phillips, Swift & Blinn 2000
Phillips, M.J., Swift Jr., L.W. & Blinn, C.R., 'Best Management Practices for Riparian Areas', in Verry, E.S., Hornbeck, J.W. & Dolloff, C.A. (eds.), *Riparian Management in Forests of the Continental Eastern United States*, Boca Raton, FL: Lewis Publishers, 2000, pp. 273–286.
Phillips, Ward & Chaffee 2008
Phillips, B., Ward, T. & Chaffee, C. (eds.), *Eco-Labelling in Fisheries: What Is It All About*, Oxford: Wiley-Blackwell, 2008.
Plagányi *et al.* 2007
Plagányi, E.E., Rademeyer, R.A., Butterworth, D.S., Cunningham, C.L. & Johnston, S.J., 'Making Management Procedures Operational – Innovations Implemented in South Africa', *ICES Journal of Marine Science: Journal du Conseil*, 2007, Vol. 64(4), pp. 626–632.
Ponte 2006
Ponte, S., Ecolabels and Fish Trade: Marine Stewardship Council Certification and the South African Hake Industry, Trade Law Center for Southern Africa, Working Paper, 2006.
Ponte 2008
Ponte, S., 'Greener than Thou: The Political Economy of Fish Ecolabeling and its Local Manifestations in South Africa', *World Development*, 2008, Vol. 36(1), pp. 159–175.
Ponte 2012
Ponte, S., 'The Marine Stewardship Council (MSC) and the Making of a Market for "Sustainable Fish"', *Journal of Agrarian Change*, 2012, Vol. 12(2–3), pp. 300–315.
Popescu & Ogushi 2013
Popescu, I. & Ogushi, T., 'Fisheries: Fisheries in Japan', in Directorate General for Internal Policies, Policy Department B: Structural and Cohesion Policies, European Parliament, 2013, pp. 15–53, available at: www.europarl.europa.eu/RegData/etudes/note/join/2014/529044/IPOL-PECH_NT(2014)529044_EN.pdf.
Prakash & Potoski 2007
Prakash, A., & Potoski, M., 'Collective Action through Voluntary Environmental Programs: A Club Theory Perspective', *The Policy Studies Journal*, 2007, Vol. 35(4), pp. 773–792.
Prasad 2003
Prasad, B.C., 'Institutional Economics and Economic Development: The Theory of Property Rights, Economic Development, Good Governance and the Environment', *International Journal of Social Economics*, 2003, Vol. 30(6), pp. 741–762.
Purnomo & Anand 2014
Purnomo, E.P. & Anand, P.B., 'The Conflict of Forest Tenure and the Emergence of Community Based Forest Management in Indonesia', *Journal of Government and Politics*, 2014, Vol. 5(1), pp. 20–31.

Purwanto 2005

Purwanto, S.A., 'Forest Resource Management and Self-Governance in Regional Autonomy Indonesia', in Erb, M., Faucher, C. & Sulistiyanto, P. (eds.), *Regionalism in Post-Suharto Indonesia*, London: Routledge, 2005, pp. 413–442.

Quevedo 2006

Quevedo, L., 'Bolivia', in Cashore, B., Gale, F., Meidinger, E. & Newsom, D. (eds.), *Confronting Sustainability: Forest Certification in Developing and Transitioning Countries*, New Haven, CT: Yale School of Forestry Environmental Studies, 2006, pp. 303–336.

Raemaekers *et al.* 2011

Raemaekers, S., Hauck, M., Bürgener, M., Mackenzie, A., Maharaj, G., Plagányi, E.E. & Britz, P.J., 'Review of the Causes of the Rise of the Illegal South African Abalone Fishery and Consequent Closure of the Rights-Based Fishery', *Ocean & Coastal Management*, 2011, Vol. 54(6), pp. 433–445.

Real 2002

Real, B., *Legal Reforms in Bolivia in the 1990s: Challenges and Opportunities for Decentralization, Indigenous Rights and Forest Management*, Doctoral thesis, University of Florida, 2002.

Redo, Millington & Hindery 2011

Redo, D., Millington, A.C. & Hindery, D., 'Deforestation Dynamics and Policy Changes in Bolivia's Post-Neoliberal Era', *Land Use Policy*, 2011, Vol. 28(1), pp. 227–241.

Rees 2005

Rees, E.B., *In What Sense a Fisheries Problem?: Negotiating Sustainable Growth in New Zealand Fisheries*, Diss., Auckland: University of Auckland, 2005.

Resosudarmo 2005

Resosudarmo, I.A.P., 'Closer to People and Trees: Will Decentralisation Work for the People and the Forests of Indonesia?', in Ribot, J.C. & Larson, A.M. (eds.), *Democratic Decentralisation through a Natural Resource Lens*, Abingdon: Routledge, 2005, pp. 110–132.

Revesz 2001

Revesz, R.L., 'Federalism and Environmental Regulation: A Public Choice Analysis', *Harvard Law Review*, 2001, Vol. 115(2), pp. 553–641.

Rhodes 2005

Rhodes, R.A.W., 'The Hollowing out of the State: The Changing Nature of the Public Service in Britain', *The Political Quarterly*, 2005, Vol. 65(2), pp. 138–151.

Richardson, Naiman & Bisson 2012

Richardson, J.S., Naiman, R.J. & Bisson, P.A., 'How Did Fixed-Width Buffers Become Standard Practice for Protecting Freshwaters and their Riparian Areas from Forest Harvest Practices?', *Freshwater Science*, 2012, Vol. 31(1), pp. 232–238.

Royo & Wells 2012

Royo, N. & Wells, A., Community Based Forest Management in Indonesia: A Review of Current Practice and Regulatory Frameworks, Background paper for the Forest Dialogue on Investing in Locally Controlled Forestry (ILCF), Yogyakarta, Indonesia, 2012.

Rudel *et al.* 2005

Rudel, T.K., Coomes, O.T., Moran, E., Achard, F., Angelsen, A., Xu, J. & Lambin, E., 'Forest Transitions: Towards a Global Understanding of Land Use Change', *Global Environmental Change*, 2005, Vol. 15(1), pp. 23–31.

Safitri 2010

Safitri, M.A., *Forest Tenure in Indonesia: The Socio-Legal Challenges of Securing Communities' Rights*, Doctoral thesis, Faculty of Law, Leiden University, 2010, available at: https://openaccess.leidenuniv.nl/handle/1887/16242.

Sahlin 2011

Sahlin, M., *Under the Cover of the Swedish Forestry Model*, Stockholm: Swedish Society for Nature Conservation (SSNC), 2011.

Sahlin 2013

Sahlin, M., Credibility at Stake: How FSC Sweden Fails to Safeguard Forest Biodiversity, Stockholm: Swedish Society for Nature Conservation, 2013, available at: www.naturskyddsforeningen.se/sites/default/files/dokument-media/rapporter/2013_engelsk_rapport_skog_credibility_at_stake.pdf.

Sahlin & Svenska Naturskyddsföreningen 2009

Sahlin, M. & Svenska Naturskyddsföreningen, *Cutting the Edge: The Loss of Natural Forests in Sweden*, Stockholm: Swedish Society for Nature Conservation (SSNC), 2009.

Saito 2008

Saito, F., 'Decentralization and Local Governance: Introduction and Overview', in Saito, F. (ed.), *Foundations for Local Governance. Decentralization in Comparative Perspective*, Heidelberg: Springer, 2008, pp. 1–24.

Sands & Peel 2012

Sands, P. & Peel, J., *Principles of International Environmental Law*, Cambridge: Cambridge University Press, 2012.

Sari 2013

Sari, I.M., *Community Forests at a Crossroads: Lessons Learned from Lubu Beringin Village Forest and Guguk Customary Forests in Jambi Province, Sumatra, Indonesia*, Doctoral thesis, University of Oslo, 2013.

Sarker *et al.* 2015

Sarker, A., Ikeda, T., Abe, T. & Inoue, K., 'Design Principles for Managing Coastal Fisheries Commons in Present-Day Japan', *Ecological Economics*, 2015, Vol. 117, pp. 32–38.

Savcor Indufor 2005

Savcor Indufor, *Effectiveness and Efficiency of FSC and PEFC Forest Certification on Pilot Areas in Nordic Countries*, Helsinki: Federation of Nordic Forest Owners' Organizations, 2005.

Schindler *et al.* 2002

Schindler, D.E., Essington, T.E., Kitchell, J.F., Boggs, C. & Hilborn, R., 'Sharks and Tunas: Fisheries Impacts on Predators with Contrasting Life Histories', *Ecological Applications*, 2002, Vol. 12(3), pp. 735–748.

Schlager & Ostrom 1992

Schlager, E. & Ostrom, E., 'Property-Rights Regimes and Natural Resources: A Conceptual Analysis', *Land Economics*, 1992, Vol. 68(3), pp. 249–262.

Schlager & Ostrom 1999

Schlager, E. & Ostrom, E., 'Property Rights Regimes and Coastal Fisheries: An Empirical Analysis', in McGinnis, M.D. (ed.), *Polycentric Governance and Development: Readings from the Workshop in Political Theory and Policy Analysis*, Ann Arbor, MI: University of Michigan Press, 1999, pp. 87–113.

Schlyter, Stjernquist & Bäckstrand 2009

Schlyter, P., Stjernquist, I. & Bäckstrand, K., 'Not Seeing the Forest for the Trees?

The Environmental Effectiveness of Forest Certification in Sweden', *Forest Policy and Economics*, 2009, Vol. 11(5–6), pp. 375–382.

Schuck *et al.* 2002

Schuck, A., Päivinen, R., Hytönen, T. & Pajari, B., Compilation of Forestry Terms and Definitions, Torikatu: European Forest Institute, Internal Report No. 6, 2002.

Scott, Cafaggi & Senden 2011

Scott, C., Cafaggi, F. & Senden, L., 'The Conceptual and Constitutional Challenge of Transnational Private Regulation', *Journal of Law and Society*, 2011, Vol. 38(1), pp. 1–19.

SCS 2001a

SCS, MSC Certification for New Zealand Commercial Hoki Fishery, 2001.

SCS 2001b

SCS, First Surveillance Report for New Zealand Hoki Fishery, 2001.

SCS 2004

Scientific Certification Systems, An MSC Assessment of the Red Rock Lobster Fishery, Baja California, Mexico, 2004.

SCS 2007

SCS, 12th Surveillance Report for New Zealand Hoki Fishery, 2007.

SCS 2011

Scientific Certification Systems, California Spiny Lobster Fishery, Baja California, Mexico, MSC Re-Certification, 2011.

SCS 2014

SCS, Baja California Lobster Fishery, Mexico, 3rd Year MSC Annual Surveillance Audit Report, 2014.

Sharpe & Hendry 2009

Sharpe, D.M.T. & Hendry, A.P., 'Life History Change in Commercially Exploited Fish Stocks: An Analysis of Trends across Studies', *Evolutionary Applications*, 2009, Vol. 2(3), pp. 260–275.

Shepard 2006

Shepard, J.P., 'Water Quality Protection in Bioenergy Production: The US System of Forestry Best Management Practices', *Biomass and Bioenergy*, 2006, Vol. 30(4), pp. 378–384.

Short 2008

Short, K., 'Case Study 4: The New Zealand Hoki, B: The WWF Perspective', in Ward, T. & Phillips, B. (eds.), *Seafood Ecolabelling, Principles and Practice*, Chichester: John Wiley & Sons, 2008, pp. 154–161.

Sinclair 1961

Sinclair, S., *License Limitation: A Method of Economic Fisheries Management*, Ottawa: Canada Department of Fisheries, 1961.

Singer 2008

Singer, B., 'Putting the National back into Forest-Related Policies: The International Forests Regime and National Policies in Brazil and Indonesia', *International Forestry Review*, 2008, Vol. 10(3), pp. 523–537.

Singer 2009

Singer, B., Indonesian Forest-Related Policies. A Multi-Sectoral Overview of Public Policies in Indonesia's Forests since 1965, 2009, available at: http://b-singer.fr/pdf/Forest_policies_in_Indonesia.pdf.

Sjöstedt & Sundström 2015

Sjöstedt, M. & Sundström, A., 'Coping with Illegal Fishing: An Institutional Account of Success and Failure in Namibia and South Africa', *Biological Conservation*, 2015, Vol. 189, pp. 78–85.

Smith & Darr 2004

Smith, W.B. & Darr, D.R., *US Forest Resource Facts and Historical Trends*, Washington, DC: USDA Forest Service, 2004.

South Africa Department of Agriculture, Forestry and Fisheries 2014

South Africa Department of Agriculture, Forestry and Fisheries, Status of the South African Marine Fishery Resources 2014, p. 1, available at: www.nda.agric.za/doaDev/sideMenu/fisheries/03_areasofwork/Resources%20Research/STATUS%20OF%20THE%20SOUTH%20AFRICAN%20MARINE%20FISHERY%20RESOURCES%202014%20WEB.pdf.

Sowman 2011

Sowman, M., 'New Perspectives in Small-Scale Fisheries Management: Challenges and Prospects for Implementation in South Africa', *South African Journal of Marine Science*, 2011, Vol. 33(2), pp. 297–311.

Sowman *et al.* 2014

Sowman, M., Sunde, J., Raemaekers, S. & Schultz, O., 'Fishing for Equality: Policy for Poverty Alleviation for South Africa's Small-Scale Fisheries', *Marine Policy*, 2014, Vol. 46, pp. 31–42.

Squires 1992

Squires, D., 'Productivity Measurement in Common Property Resource Industries: An Application to the Pacific Coast Trawl Fishery', *Rand Journal of Economics*, 1992, Vol. 23(2), pp. 221–236.

Squires *et al.* 1998

Squires, D., Campbell, H., Cunningham, S., Dewees, C., Grafton, R.Q., Herrick, S.F., Kirkley, J., Pascoe, S., Salvanes, K., Shallard, B. & Turris, B., 'Individual Transferable Quotas in Multispecies Fisheries', *Marine Policy*, 1998, Vol. 22(2), pp. 135–159.

Srivastava 2011

Srivastava, N., 'Changing Dynamics of Forest Regulation: Coming Full Circle?', *Review of European Community & International Environmental Law*, 2011, Vol. 20(2), pp. 113–122.

Steering Committee of the State-of-Knowledge Assessment of Standards and Certification 2012

Steering Committee of the State-of-Knowledge Assessment of Standards and Certification, *Toward Sustainability: The Roles and Limitations of Certification*, Washington, DC: RESOLVE Inc., 2012.

Stewart & Callagher 2003

Stewart, J.M. & Callagher, P.D., 'New Zealand Fisheries Management: Changes in Property Rights Structure and Implications for Sustainability', *Sustainable Development*, 2003, Vol. 11(2), pp. 69–76.

Stewart, Walshe & Moodie 2006

Stewart, J., Walshe, K. & Moodie, B., 'The Demise of the Small Fisher? A Profile of Exiters from the New Zealand Fishery', *Marine Policy*, 2006, Vol. 30(4), pp. 328–340.

Stokke 2012

Stokke, O.S., 'International Fisheries Politics: from Sustainability to Precaution', in Andresen, S., Lerum Boasson, E. & Hønneland, G. (eds.), *International Environmental Agreements: An Introduction*, London/New York: Routledge, 2012, pp. 97–116.

Stora Enso 2014
Stora Enso, Global Responsibility Performance, Part of Stora Enso's Annual Report 2014, available at: http://assets.storaenso.com/se/com/DownloadCenter Documents/Global_Responsibility_Performance_2014.pdf.
Stroup & Baden 1983
Stroup, R.L. & Baden, J.A., *Natural Resources: Bureaucratic Myths and Environmental Management*, Cambridge: Ballinger Publishing Company, 1983.
Stroup & Goodman 1992
Stroup, R.L. & Goodman, S.L., 'Property Rights, Environmental Resources, and the Future', *Harvard Journal of Law and Public Policy*, 1992, Vol. 15, pp. 427–454.
Suenaga 2008
Suenaga, S., 'Sandfish Resource Co-management in Akita Prefecture, Japan', in Townsend, R.E., Shotton, R. & Uchida, H. (eds.), *Case Studies in Fisheries Self-Governance*, Rome: FAO, 2008, pp. 191–200.
Sunderlin, Hatcher & Liddle 2008
Sunderlin, W.D., Hatcher, J. & Liddle, M., *From Exclusion to Ownership? Challenges and Opportunities in Advancing Forest Tenure Reform*, Washington DC: Rights and Resources Initiative 2008.
Sundström 2013
Sundström, A., 'Corruption in the Commons: Why Bribery Hampers Enforcement of Environmental Regulations in South African Fisheries', *International Journal of the Commons*, 2013, Vol. 7(2), pp. 454–472.
Swedish Forest Agency 2005
Swedish Forest Agency, *Quantitative Targets of Swedish Forest Policy*, Jönköping: Skogsstyrelsen, 2005.
Swedish Forest Agency 2012
Swedish Forest Agency, Swedish Statistical Yearbooks of Forestry 2012, available at: www.skogsstyrelsen.se/globalassets/statistik/historisk-statistik/skogsstatistisk-arsbok-2010-2014/skogsstatistisk-arsbok-2012.pdf.
Swedish Forest Agency 2014
Swedish Forest Agency, *Swedish Statistical Yearbooks of Forestry*, Jönköping: Skogsstyrelsen, 2014.
Sweeney *et al.* 2004
Sweeney, B.W., Bott, T.L., Jackson, J.K., Kaplan, L.A., Newbold, J.D., Standley, L.J., Hession, W.C. & Horwitz, R.J., 'Riparian Deforestation, Stream Narrowing, and Loss of Stream Ecosystem Services', *Proceedings of the National Academy of Sciences of the United States of America*, 2004, Vol. 101(39), pp. 14132–14137.
Symes & Crean 1995
Symes, D. & Crean, K., 'Privatisation of the Commons: The Introduction of Individual Transferable Quotas in Developed Fisheries', *Geoforum*, 1995, Vol. 26(2), pp. 175–185.
Szczepanski 2002
Szczepanski, K., 'Land Policy and Adat Law in Indonesia's Forests', *Pacific Rim Law & Policy Journal*, 2002, Vol. 11(1), pp. 231–255.
Tacconi 2007
Tacconi, L., 'Verification and Certification of Forest Products and Illegal Logging in Indonesia', in Tacconi, L. (ed.), *Illegal Logging: Law Enforcement, Livelihoods and the Timber Trade*, London: Earthscan, 2007, pp. 251–274.

Tanoue 2015

Tanoue, W., *Japan's Total Allowable Catch Systems in Fishery Resource Management*, Master's thesis, Washington: University of Washington, 2015.

Thrane, Ziegler & Sonesson 2009

Thrane, M., Ziegler, F. & Sonesson, U., 'Eco-Labelling of Wild-Caught Seafood Products', *Journal of Cleaner Production*, 2009, Vol. 17(3), pp. 416–423.

Townsend 1990

Townsend, R.E., 'Entry Restrictions in the Fishery: A Survey of the Evidence', *Land Economics*, 1990, Vol. 66(4), pp. 359–378.

Townsend 2010

Townsend, R.E., 'Transactions Costs as an Obstacle to Fisheries Self-Governance in New Zealand', *Australian Journal of Agricultural and Resource Economics*, 2010, Vol. 54(3), pp. 301–320.

Uberhuaga, Larsen & Treue 2011

Uberhuaga, P., Larsen, H.O. & Treue, T., 'Indigenous Forest Management in Bolivia: Potentials for Livelihood Improvement', *International Forestry Review*, 2011, Vol. 13(1), pp. 80–95.

Uchida 2010

Uchida, H., 'Fishing Effort Coordination and Revenue Pooling Arrangement in Fishery Co-management: Evidence from Japan', *KMI International Journal of Maritime Affairs and Fisheries*, 2010, Vol. 1(2), pp. 1–26.

Uchida & Baba 2008

Uchida, H. & Baba, O., 'Fishery Management and the Pooling Arrangement in the Sakuraebi Fishery in Japan', in Townsend, R.E., Shotton, R. & Uchida, H. (eds.), *Case Studies in Fisheries Self-Governance*, Rome: FAO, 2008, pp. 175–189.

Uchida & Makino 2008

Uchida, H. & Makino, M., 'Japanese Coastal Fishery Co-management: An Overview', in Townsend, R.E., Shotton, R. & Uchida, H. (eds.), *Case Studies in Fisheries Self-Governance*, Rome: FAO, 2008, pp. 221–229.

Uchida & Watanobe 2008

Uchida, H. & Watanobe, M., 'Walleye Pollack (Suketoudara) Fishery Management in the Hiyama Region of Hokkaido, Japan', in Townsend, R.E., Shotton, R. & Uchida, H. (eds.), *Case Studies in Fisheries Self-Governance*, Rome: FAO, 2008, pp. 163–174.

Uggla, Forsberg & Larsson 2016

Uggla, Y., Forsberg, M. & Larsson, S., 'Dissimilar Framings of Forest Biodiversity Preservation: Uncertainty and Legal Ambiguity as Contributing Factors', *Forest Policy and Economics*, 2016, Vol. 62, pp. 36–42.

United States Department of Agriculture 2011

United States Department of Agriculture, National Report on Sustainable Forests 2010, available at: https://www.fs.fed.us/research/sustain/docs/national-reports/2010/2010-sustainability-report.pdf (2011).

Urano 2014

Urano, M., 'Impacts of Newly Liberalised Policies on Customary Land Rights of Forest-Dwelling Populations: A Case Study from East Kalimantan, Indonesia', *Asia Pacific Viewpoint*, 2014, Vol. 55(1), pp. 6–23.

Van Sittert *et al.* 2006

Van Sittert, L., Branch, G., Hauck, M. & Sowman, M., 'Benchmarking the First Decade of Post-Apartheid Fisheries Reform in South Africa', *Marine Policy*, 2006, Vol. 30(1), pp. 96–110.

Verry & Dolloff 2000

Verry, E.S. & Dolloff, C.A., 'The Challenge of Managing for Healthy Riparian Areas', in Verry, E.S., Hornbeck, J.W. & Dolloff, C.A. (eds.), *Riparian Management in Forests of the Continental Eastern United States*, Boca Raton, FL: Lewis Publishers, CRC Press LLC, 2000, pp. 1–22.

Visseren-Hamakers & Pattberg 2013

Visseren-Hamakers, I.J. & Pattberg, P., 'We Can't See the Forest for the Trees. The Environmental Impact of Global Forest Certification Is Unknown', *Gaia – Ecological Perspectives for Science and Society*, 2013, Vol. 22(1), pp. 25–28.

Wakamatsu *et al.* 2010

Wakamatsu, H., Uchida, H., Roheim, C.A., Anderson, C.M. & Managi, S., 'Consumer Preferences for Eco-Labeled Seafood in Japan and the Influence of Information on the Preferences', in IIFET Montpellier Proceeding 2010, pp. 1–12, available at: http://ir/library.oregonstate.edu/xmlui/handle/1957/39233.

Walker *et al.* 2013

Walker, N., Patel, S., Davis, F., Milledge, S. & Hulse, J., *Demand-Side Interventions to Reduce Deforestation and Forest Degradation*, London: International Institute for Environment and Development (IIED), 2013.

Walshe 2010

Walshe, K.A.R., *The Fisheries' Trinity: Re-Conceptualising New Zealand's Inshore Fisheries Management*, Diss., Auckland: University of Auckland, 2010.

Ward 2008

Ward, T.J., 'Barriers to Biodiversity Conservation in Marine Fishery Certification', *Fish and Fisheries*, 2008, Vol. 9(2), pp. 169–177.

Warner, Gjerde & Freestone 2014

Warner, R., Gjerde, K. & Freestone, D., 'Regional Governance for Fisheries and Biodiversity', in Garcia, S.M., Rice, J. & Charles, A. (eds.), *Governance of Marine Fisheries and Biodiversity Conservation: Interaction and Coevolution*, Chichester: John Wiley & Sons, 2014, pp. 211–224.

Washington & Ababouch 2011

Washington, S. & Ababouch, L., *Private Standards and Certification in Fisheries and Aquaculture: Current Practice and Emerging Issues*, Rome: Food and Agriculture Organization of the United Nations, 2011.

Webb & Shivakoti 2008

Webb, E.L. & Shivakoti, G., *Decentralization, Forests and Rural Communities: Policy Outcomes in Southeast Asia*, Los Angeles/London/New Delhi/Singapore: Sage Publications India, 2008.

Weeber & Wallace 2008

Weeber, B. & Wallace, C., 'The New Zealand Best Fish Guide', in Ward, T. & Phillips, B. (eds.), *Seafood Ecolabelling: Principles and Practice*, Chichester: John Wiley & Sons, 2008, pp. 367–388.

Wells 2008

Wells, A., 'Verification of Legal Compliance in Indonesia', in Brown D., Schreckenberg, K., Bird, N., Cerutti, P., Del Gatto, F., Diaw, C., Fomété, T., Luttrell, C., Navaroo, G., Oberndorf, R., Thiel, H. & Wells, A. (eds.), *Legal Timber: Verification and Governance in the Forest Sector*, London: ODI, 2008, pp. 173–186.

White & Martin 2002

White, A. & Martin, A., *Who Owns the World's Forests? Forest Tenure and Public Forests in Transition*, Washington, DC: Forest Trends, 2002.

Wijaya *et al.* 2014

Wijaya, I.M. Rosyidie, A., & Onimaru, T., 'The Implementation of Mandatory Forest Certification: A Case Study on Forest Concession in Indonesia', *Jurnal Perencanaan Wilayah Dan Kota B*, 2014, pp. 655–666.

Wilen, Cancino & Uchida 2012

Wilen, J.E., Cancino, J. & Uchida, H., 'The Economics of Territorial Use Rights Fisheries, or TURFs', *Review of Environmental Economics and Policy*, 2012, Vol. 6(2), pp. 237–257.

Wilson *et al.* 2006

Wilson, D.C., Ahmed, M., Siar, S.V. & Kanagaratnam, U., 'Cross-Scale Linkages and Adaptive Management: Fisheries Co-management in Asia', *Marine Policy*, 2006, Vol. 30(5), pp. 523–533.

Witbooi 2006

Witbooi, E., 'Law and Fisheries Reform: Legislative and Policy Developments in South African Fisheries over the Decade 1994–2004', *Marine Policy*, 2006, Vol. 30(1), pp. 30–42.

Wood 2009

Wood, P.J., *Public Forests, Private Governance: The Role of Provincial Governments in FSC Forest Certification*, Doctoral thesis, University of Toronto, 2009.

Worm *et al.* 2006

Worm, B., Barbier, E.B., Beaumont, N., Duffy, J.E., Folke, C., Halpern, B.S., Jackson, J.B.C., Lotze, H.K., Micheli, F., Palumbi, S.R., Sala, E., Selkoe, K.A., Stachowicz, J.J. & Watson, R., 'Impacts of Biodiversity Loss on Ocean Ecosystem Services', *Science*, 2006, Vol. 314(5800), pp. 787–790.

Wulder, White & Coops 2011

Wulder, M.A., White, J.C. & Coops, N.C., 'Fragmentation Regimes of Canada's Forests', *The Canadian Geographer/Le Géographe Canadien*, 2011, Vol. 55(3), pp. 288–300.

WWF 2011

WWF, Fisheries: Facts and Trends South Africa, 2011, available at: http://awsassets.wwf.org.za/downloads/wwf_a4_fish_facts_report_lr.pdf.

Yamamoto 1995

Yamamoto, T., 'Development of a Community-Based Fishery Management System in Japan', *Marine Resource Economics*, 1995, Vol. 10(1), pp. 21–34.

Yandle 2006

Yandle, T., 'Sharing Natural Resource Management Responsibility: Examining the New Zealand Rock Lobster Co-management Experience', *Policy Sciences*, 2006, Vol. 39(3), pp. 249–278.

Yandle 2007

Yandle, T., 'Understanding the Consequences of Property Rights Mismatches: A Case Study of New Zealand's Marine Resources', *Ecology and Society*, 2007, Vol. 12(2), pp. 1–34, available at: www.ecologyandsociety.org/vol12/iss2/art27/.

Yandle 2008

Yandle, T., 'The Promise and Perils of Building a Co-management Regime: An Institutional Assessment of New Zealand Fisheries Management between 1999 and 2005', *Marine Policy*, 2008, Vol. 32(1), pp. 132–141.

Yandle & Dewees 2008

Yandle, T. & Dewees, C.M., 'Consolidation in an Individual Transferable Quota Regime: Lessons from New Zealand, 1986–1999', *Environmental Management*, 2008, Vol. 41(6), pp. 915–928.

Young 2001
Young, E., 'State Intervention and Abuse of the Commons: Fisheries Development in Baja California Sur, Mexico', *Annals of the Association of American Geographers*, 2001, Vol. 91(2), pp. 283–306.
Young 2013
Young, M., 'Achieving Equity in the Fishing Industry: The Fate of Informal Fishers in the Context of the Policy for the Small-Scale Fisheries Sector in South Africa', *Potchefstroomse Elekroniese Regsblad*, 2013, Vol. 16(5), pp. 288–326.
Zehavi 2012
Zehavi, A., 'New Governance and Policy Instruments: are Governments Going "SOFT"?', in Levi-Faur, D. (ed.), *The Oxford Handbook of Governance*, Oxford: Oxford University Press, 2012, pp. 242–254.

Index

For Product Safety Concerns and Information please contact our EU representative GPSR@taylorandfrancis.com Taylor & Francis Verlag GmbH, Kaufingerstraße 24, 80331 München, Germany

Batch number: 10725973

Printed by Printforce, the Netherlands